Eclogae Geologicae Helvetiae

Swiss Journal of Geosciences

Supplement 1:
Radiolaria
Siliceous Plankton through Time

Zeitschrift der / Revue de la
Schweizerischen Geologischen Gesellschaft / Société Géologique Suisse
Schweizerischen Paläontologischen Gesellschaft / Société Paléontologique Suisse
Schweizerischen Fachgruppe für Geophysik / Groupement Suisse de Géophysique
Schweizerischen Fachgruppe für Sedimentologie (SwissSed) / Groupement Suisse de Sédimentologie (SwissSed)
Schweizerischen Fachgruppe für Tektonik (Swiss Tectonic Studies Group) / Groupement Suisse des Téctoniciens (Swiss Tectonic Studies Group)

Contents

0012-9402/06/01SIII-V
DOI 10.1007/s00015-006-0611-4
Birkhäuser Verlag, Basel, 2006

Eclogae geol. Helv. 99 (2006) Supplement 1, SIII–SV

UNIL | Université de Lausanne

Radiolaria – Siliceous plankton through time

Proceedings of the Tenth Meeting of the International Association of Radiolarian Palaeontologists

INTERRAD X, held at the University of Lausanne, Switzerland, September 2003

Preface

Peter O. Baumgartner

Radiolaria are a very diverse marine siliceous microplankton group that have existed at least since the Cambrian to the Recent. Fossil Radiolaria were known since the middle of the 19th Century and studied by many scientists at the turn of the 19th/20th Century, in part as a consequence of the discoveries made in samples collected be the HMS Challenger expeditions. Intense biostratigraphic radiolarian research was spurted again in the early 1970's by the Deep Sea Drilling Project. Biochronologic zonations based on samples recovered by Ocean Drilling were first proposed for the Tertiary (Riedel & Sanfilippo 1978) and the Cretaceous (Foreman 1973, 1975). However, the radiolarites, cherty oceanic sediments associated with ophiolites, had so far only been studied in thin sections. This changed when radiolarian workers discovered the use of hydrofluoric acid to extract Radiolaria from chert and other siliceous rocks (Dumitrica 1970, Pessagno & Newport 1972) The new method allowed to work on land samples from mountain ranges that had undergone burial diagenesis or even metamorphosis. As a consequence, radiolarian biochronology was rapidly extended into the Jurassic, Triassic and the Palaeozoic (Pessagno 1977, Pessagno et al. 1979, Yao et al. 1980, Nakaseko & Nishimura 1979, Holdsworth & Jones 1980). A major obstacle to the elaboration of detailed biochronozones is the discontinuous nature of the radiolarian record, due to spotty preservation. Biogenous, opaline silica is unstable in the ocean and in the diagenetic environment. Large amounts of the more delicate radiolarian tests become dissolved already during their descent in the water column and in the bottom sediments (Takahashi & Honjo 1981). During early diagenesis, opaline silica dissolves and radiolarians become preserved either as casts of the mould that leaves the dissolved test, or as replacement by quartz, pyrite or other minerals. In Palaeozoic and Mesozoic samples, usually only robust forms are preserved. However, exceptionally well preserved samples from special diagenetic environments, such as Middle Jurassic manganese carbonate nodules (Yao 1997), or euxinic environments like the Late Jurassic Solnhofen Limestone (Zügel 1997, Dumitrica & Zügel 2002) show several hundred morphotypes, a diversity that rivals with the diversity of living radiolarians. These samples teach us, that the average radiolarian assemblage extracted by harsh chemical treatments from rocks is a poor residue of dissolution resistant forms. The consequence of this unavoidable fact is the discontinuous record of radiolarian ranges, which greatly hampers biochronologic correlation. A major

Institut de Géologie et Paléontologie, Université de Lausanne, Anthropole, CH-1015 Lausanne, Switzerland. E-mail: Peter.Baumgartner@unil.ch

step in resolving this problem has been the use of the Unitary Associations method (Guex 1977, 1991). Already in early attempts for a Jurassic-Cretaceous radiolarian zonation (Baumgartner et al. 1980), we realised that this method allows integration of the data on mutual co-occurrence of all considered taxa from all studied sections, thus overcoming local gaps of preservation. Unitary Associations represent maximal sets of co-occurring taxa that are put in a stratigraphic sequence by their identification in all studied sections. Starting in 1984 (Baumgartner 1984a,b) computer programs (Guex and Davaud 1982) allowed the calculation of Unitary Associations from large data sets. The largest data set so far treated was the MRD database on Middle Jurassic to Early Cretaceous radiolarians (Baumgartner et al. 1995). A radiolarian biozonation including 22 Unitary Association Zones for the Middle Jurassic to Early Cretaceous interval was calculated with the Biograph software (Savary & Guex 1991). Today the majority of Mesozoic radiolarian biozonations are based on Unitary Associations. An excellent example is presented in this volume (Beccaro, this volume). A new software, UA-Graph (Hammer et al. 2001) has been developed.

Dating pelagic sediments by radiolarian biochronology has totally changed our understanding of the geology of Tethyan and Circumpacific mountain ranges. Suture zones and ophiolites, interpreted in the frame of plate tectonics as remnants of ancient oceans, needed to be dated. A vast campaign of sampling and dating radiolarites started in the 1980's and is still going on.

Today, Radiolaria are an important microfossil group used worldwide to date mostly basinal sediments of the entire Phanerozoic.

The spectacular diversity of Radiolaria has spurted biogeographic and palaeo-oceanographic studies based on radiolarian assemblages from Cenozoic (Hollis, this volume, Moore 1978) to recent samples (Yamashita et al. 2002) from all oceans. In the modern ocean about 400 known species occupy almost all ecologic niches from shallow water to the deep ocean. Many species are endemic to certain water masses and can therefore be used as oceanographic tracers (Casey 1971a,b). The distribution of Late Tertiary to Holocene Radiolaria in piston cores from the world's oceans have been widely used to trace paleooceanographic features, such as the upwelling zones (Caulet et al. 1992), the growth and vane of polar water masses through the Ice Ages (Romine and Moore 1981, Vénec-Peyré et al. 1997) as well as the major current systems, such as the Humbolt Current (Molina-Cruz 1977).

This volume gives a representative view of research topics discussed at the 10th International Meeting of Radiolarian Palaeontologists, held at the University of Lausanne in 2003. A large effort is currently made to refine and better understand Palaeozoic and early Mesozoic radiolarian biochronology (Feng et al. this volume, Marquez et al. this volume, Wonganan & Caridroit this volume). Cherts associated with basalts and other ophiolitic rocks are often the only remnants of Palaeozoic and Mesozoic ocean basins. Their reconstruction is essentially based on biochronologic ages and palaeobiogeographic affinities of Radiolaria. Much work has dealt with radiolarian faunal changes related to major geologic boundaries in the Palaeozoic (Wang & Luo this volume), at the Permo-Triassic boundary (DeWever et al. this volume), and in the Cenozoic (Hollis, this volume). Mesozoic radiolarian biochronology continues to be improved and better calibrated (Danelian this volume, Beccaro this volume, Bandini et al. this volume).

For a number of years, the radiolarian community is concerned with the development of databases. With the rapid development of computer technologies, databases have become more and more sophisticated and for the last few years only, have become available online. There has been a common agreement among radiolarian specialists, that *Radiolaria.org* is the portal that provides links to most radiolarian databases and hosts a wealth of information on Radiolaria and radiolarian workers (Dolven, this volume).

REFERENCES

Baumgartner, P. O. 1984a: Comparison of Unitary Associations and probabilistic ranking and scaling as applied to Mesozoic radiolarians. Computer & Geosciences 10(1), 167-184.

Baumgartner, P. O. 1984b: A Middle Jurassic-Early Cretaceous low-latitude radiolarian zonation based on unitary associations and age of Tethyan radiolarites. Eclogae Geologicae Helvetiae 77(3), 729-837.

Baumgartner, P. O., Bartolini, A., Carter, E., Conti, M., Cortese, G., Danelain, T., Dumitrica-Jud, R., Gorican, S., Guex, J., Hull, D., Kito, N., Marcucci, M., Matsuoka, A., Murchey, B., O'Dogherty, L., Savary, J., Vishnevskaya, V., Widz, D. & Yao, A. 1995: Middle Jurassic to Early Cretaceous Radiolarian Biochronology of Tethys Based on Unitary Associations. In: Baumgartner, P. O., O'Dogherty, L., Gorican, S., Urquhart, E., Pillevuit A., & De Wever, P. (Eds): Middle Jurassic to Lower Cretaceous Radiolaria of Tethys: Occurrences, Systematics, Biochronology, Mémoires de Géologie (Lausanne), 23, 1013-1043.

Baumgartner, P. O., de Wever, P. & Kocher, R. 1980: Correlation of Tethyan Late Jurassic Early Cretaceous radiolarian events. Cahiers de Micropal. C.R.N.S., 1980/2, 23-72.

Casey, R. E. 1971a: Distribution of polycystine Radiolaria in the oceans in relation to physical and chemical conditions. In: Funnell, B. M., & Riedel, W. R. (Eds.): The Micropalaeontology of Oceans, Cambridge, UK: Cambridge University Press, 151-159.

Casey, R. E. 1971b: Radiolarians as indicators of past and present water-masses In: Funnell, B. M., & Riedel, W. R. (Eds.): The Micropalaeontology of Oceans, Cambridge, UK: Cambridge University Press, 331-341.

Caulet, J. P., Vènec-Peyré, M.-T., Vergnaud-Grazzini, C. & Nigrini, C. 1992: Variation of South Somalian upwelling during the last 160 ka: radiolarian and foraminifera records in core MD 85674. Geological Society Special Publication.

Dumitrica, P. 1970: Cryptocephalic and cryptothoracic Nassellaria in some Mesozoic deposits of Romania. Revue roumaine de Géologie, Géophysique et Géographie (série Géologie) 14(1), 45-124.

Dumitrica, P. & Zügel, P. 2002: *Mendacastrum* n. gen. and *Domuzdagia* n. gen., two Jurassic spherical Spumellaria (Radiolaria) with hagiastrid medullary shell. Micropaleontology 48(1), 23-34.

Forman, H. P., 1973: Radiolaria of Leg 10 with systematics and ranges for the families Amphipyndacidae, Artostrobiidae, and Theoperidae. In: Worzel, J.L., Bryant, W., et al. (Eds): Initial Reports of the Deep Sea Drilling Project, 10, U.S. Government Printing Office, Washington DC, 407–474.

FORMAN, H. P. 1975: Radiolaria from the North Pacific, Deep Sea Drilling Project, Leg 32. In: Larson, R. L., Moberly, R. et al. (Eds): Initial Reports of the Deep Sea Drilling Project, 32. U. S. Government Printing Office, Washington, D.C., 579-676.

GUEX, J. 1977: Une nouvelle méthode d'analyse biochronologique. Bull. Géol. Lausanne, 224, 309–322. Bull. Géol. Lausanne, 313, 317–340.

GUEX, J. 1991: Biochronologic Correlations. Springer-Verlag, 1–252.

GUEX, J. & DAVAUD, E. 1982: Recherche automatique des associations unitaires en biochronologie. Bull. Soc. Vaud. Sci. Nat., 76, no. 361, 53–69.

HOLDSWORTH, B. K. & JONES, D. L.1980: Preliminary radiolarian zonation for late Devonian through Permian time. Geology 8(6), 281-285.

MOLINA CRUZ, A. 1977: Radiolarian assemblages and their relationship to the oceanography of the subtropical southeastern Pacific. Marine Micropaleontology 2(4), 315-352.

MOORE, T. C. 1978: The distribution of radiolarian assemblages in the modern and ice-age Pacific. Marine Micropaleontology 3(3), 229-266.

NAKASEKO, K. & NISHIMURA, A. 1979: Upper Triassic Radiolaria from southwest Japan. Science Reports, College of General Education, Osaka University, 28, 2 (2), 61-109.

PESSAGNO, E. A., JR. 1977: Upper Jurassic Radiolaria and radiolarian biostratigraphy of the California Coast Ranges. Micropaleontology 23(1), 56-113.

PESSAGNO, E. A., JR., FINCH, W. & ABBOTT, P. L. 1979: Upper Triassic Radiolaria from the San Hipolito Formation, Baja California. Micropaleontology 25(2), 160-197.

PESSAGNO, E. A. & NEWPORT, R. L. 1972: A technique for extracting Radiolaria from radiolarian cherts. Micropaleontology 18(2), 231-234.

RIEDEL, W. R. & SANFILIPPO, A. 1978: Stratigraphy and evolution of tropical Cenozoic radiolarians. Micropaleontology 24(1), 61-96.

ROMINE, K. & MOORE, T. C., JR. 1981: Radiolarian assemblage distributions and paleoceanography of the eastern equatorial Pacific Ocean during the last 127,000 years.; CLIMAP's regional ocean dynamics. Palaeogeography, Palaeoclimatology, Palaeoecology 35(2-4), 281-314.

SAVARY, J. & GUEX, J. 1991: BioGraph: un nouveau programme de construction des corrélations biochronologiques basées sur les associations unitaires. Bull. Géol. Lausanne, 313, 317–340.

TAKAHASHI, K. & HONJO, S. 1981: Vertical flux of Radiolaria; a taxon-quantitative sediment trap study from the western tropical Atlantic. Micropaleontology 27(2), 140-190.

HAMMER, Ø., HARPER, D.A.T. & RYAN, P. D. 2001: PAST: Paleontological Statistics Software Package for Education and Data Analysis. Palaeontologia Electronica 4(1): 1-99, http://palaeoelectronica.org/2001_1/past/issue1_01.htm

VENEC PEYRE MARIE, T., CAULET, J. P. & VERGNAUD GRAZZINI, C. 1997: Glacial/ interglacial changes in the equatorial part of the Somali Basin (NW Indian Ocean) during the last 355 kyr. Paleoceanography 12(5), 640-648.

YAO, A. 1997: Faunal change of Early - middle Jurassic radiolarians. News of Osaka Micropaleontologists 10, 155-182.

YAO, A., MATSUDA, T. & ISOZAKI, Y. 1980: Triassic and Jurassic radiolarians from the Inuyama area, central Japan. Journal of Geosciences, Osaka City University 23, 135-154.

YAMASHITA, H., TAKAHASHI, K. & FUJITANI, N. 2002: Zonal and vertical distribution of radiolarians in the western and central Equatorial Pacific in January 1999. Deep-Sea Research 11, 49, 2823-2862.

ZÜGEL, P. 1997: Discovery of a radiolarian fauna from the Tithonian of the Solnhofen area (Southern Franconian Alb, southern Germany). Paläontologische Zeitschrift 71, 197-209.

0012-9402/06/01S0001-20
DOI 10.1007/s00015-006-0605-2
Birkhäuser Verlag, Basel, 2006

Eclogae geol. Helv. 99 (2006) Supplement 1, S1–S20

Turonian Radiolarians from Karnezeika, Argolis Peninsula, Peloponnesus (Greece)

ALEXANDRE N. BANDINI[1], PETER O. BAUMGARTNER[1] & MICHÈLE CARON[2]

Key words: Mesozoic, Upper Cretaceous, Turonian, Radiolaria, Foraminifera, Tethys, Pelagonian Zone, Argolis Peninsula, Greece

ABSTRACT

Near Karnezeika a roughly 140 m thick Upper Cretaceous section consists of interbedded pelagic limestones, cherts and coarse polymict breccias including ophiolites and shallow water limestones. At the base, pink pelagic limestones rest on deeply altered and fractured Lower Jurassic Pantokrator Limestone. This first pelagic facies is dated as middle Turonian, based on planktonic Foraminifera. Over 100 m of coarse ophiolite-carbonate breccias, interpreted as a channel or canyon fill in a pelagic environment, document the erosion of the Late Jurassic nappe edifice along the Cretaceous Pelagonian margin. Above these breccias, we mesured 16 m of principally pink and red pelagic limestones and radiolarian cherts, in which we recovered well-preserved radiolarians discussed here. In this interval, the presence of planktonic Foraminifera allows to state a late Turonian to Coniacian age. More than 40 radiolarian species are described and figured in this work. The radiolarian chronostratigraphy established by 10 different authors in 11 publications was compared for this study and used to establish radiolarian ranges. This exercise shows major discrepancies between authors for the radiolarian ranges of the studied assemblage. Nevertheless, a Turonian age can be stated based on a synthesis of cited radiolarian ranges. This age is consistent with the age based on planktonic foraminifera. In combining the ages of both Radiolaria and planktonic Foraminifera, the studied samples can be restricted to the late Turonian. However, the discrepancies of published radiolarian ranges call for an urgent, major revision of the Late Cretaceous radiolarian biochronology. The integration of planktonic foraminifera with radiolarians may greatly enhance biochronologic resolution in sections where both groups occur.

Introduction

Relatively few studies exist of Upper Cretaceous Radiolaria from Greece, except for some occurences in the Pindos-Olonos zone (De Wever & Thiébault 1981; Thiébault et al. 1981; De Wever & Origlia-Devos 1982; Neumann 2003). In the Hellenides, ongoing tectonic activity is reflected in a small-scaled puzzle of shallow, detrital and pelagic facies in palaeogeographic realms such as the Pelagonian (Vrielynck 1981). Upper Cretaceous radiolarian occurrences are therefore restricted to times of high silica producticity in pelagic palaeoenvironments. One of the objectives of this study was to compare published radiolarian ranges given by different authors for the Upper Cretaceous and to try to establish an acceptable radiolarian age in spite of the differences in radiolarian ranges given by the various authors.

For this preliminary work we collected 17 samples in a total of 140 m of section for the study of planktonic Foraminifera and Radiolaria. 5 samples in a 2.6 m interval above coarse ophiolite-carbonate breccias yielded well-preserved radiolarians. From these samples 41 radiolarian species belonging to 18 genera are described and figured in this work. Overall, the assemblage resembles those described by O'Dogherty (1994) from the lower Turonian. The radiolarian biostratigraphy established by the following 11 publications (given with regions) was compared for this study: Dumitrica (1975, Romania), Foreman (1975, Pacific and 1977, Atlantic), O'Dogherty (1994, Italy and Spain), Pessagno (1976, California), Riedel & Sanfilippo (1974, Composite), Sanfilippo & Riedel (1985, Composite), Schaaf (1985, Composite), Taketani (1982, Japan), Thurow (1988, Atlantic), Vishnevskaya (2001, Russia).

The presence of abundant planktonic Foraminifera both at the base of the studied Cretaceous section and immediately above the radiolarian samples allow the comparison between the ages determined by radiolarian and foraminiferal biostratigraphy (Caron 1985).

[1] Institut de géologie et de paléontologie, Anthropole-Dorigny, Université de Lausanne, 1015 Lausanne, Switzerland.
Email: AlexandreNicolas.Bondini@unil.ch, Peter.Baumgartner@unil.ch
[2] Institut de géologie et de paléotonlogie, Université de Fribourg, Pérolles, 1700 Fribourg, Switzerland. Email: Michele.Caron@unifr.ch

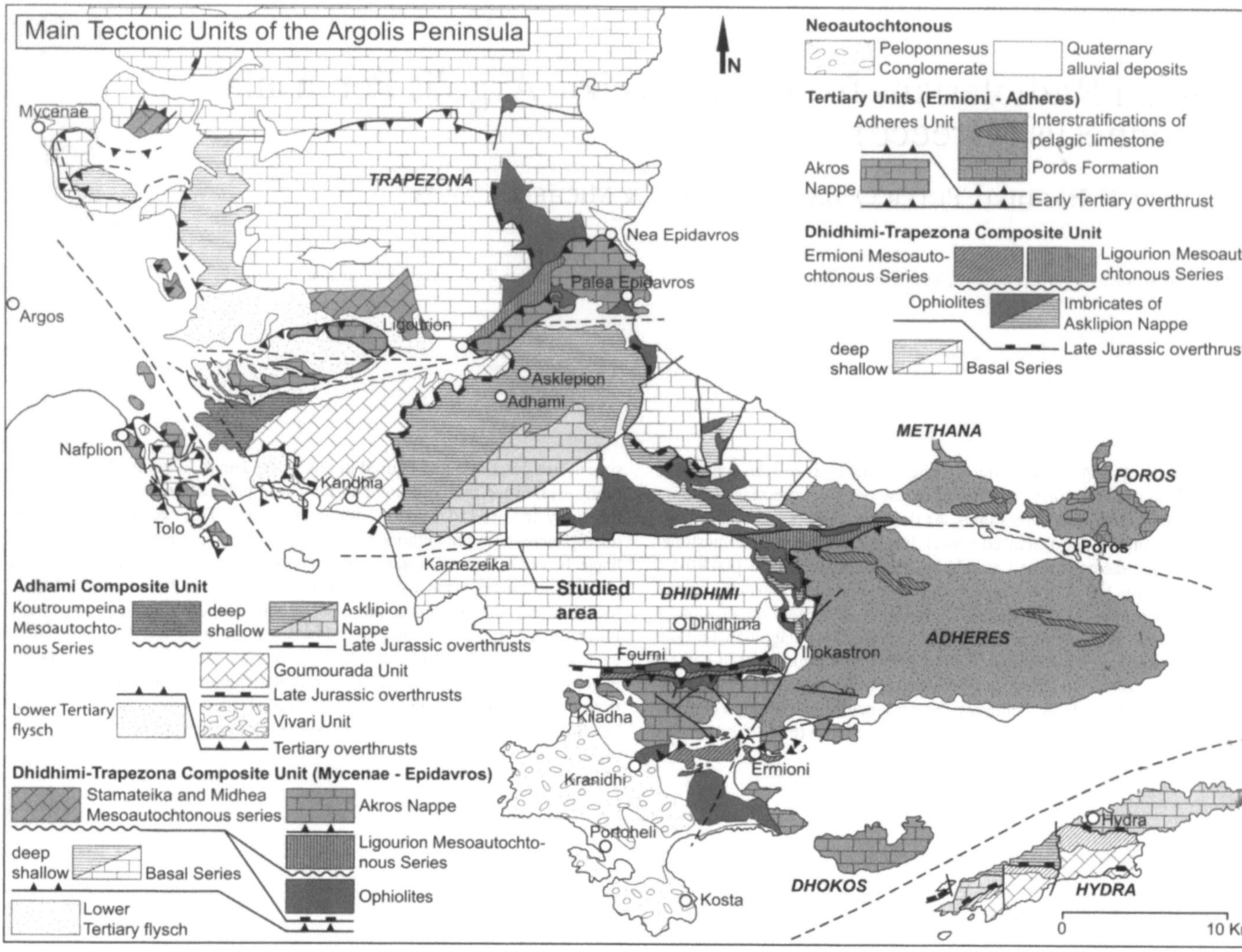

Fig. 1. Overview of the tectono-stratigraphy of the Argolis Peninsula. Based on: Baumgartner (1985), Vrielynck (1981), Clift (Poros Island) and several unpublished diploma thesis University of Lausanne (Bandini 2004; Giraud 2005; Glassey 2005).

Geological Setting

In the Argolis Peninsula, the Jurassic nappe edifice of the eastern Pelagonian margin is unconformably overlain by several different Cretaceous series (Vrielynck 1981). The studied section is located in the *Depression of Karnezeika-Stavropodhi*, a complex nappe syncline affected by neotectonic E-W trending subvertical faults (Fig. 1). The northern edge of this zone is built by Upper Triassic to Lower Jurassic *Pantokrator Limestone* (Fig. 2) belonging to the *Basal Series* of the *Dhidhimi-Trapezona Composite Unit* (Baumgartner 1985). These series became overthrust during the late Jurassic by nappes including ophiolites and then the area became deeply eroded during latest Jurassic and Early Cretaceous times.

Upper Cretaceous pelagic and coarse clastic sediments unconformably overlie the Pantokrator limestone of the basal series and contain disorganised boulder breccias of basalts and shallow water limestones that document the ongoing erosion of the Late Jurassic Pelagonian nappe edifice in a high-relief, deeper marine environment.

Stratigraphy and sedimentology

Near Karnezeika a roughly 140 m thick section consists of interbedded pelagic limestones, cherts and boulder breccias containing abundant ophiolite clasts along with boulders of the underlying Pantokrator Limestone (Fig. 3). At the base, pink pelagic limestones rest on deeply altered and fractured Pantokrator Limestone. Over 100 m of coarse ophiolite-carbonate breccias represent a channel or canyon fill in a pelagic environment. The following 16 m are principally pink and red pelagic limestones and radiolarian cherts, from which we recovered well-preserved radiolarians described here.

The studied section rests conformably on the breccias de-

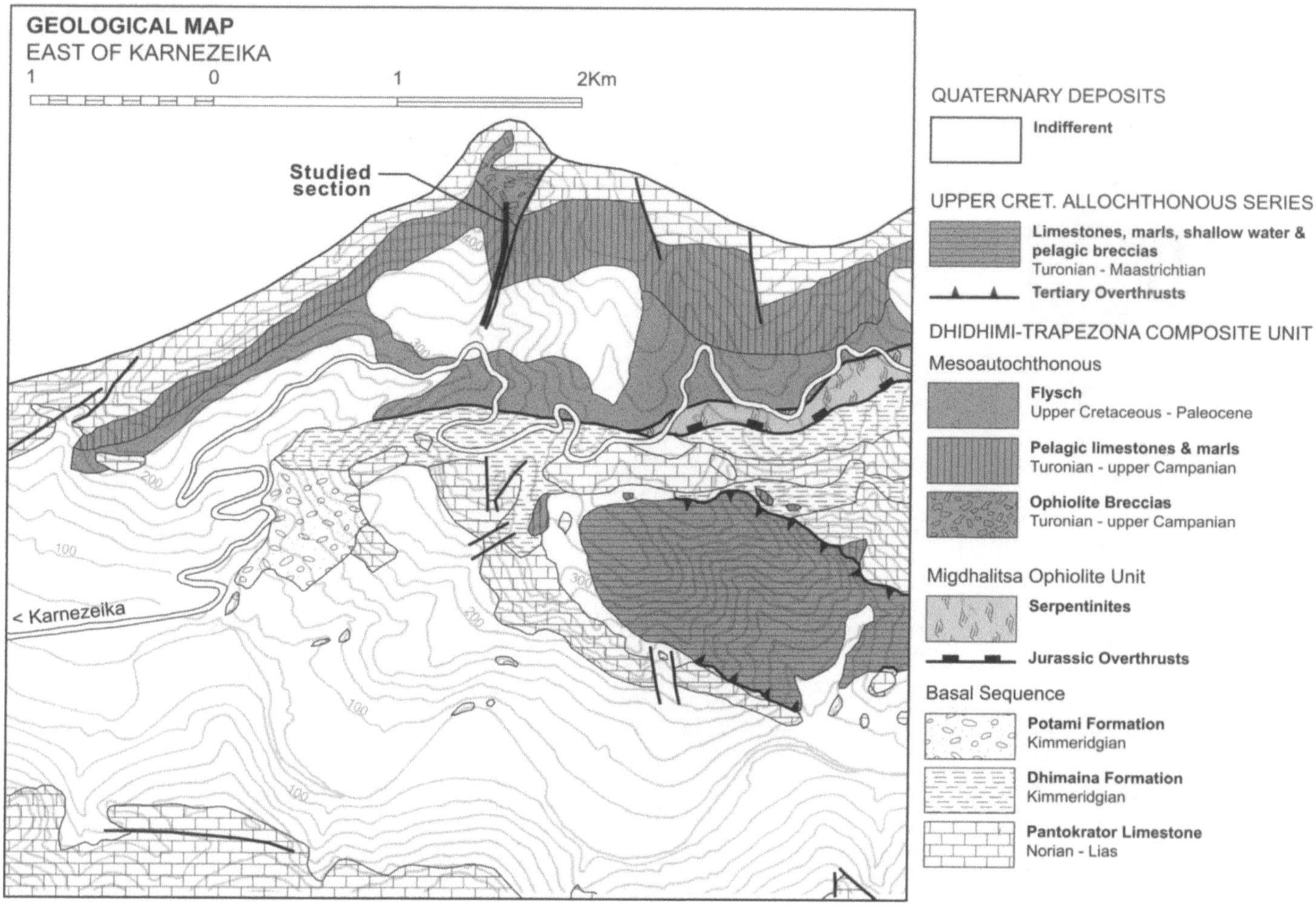

Fig. 2. Geological Map of the area east of Karnezeika (modified after Vernez 1990)

scribed above. It is a 16 m thick alternation of marls and red radiolarian cherts with detrital levels evolving progressively into a pelagic limestone rich in Globotruncanidae. Up section follow 10 m of disorganised boulder breccias with a limestone interbed, which are in turn unconformably covered by quaternary breccias.

Radiolarian biochronology

Today no standard radiolarian biochronology is available for the Late Cretaceous. However, several local and regional radiolarian zonations have been proposed in the past (see citations above).

In comparing the published range charts it becomes evident that the chronostratigraphic range of any given taxon proposed by different authors shows major discrepancies for the studied assemblage (Figs. 4a–b–c–d) from one pubication to the other. This may be due to uncertainties in the chronostratigraphic calibration of radiolarian occurrences. However, most radiolarian zonations of the Late Cretaceous are rather well calibrated by means of plaktonic foraminifera and nannofossils. We rather believe that the differences result from locally inclompete ranges of radiolarian taxa, either due to plaeobiogeographic or paleoecologic exclusions, or due to preservational (diagenetic) biases. The only way to use these radiolarian zonations is to maximise the range of each taxon by stacking the "spartial" ranges expressed in each publication. In principle, the best way of doing this is to create Unitary Associations (Guex 1977–1991: Baumgartner et al. 1980; Guex & Davaud 1982, 1984; Baumgartner 1984 and Savary & Guex 1991) in using the occurrence data of well-defined taxa only, detached from the chronostratigraphic calibrations (Baumgartner et al. 1995). In such a way, we can construct a range chart for the Late Cretaceous that reflects maximal ranges of each taxon with respect to the maximal ranges of all other taxa. This work is in progress, but not completed (Jackett et al. 2002 and Diserens et al. 2003). For this paper the comparison of ranges is based on the chronostratigraphic range of each taxon expessed by each author. We have simply stacked these chronostratigraphic ranges to obtain a minimum and a maximum age for the existence of each taxon. We are aware of the possible errors that may arise

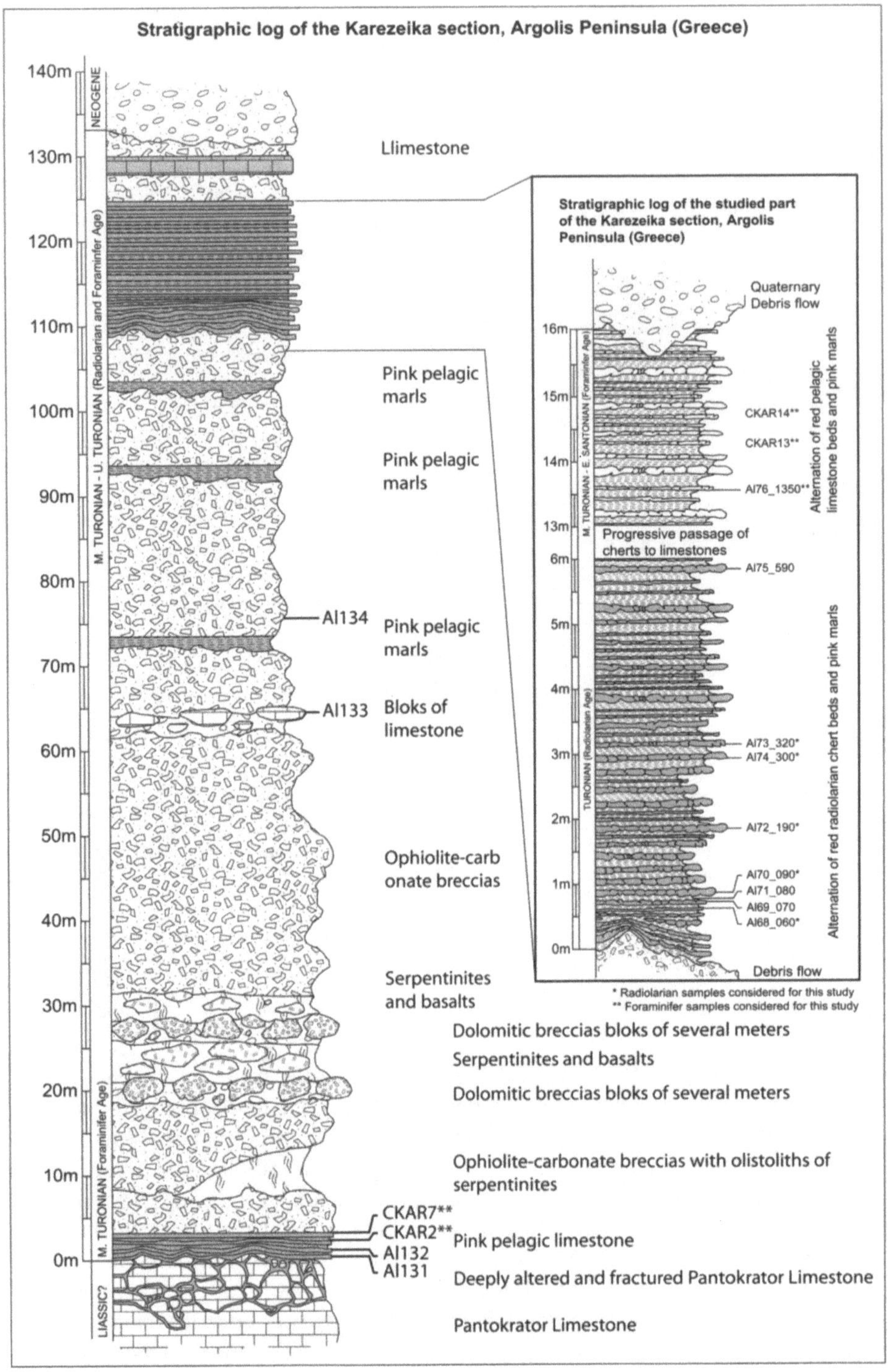

Fig. 3. Stratigraphic log of the Karnezeika section. The inset at right shows the pelagic interval studied for radiolarians and planktonic foraminifera with the samples.

using this procedure. We are, however, in good company, since this procedure was practiced in a series of papers on Mesozoic Radiolaria (e.g. De Wever et al. 1986).

Despite the major discrepancies between the ranges of each author, a Turonian age can be stated using the procedure described above. This age is mainly based on the presence of *Patellula ecliptica*, *Patellula heroica*, *Praeconocaryomma californiaensis* and *Afens liriodes* (have not been cited from earlier than early Turonian), found together with *Pseudoaulophacus putahensis* (has not been cited from later than late Turonian). Moreover, *Dictyomitra urakawensis, Stichomitra communis* and *Pseudodictyomitra pseudomacrocephala* have not been cited from later than Coniacian, *Crucella messinae* have not been cited from later than early Santonian (Fig. 5).

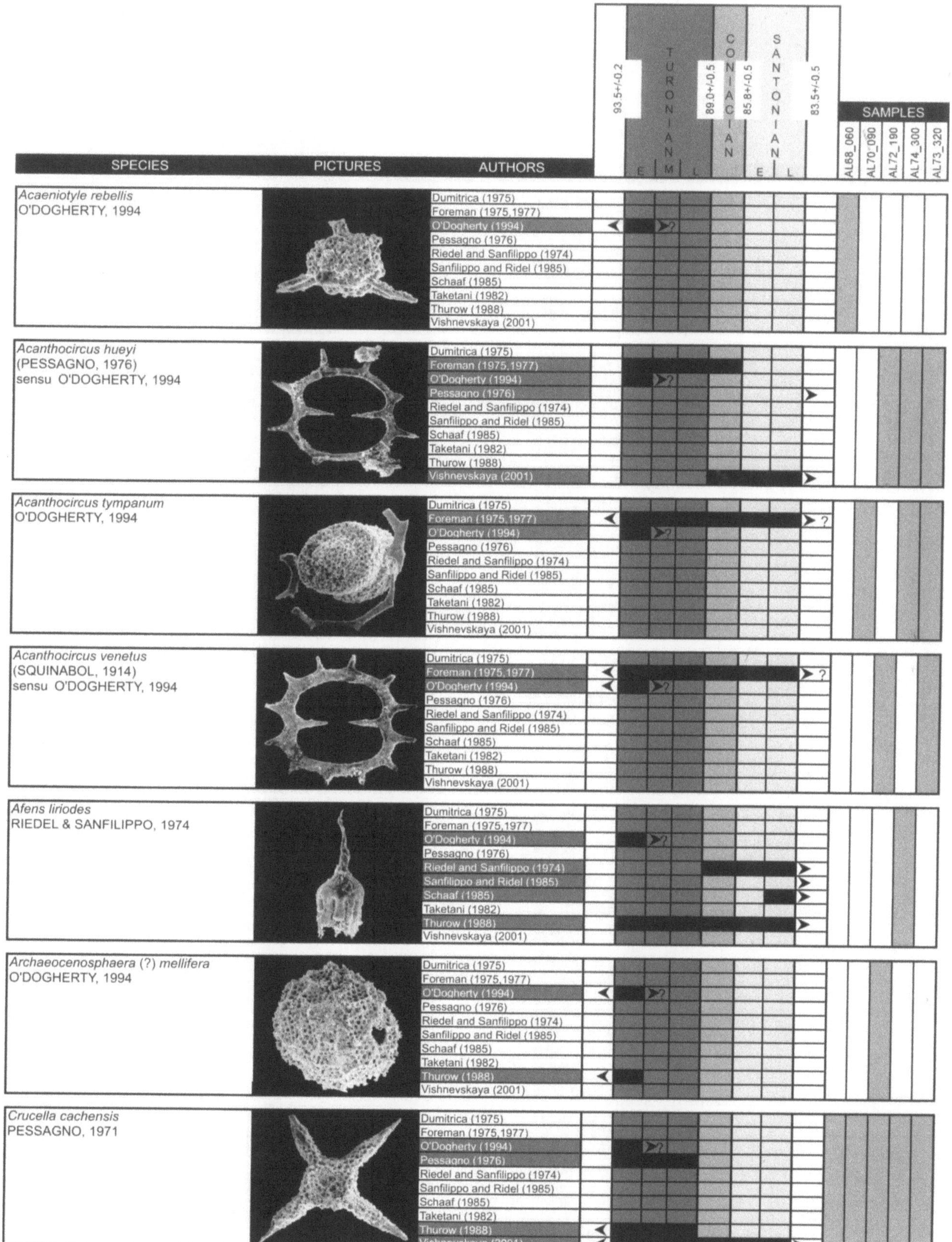

Fig. 4a. Selected Turonian-Santonian radiolarian ranges according to 10 authors as cited in the figures. Note the major discrepencies between authors.

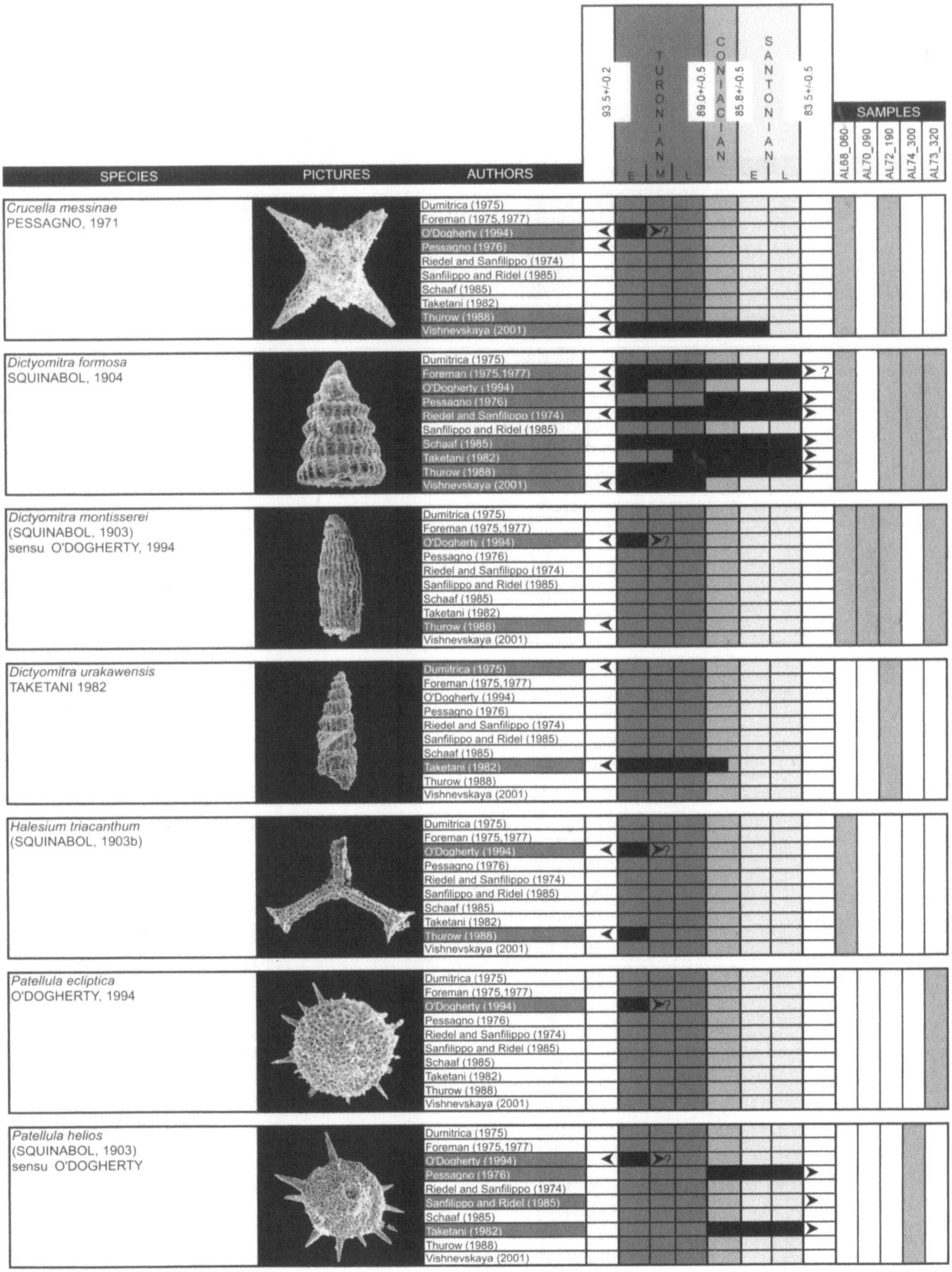

Fig. 4b. Selected Turonian-Santonian radiolarian ranges according to 10 authors as cited in the figures. Note the major discrepencies between authors.

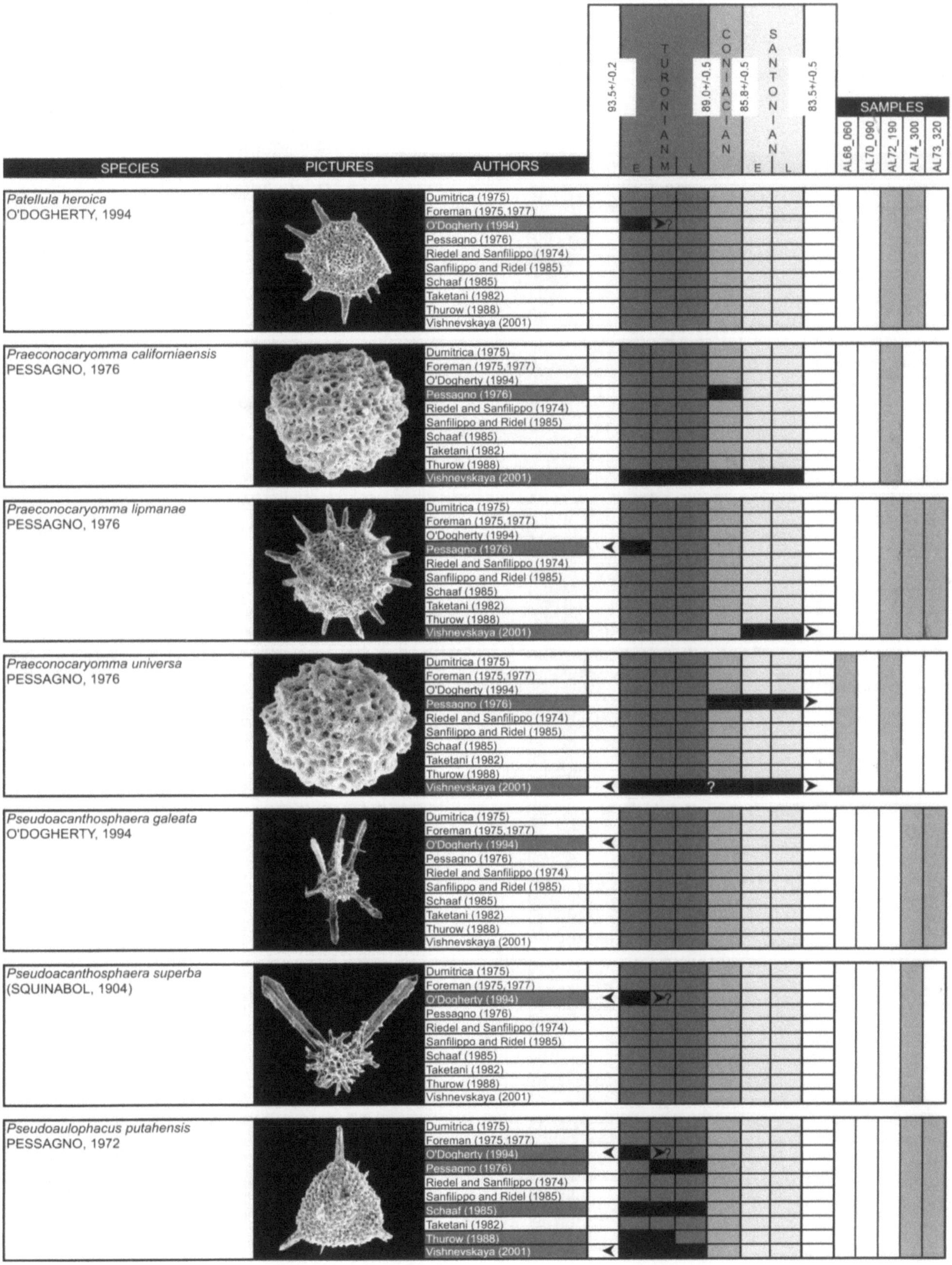

Fig. 4c. Selected Turonian-Santonian radiolarian ranges according to 10 authors as cited in the figures. Note the major discrepencies between authors.

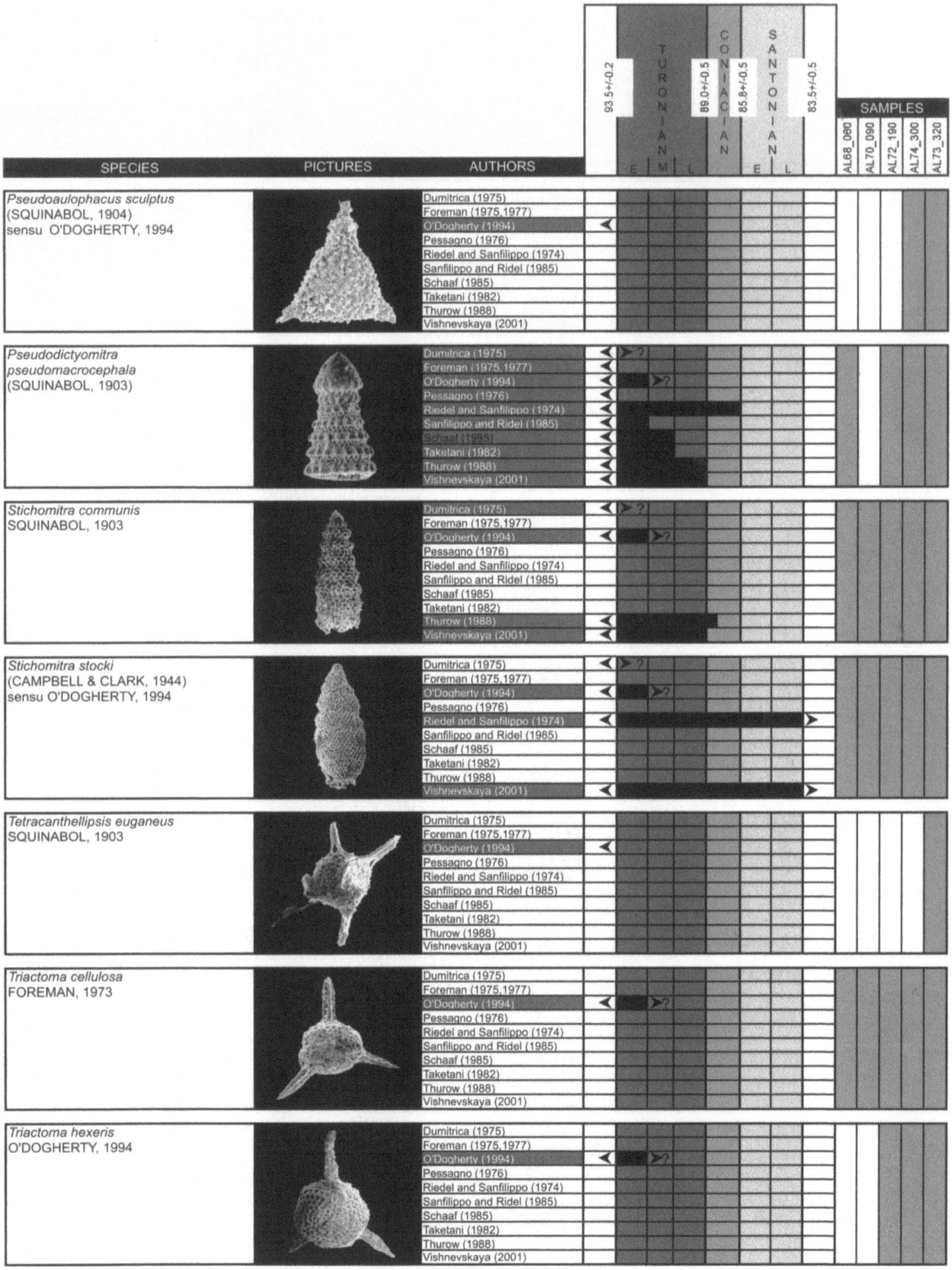

Fig. 4d. Selected Turonian-Santonian radiolarian ranges according to 10 authors as cited in the figures. Note the major discrepencies between authors.

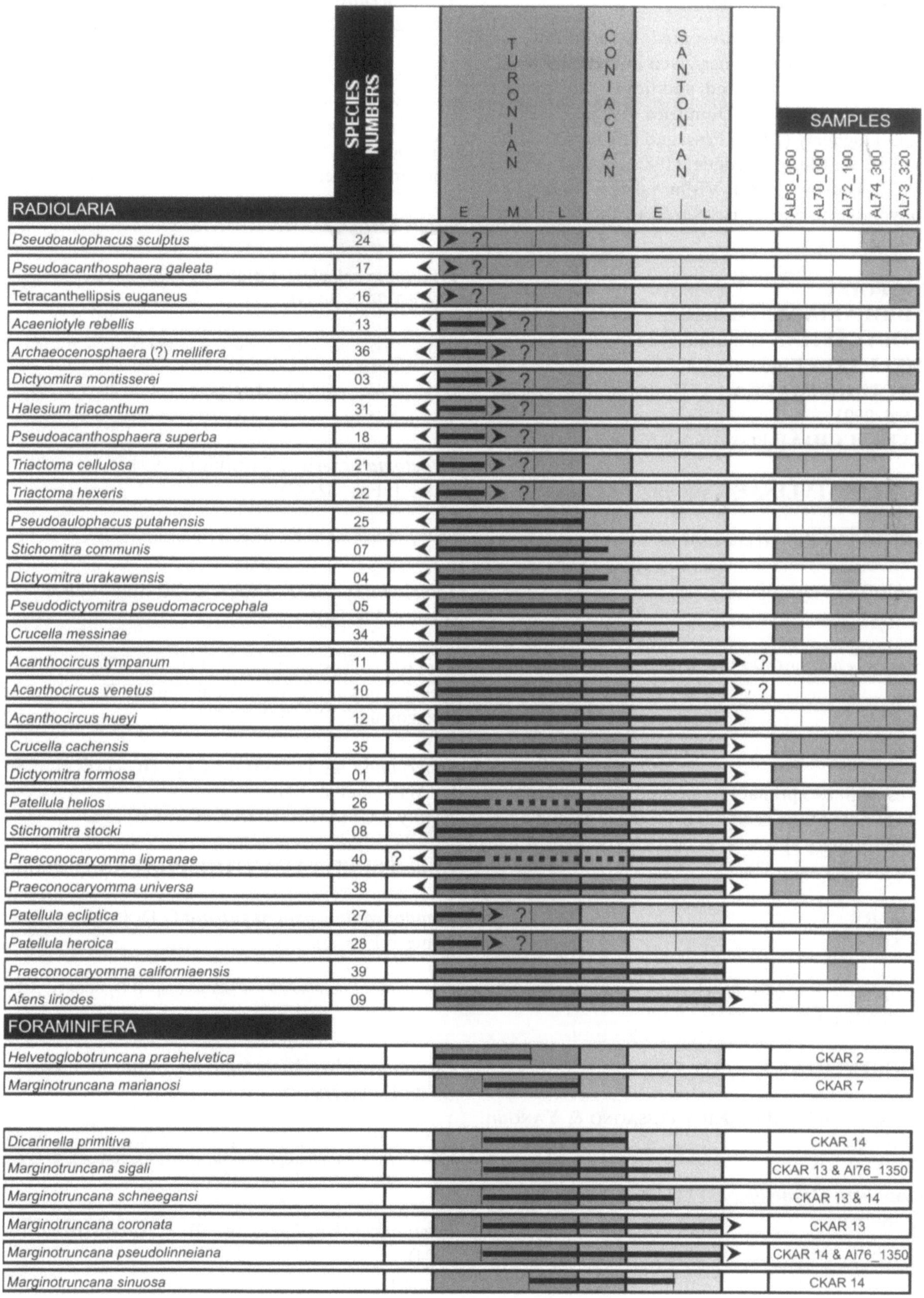

Fig. 5. Summary of Turonian to Santonian radiolarian and foraminfer ages obtained for the studied samples. The radiolarian ranges represent maximum ranges according to the 10 authors cited in figure 4.

Systematic Paleontology of Radiolaria

The suprageneric systematics presented here follow De Wever et al. (2001). The synonymies given include the original description of each taxon and additional synonymies from the following publications: Dumitrica (1975), Foreman (1975, 1977), O'Dogherty (1994), Pessagno (1976), Riedel & Sanfilippo (1974), Sanfilippo & Riedel (1985), Schaaf (1985), Taketani (1982), Thurow (1988), Vishnevskaya (2001). For further synonymies the reader is refered to these publications.

Class **ACTINOPODA**
Subclass **RADIOLARIA** MÜLLER 1858
Superorder **POLYCYSTINA** EHRENBERG 1838
Order **SPUMELLARIA** EHRENBERG 1875
Superfamily **ACTINOMMACEA (A)** HAECKEL 1862, *emend.* DE WEVER et al. 2001
Family **PARVIVACCIDAE** PESSAGNO & YANG *in* PESSAGNO et al. 1989
Subfamily **ACAENIOTYLINAE** YANG 1993

Genus **ACAENIOTYLE** FOREMAN 1973

Acaeniotyle rebellis O'DOGHERTY 1994
(Plate 1, Figs. 21–22)

1994 *Acaeniotyle rebellis* n. sp. O'DOGHERTY, p. 287–288, pl. 51, figs. 5–10.

Acaeniotyle* sp. *A
(Plate 1, Fig. 23)

Remarks. – Test with two primary spines well preserved and at the bottom right part of the shell a third spine. The angle between the three primary spines is 90 degrees (perhaps it had originally four primary spines?).

Acaeniotyle* sp. *B
(Plate 1, Fig. 24)

Remarks. – Specimen not very well preserved with only one spine. Test with tubercles.

Family **XIPHOSTYLIDAE** HAECKEL 1881

Genus **ARCHAEOCENOSPHAERA** PESSAGNO & YANG *in* PESSAGNO et al. 1989

Archaeocenosphaera* (?) *mellifera O'DOGHERTY 1994
(Plate 2, Fig. 18–19)

1988 *Hemicryptocapsa polyhedra* DUMITRICA. – THUROW,, p. 401, pl. 1, fig. 1.
1988 *Hemicryptocapsa* sp. cf. *H. polyhedra* DUMITRICA. – THUROW, p. 401, pl. 5, fig. 2.
1994 *Archaeocenosphaera*? *mellifera* n. sp. O'DOGHERTY, p. 375–376, pl. 74, figs. 1–5.

***Archaeocenosphaera* (?) sp.**
(Plate 2, Figs. 20–21)

Remarks. – Cortical shell very large, spherical with symmetrical meshwork. Cortical shell with very large polygonal pores. A medullary spongy cortical shell can be seen through the large pores of the first one. Very long spines occur at pore junctions. Spines with circular cross-section.

Genus **TRIACTOMA** RÜST 1885

Triactoma cellulosa FOREMAN 1973
(Plate 1, Figs. 30–31)

1973 *Triactoma cellulosa* new species FOREMAN, p. 259, pl. 2, figs 9–10; pl. 16, fig. 9.
1994 *Triactoma cellulosa* FOREMAN. – O'DOGHERTY, p. 300–301, pl. 54, figs 19–23.

Triactoma hexeris O'DOGHERTY 1994
(Plate 1, Figs. 32–33)

1994 *Triactoma hexeris* n. sp. O'DOGHERTY, p. 303, pl. 55, figs. 14–21.

Triactoma* sp. aff. *T. hexeris O'DOGHERTY 1994
(Plate 1, Fig. 34))

Remarks. – Test with cortical shell less rouded then previous species, but more hexagonal thiner. Pores are not clearly hexagonal.

Superfamily **ACTINOMMACEA (B)** HAECKEL 1862, *emend.* DE WEVER et al. 2001
Family **ACTINOMMIDAE** HAECKEL 1862

Genus **PSEUDOACANTHOSPHAERA** O'DOGHERTY 1994

Pseudoacanthosphaera galeata O'DOGHERTY 1994
(Plate 1, Fig. 26)

1994 *Pseudoacanthosphaera galeata* n. sp. O'DOGHERTY, p. 297, pl. 53, figs. 16–19.

Pseudoacanthosphaera superba (SQUINABOL 1904)
(Plate 1, Fig. 27)

1904 *Trisphaera superba* n. sp. SQUINABOL, p. 190, pl. 2, fig. 13.
1994 *Pseudoacanthosphaera superba* (SQUINABOL). – O'DOGHERTY, p. 298–299, pl. 54, figs. 5–10.

Pseudoacanthosphaera* sp. aff. *P. spinosissima (SQUINABOL 1904)
(Plate 1, Fig. 28)

Remarks. – Test with a small spinose cortical shell and two long three-bladed primary spines.

***Pseudoacanthosphaera* (?) sp.**
(Plate 1, Fig. 29)

Remarks. – Test with ellipsoidal cortical shell and three, maybe four primary spines (the fourth one is not visible). Primary spines three-bladed. Meshwork developing small secondary spines at pore vertices.

Genus **TETRACANTHELLIPSIS** SQUINABOL 1903

Tetracanthellipsis euganeus SQUINABOL 1903
(Plate 1, Fig. 25)

1903 *Tetracanthellipsis euganeus* n. sp. SQUINABOL, p. 117, pl. 8, fig. 9.
1994 *Tetracanthellipsis euganeus* SQUINABOL. – O'DOGHERTY, p. 295–296, pl. 53, figs. 8–10.

Superfamily **ACTINOMMACEA (C)** HAECKEL 1862, *emend.* DE WEVER et al. 2001
Family **CONOCARYOMMIDAE** LIPMAN 1969

Genus **PRAECONOCARYOMMA** PESSAGNO 1976

Praeconocaryomma universa PESSAGNO 1976
(Plate 2, Figs. 22–23)

1976 *Praeconocaryomma universa* n. sp. PESSAGNO, p. 42, pl. 6, fig. 14–16.
1982 *Praeconocaryomma universa* PESSAGNO. – TAKETANI, p. 47, pl. 1, figs. 3a–b, 4; pl. 9, fig. 4.
2001 *Praeconocaryomma universa* PESSAGNO. – VISHNEVSKAYA, p. 179, pl. 21, fig. 3; pl. 24, fig. 1; pl. 97, fig. 1; pl. 113, fig. 5; pl. 125, fig. 1–2; pl. 126, fig. 1.
2001 *Praeconocaryomma? universa* PESSAGNO. – VISHNEVSKAYA, p. 179, pl. 21, figs. 11.
2001 *Praeconocaryomma* ex gr. *universa* PESSAGNO. – VISHNEVSKAYA, p. 179, pl. 80, figs. 4; pl. 81, fig. 1.

Praeconocaryomma californiaensis PESSAGNO 1976
(Plate 2, Figs. 24–25)

1976 *Praeconocaryomma californiaensis* n. sp. PESSAGNO, p. 41, pl. 7, fig. 1–8.
1982 *Praeconocaryomma californiaensis* PESSAGNO – TAKETANI, p. 47, pl. 1, figs. 2a–c; pl. 9, figs. 1–2.

Praeconocaryomma lipmanae PESSAGNO 1976
(Plate 2, Figs. 26–27)

1976 *Praeconocaryomma lipmanae* n. sp. PESSAGNO, p. 41–42, pl. 4, fig. 12–13.
1982 *Praeconocaryomma lipmanae* PESSAGNO. – TAKETANI, p. 47, pl. 9, fig. 3.

***Praeconocaryomma* sp.**
(Plate 2, Figs. 28–29)

Remarks. – Elliptical cortical shell in outline with numerous equally spaced, cone-like mammae. Cortical shell with or without radial spines circular in cross-section projecting from center of each mamma. Cortical shell with pore frames of uniform size. Cortical shell with about ten elliptical pore frames on each mamma.

Superfamily **PYLONIACEA** HAECKEL 1881
Subsuperfamily **DACTYLIOSPHAERILAE** SQUINABOL 1904
Family **HAGIASTRIDAE** RIEDEL 1971

Genus **CRUCELLA** PESSAGNO 1971

Crucella messinae PESSAGNO 1971
(Plate 2, Figs. 14–15)

1971 *Crucella messinae* n. sp. PESSAGNO, p. 56, pl. 6, figs. 1–3.
1975 *Crucella* sp. FOREMAN, p. 612, pl. 1D, fig. 7; pl. 2D, fig. 9.
1976 *Crucella messinae* PESSAGNO. – PESSAGNO, p. 32, pl. 1, figs. 4.
non 1982 *Crucella messinae* PESSAGNO. – TAKETANI, p. 50, pl. 9, fig. 17.
1988 *Crucella messinae* PESSAGNO. – THUROW,, p. 399, pl. 5, fig. 22.
? 1988 *Crucella* sp. *B* THUROW,, p. 399, pl. 2, fig. 15.
1994 *Crucella messinae* PESSAGNO. – O'DOGHERTY, p. 368, pl. 70, figs. 21–24, pl. 71, figs. 1–6.
2001 *Crucella* cf. *messinae* PESSAGNO. – VISHNEVSKAYA, p. 158, pl. 114, fig. 10.

Crucella cachensis PESSAGNO 1971
(Plate 2, Figs. 16–17)

1971 *Crucella cachensis* n. sp. PESSAGNO, p. 53, pl. 9, figs. 1–3.
1976 *Crucella cachensis* PESSAGNO. – PESSAGNO, p. 31, pl. 3, figs. 14–15.
1982 *Crucella cachensis* PESSAGNO. – TAKETANI, p. 50, pl. 9, fig. 16.
1988 *Crucella cachensis* PESSAGNO. – THUROW,, p. 399, pl. 2, fig. 13.
1994 *Crucella cachensis* PESSAGNO. – O'DOGHERTY, p. 370, pl. 71, figs. 15–22.
2001 *Crucella cachensis* PESSAGNO. – VISHNEVSKAYA, p. 158, pl. 95, fig. 5; pl. 125, fig. 12; pl. 129, fig. 3.

Subsuperfamily **PATULIBRACCHIILAE** PESSAGNO 1971
Family **PATULIBRACCHIIDAE** PESSAGNO 1971

Genus **HALESIUM** PESSAGNO 1971

Halesium triacanthum (SQUINABOL 1903) **sensu** O'DOGHERTY, 1994
(Plate 2, Fig. 10)

1903 *Dictyastrum triacanthos* n. sp. SQUINABOL, p. 121, pl. 9, fig. 28.
? 1988 *Halesium quadratum* PESSAGNO. – THUROW,, p. 401, pl. 2, fig. 10.
1994 *Halesium triacanthum* (SQUINABOL). – O'DOGHERTY, p. 350–351, pl. 65, figs. 9–14.

***Halesium* sp.**
(Plate 2, Figs. 11–12)

Remarks. – Test with 3 relatively thick and small rays.

Genus **PESSAGNOBRACCHIA** KOZUR & MOSTLER 1978

***Pessagnobracchia* sp.**
(Plate 2, Fig. 13)

Remarks. – Spongy three-rayed test with irregular arrangement of pores on rays.

Family **PSEUDOAULOPHACIDAE** RIEDEL 1967
Subfamily **PSEUDOAULOPHACINAE** RIEDEL 1967

Genus **DACTILYODISCUS** SQUINABOL 1903

***Dactyliodiscus* sp.**
(Plate 2, Figs. 8–9)

Remarks. – Test is disc-shaped and circular in outline with a variable number of equatorial spines. Meshwork spongy with irregularly pore frames. Poorly defined central raised area. Upper and lower surfaces of the test with small tubercles.

Genus **PSEUDOAULOPHACUS** PESSAGNO 1963

Pseudoaulophacus sculptus (SQUINABOL 1904) **sensu** O'DOGHERTY 1994
(Plate 1, Figs. 35–36)

1904 *Theodiscus sculptus* n. sp. SQUINABOL, p. 200, pl. 4, fig. 9.
1988 *Alievium superbum* « Cenomanian » form (SQUINABOL). – THUROW,, p. 397, pl. 5, fig. 11.
1988 *Alievium* sp. *A*. THUROW, p. 397, pl. 5, fig. 12.
1994 *Pseudoaulophacus sculptus* (SQUINABOL). – O'DOGHERTY, p. 319–320, pl. 59, figs. 1–4.

Pseudoaulophacus putahensis PESSAGNO 1972
(Plate 1, Figs. 37–38)

1972 *Pseudoaulophacus putahensis* n. sp. PESSAGNO, p. 301, pl. 27, fig. 1.
1976 *Pseudoaulophacus putahensis* PESSAGNO. – PESSAGNO, p. 28, pl. 3, fig. 13.
1988 *Pseudoaulophacus putahensis* PESSAGNO. – THUROW,, p. 404, pl. 2, fig. 4.
1994 *Pseudoaulophacus putahensis* PESSAGNO. – O'DOGHERTY, p. 320–321, pl. 59, figs. 5–13.
2001 *Pseudoaulophacus putahensis* PESSAGNO. – VISHNEVSKAYA, p. 181, pl. 130, fig. 7.

Superfamily **SPONGURACEA** HAECKEL 1862
Family **SPONGURIDAE** HAECKEL 1862

Genus **PATELLULA** KOZLOVA 1972

Patellula helios (SQUINABOL 1903) **sensu** O'DOGHERTY 1994
(Plate 2, Figs. 1–2)

1903 *Stylotrochus helios* n. sp. SQUINABOL, p. 124, pl. 10, figs. 23–23a.
1976 *Pseudoaulophacus lenticulatus* (WHITE). – PESSAGNO, p. 28, pl. 9, figs. 11–12.
1982 *Pseudoaulophacus lenticulatus* (WHITE). – TAKETANI, p. 51, pl. 10, fig. 11.
1985 *Pseudoaulophacus lenticulatus* (WHITE). – SANFILIPPO & RIEDEL, p. 596, text-figs. 6.4a–b.
1994 *Patellula helios* (SQUINABOL). – O'DOGHERTY, p. 327–328, pl. 60, figs. 19–24.

Patellula ecliptica O'DOGHERTY 1994
(Plate 2, Figs. 3–4)

1994 *Patellula ecliptica* n. sp. O'DOGHERTY, p. 329, pl. 61, figs. 1–5.

Patellula heroica O'DOGHERTY 1994
(Plate 2, Figs. 5–6)

1994 *Patellula heroica* n. sp. O'DOGHERTY, p. 330, pl. 61, figs. 6–11.

***Patellula* sp.**
(Plate 2, Fig. 7)

Remarks. – Test flattened ellipsoidal with fifteen spines radiating in the same equatorial plane.

Order **ENTACTINARIA** KOZUR & MOSTLER 1982
Family **SATURNALIDAE** DEFLANDRE 1953
Subfamily **SATURNALINAE** DEFLANDRE 1953

Genus **ACANTHOCIRCUS** SQUINABOL 1903

Acanthocircus venetus (SQUINABOL 1914) **sensu** O'DOGHERTY 1994
(Plate 1, Figs. 15–16)

1914 *Saturnalis venetus* n. f. SQUINABOL, p. 269, 299, pl. 20[1], fig. 2; pl. 24[5], fig. 1.
? 1975 *Spongosaturnalis horridus* (SQUINABOL). – FOREMAN, p. 610, pl. 4, fig. 3.
1975 *Spongosaturnalis hueyi* group (PESSAGNO). – FOREMAN, p. 611, pl. 1B, figs. 1–3; non pl. 1A, figs. 7–8 (=*A. tympanum*?).
1975 *Spongosaturnalis* (?) *preclarus* new species FOREMAN, p. 611, pl. 1A, figs. 4–5; pl. 4, fig. 8.
1994 *Acanthocircus venetus* (SQUINABOL). – O'DOGHERTY, p. 256, pl. 45, figs. 1–8.

Acanthocircus tympanum O'DOGHERTY 1994
(Plate 1, Figs. 17–18)

pars 1975 *Spongosaturnalis hueyi* group (PESSAGNO). – FOREMAN, p. 611, pl. 1A, figs. 7–8; non pl. 1B, figs. 1–3 (=*A. venetus*?)
1994 *Acanthocircus tympanum* n. sp. O'DOGHERTY, p. 259–260, pl. 45, figs. 17–24.

Acanthocircus hueyi (PESSAGNO 1976) **sensu** O'DOGHERTY 1994
(Plate 1, Figs. 19–20)

1975 *Spongosaturnalis hueyi* (PESSAGNO). – FOREMAN, p. 611, pl. 1A, fig. 6; pl. 4, fig. 10.
1976 *Spongosaturnalis hueyi* n. sp. PESSAGNO, p. 39, pl. 12, fig. 1.
1994 *Acanthocircus hueyi* (PESSAGNO). – O'DOGHERTY, p. 260–261, pl. 46, figs. 1–5.
2001 *Spongosaturnalis hueyi* (PESSAGNO). – VISHNEVSKAYA, p. 186, pl. 122, fig. 2.
2001 *Spongosaturnalis* ex. gr. *hueyi* (PESSAGNO). – VISHNEVSKAYA, p. 186, pl. 92, figs. 7–8, pl. 95, figs. 1–3.

Order **NASSELARIA** EHRENBERG 1875
Superfamily **ARCHAEDICTYOMITRACEA** PESSAGNO 1976

Family **ARCHAEDICTYOMITRIDAE** PESSAGNO 1976

Genus **DICTYOMITRA** ZITTEL 1876

Dictyomitra formosa SQUINABOL 1904
(Plate 1, Figs. 1–2)

1904 *Dictyomitra formosa* n. sp. SQUINABOL, p. 232, pl. 10, fig. 4.
1974 *Dictyomitra torquata* FOREMAN. – RIEDEL & SANFILIPPO, p. 778, pl. 5, figs. 1, 2 and 4.
1975 *Dictyomitra duodecimcostata* (SQUINABOL). – FOREMAN, p. 614, pl. 7, fig. 8; pl. 1G, fig. 5.
1976 *Dictyomitra formosa* SQUINABOL. – PESSAGNO, p. 51, pl. 8, figs. 10–12.
1982 *Dictyomitra formosa* SQUINABOL. – TAKETANI, p. 58, *pl.* 4, figs. 6a–b; pl. 11, fig. 13.
1985 *Dictyomitra formosa* SQUINABOL. – SCHAAF,, text-fig.11, p. 250.
1988 *Dictyomitra formosa* SQUINABOL. – THUROW,, p. 400, pl. 1, figs. 23 and 25.
1994 *Dictyomitra formosa* SQUINABOL. – O'DOGHERTY, p. 80, pl. 4, figs. 8–12.
2001 *Dictyomitra formosa* SQUINABOL. – VISHNEVSKAYA, p. 160, pl. 25, fig. 10.

Dictyomitra* sp. cf. *D. formosa SQUINABOL 1904
(Plate 1, Fig. 3)
(species number 2)
Dictyomitra montisserei (SQUINABOL 1903) **sensu** O'DOGHERTY 1994
(Plate 1, Figs. 4–5)

1903 *Stichophormis Montis Serei* n. sp. SQUINABOL, p. 137, pl. 8, fig. 38.
1975 *Dictyomitra* sp. *A*. FOREMAN, p. 615, pl. 1G, fig. 7; pl. 2G, figs. 18 and 20.
1982 *Archaedictyomitra* sp. *A*. TAKETANI, p. 58, pl. 4, figs. 5a–b.
? 1988 *Archaedictyomitra lacrimula* (FOREMAN). – THUROW,, p. 397, pl. 3, fig. 8.
1988 *Archaedictyomitra simplex* PESSAGNO. – THUROW,, p. 398, pl. 3, fig. 9.
1994 *Dictyomitra montisserei* (SQUINABOL 1903b). – O'DOGHERTY, p. 77, pl. 3, figs. 1–29.

Dictyomitra urakawensis (TAKETANI 1982)
(Plate 1, Fig. 6)

? 1975 *Dictyomitra* sp. – DUMITRICA,, text-fig.2; 8.
1982 *Dictyomitra urakawensis* n. sp. TAKETANI, p. 59, pl. 4, figs. 8a–b; pl. 11, fig. 16.

Superfamily **AMPHIPYNDACEA** RIEDEL 1967
Family **SPONGOCAPSULIDAE** PESSAGNO1977

Genus **TORCULUM** O'DOGHERTY 1995

Torculum coronatum (SQUINABOL1904)
(Plate 1, Fig. 9)

1904 *Theoconus coronatus* n. sp. SQUINABOL, p. 220, pl. 8, fig. 3.
1976 *Stichomitra* (?) *zamoraensis* n. sp. PESSAGNO, p. 54, pl. 3, figs. 7–9.
1982 *Spongocapsula* (?) *zamoraensis* (PESSAGNO). – TAKETANI, p. 62, pl. 5, figs. 6a–b; pl. 12, figs. 12–13.
1988 *Theoconus coronatus* Group SQUINABOL. – THUROW,, p. 407, pl. 4, figs. 2.
1988 *Theoconus* sp. *A* cf. *T. coronatus* SQUINABOL. – THUROW,, p. 407, pl. 4, figs. 3–4.
2001 *Spongocapsula* aff. *zamoraensis* (PESSAGNO). – VISHNEVSKAYA, p. 186, pl. 77, fig. 5; pl. 90, figs. 5–6.

Superfamily **EUCYRTIDIACEA** EHRENBERG 1847
Family **PSEUDODICTYOMITRIDAE** PESSAGNO 1977

Genus **PSEUDODICTYOMITRA** PESSAGNO 1977

Pseudodictyomitra pseudomacrocephala (SQUINABOL 1903)
(Plate 1, Figs. 7–8)

1903 *Dictyomitra pseudomacrocephala* n. sp. SQUINABOL, p. 139, pl. 10, fig. 2.
1974 *Dictyomitra macrocephala* SQUINABOL. – RIEDEL & SANFILIPPO, p. 778, pl. 4, figs. 10–11; pl. 14, fig. 11.
1975 *Dictyomitra pseudomacrocephala* SQUINABOL. – DUMITRICA,, text-fig.2, 19.
1975 *Dictyomitra pseudomacrocephala* SQUINABOL. – FOREMAN, p. 614, pl. 7, fig. 10.
1976 *Dictyomitra (?) pseudomacrocephala* SQUINABOL. – PESSAGNO, p. 53, pl. 3, figs. 2–3.
1982 *Pseudodictyomitra pseudomacrocephala* (SQUINABOL). – TAKETANI, p. 61, pl. 5, figs. 4a–b; pl. 12, figs. 7–8.
1985 *Pseudodictyomitra pseudomacrocephala* (SQUINABOL). – SANFILIPPO & RIEDEL, p. 608, text-figs. 10. 1a–b.
1985 *Pseudodictyomitra pseudomacrocephala* (SQUINABOL). – SCHAAF, text-fig.11, p. 250
1988 *Pseudodictyomitra pseudomacrocephala* (SQUINABOL). – THUROW,, p. 405, pl. 1, fig. 13; pl. 3, figs. 11–16.
1994 *Pseudodictyomitra pseudomacrocephala* (SQUINABOL). – O'DOGHERTY, p. 108–109, pl. 8, figs. 5–8.
2001 *Pseudodictyomitra pseudomacrocephala* (SQUINABOL). – VISHNEVSKAYA, p. 183–184, pl. 20, fig. 6; pl. 24, fig. 10; pl. 100, fig. 3; pl. 129, figs. 5, 9 and 10.

Family **EUCYRTIDIIAE** EHRENBERG 1847

Genus **STICHOMITRA** CAYEUX 1897

Stichomitra communis SQUINABOL 1903
(Plate 1, Figs. 10–11)

1903 *Stichomitra communis* n. sp. SQUINABOL, p. 141, pl. 8, fig. 40.
1975 *Stichomitra* sp. DUMITRICA,, text-fig.2, 21.
1975 *Stichomitra* spp. cf. *D. tekschaensis* ALIEV. – FOREMAN, p. 615, pl. 2H, fig1.
1982 *Stichomitra communis* SQUINABOL. – TAKETANI, p. 54, pl. 3, fig. 9; pl. 11, fig. 5.
1988 *Stichomitra communis* SQUINABOL. – THUROW,, p. 406, pl. 4, fig. 10.
1988 *Stichomitra* sp. cf. *S. communis* SQUINABOL. – THUROW,, p. 406, pl. 4, fig. 9.
1994 *Stichomitra communis* SQUINABOL. – O'DOGHERTY, p. 144–145, pl. 17, figs. 6–16.
2001 *Stichomitra communis* SQUINABOL. – VISHNEVSKAYA, p. 188, pl. 23, fig. 8; pl. 79, fig. 3; pl. 129, fig. 8.

Stichomitra stocki (CAMPBELL & CLARK 1944) **sensu** O'-DOGHERTY 1994
(Plate 1, Figs. 12–13)

1944 *Stichocapsa* (?) *stocki* n. sp. CAMPBELL & CLARK, p. 44, pl. 8, figs. 31–33.

1974 *Amphipyndax stocki* (CAMPBELL & CLARK). – RIEDEL & SANFILIPPO, p. 775, pl. 15, fig. 11; pl. 11, figs. 1–3.
1975 *Amphipyndax stocki* (CAMPBELL & CLARK). – DUMITRICA, text-fig.2.23.
1982 *Amphipyndax stocki* (CAMPBELL & CLARK). – TAKETANI, p. 52, pl. 2, figs. 9a–b; pl. 10, figs. 13–14.
1982 *Amphipyndax* sp. TAKETANI, p. 52, pl. 10, fig. 16.
1988 *Stichomitra* (?) sp. *A*. THUROW,, p. 406, pl. 1, fig. 17.
1994 *Stichomitra stocki* (CAMPBELL & CLARK). O'DOGHERTY, p. 147–148 and 150, pl. 18, figs. 9–15.
2001 *Amphipyndax stocki* (CAMPBELL & CLARK). – VISHNEVSKAYA, p. 146, pl. 1, fig. 13; pl. 4, fig. 11–13; pl. 6, fig. 12; pl. 16, fig. 1; pl. 93, fig. 4–5; pl. 94, fig. 8 and 10; pl. 99, fig. 1–3 and 8; pl. 114, fig. 12.
2001 *Amphipyndax stocki* (CAMPBELL & CLARK) var. A VISHNEVSKAYA. – VISHNEVSKAYA, p. 146–147, pl. 16, fig. 2–6; pl. 26, fig. 6; pl. 100, fig. 4; pl. 123, fig. 16–21 and 23.
2001 *Amphipyndax stocki* (CAMPBELL & CLARK) var. B VISHNEVSKAYA. – VISHNEVSKAYA, p147, pl. 3, fig. 6; pl. 12, fig. 3 and 5; pl. 15, fig. 1–5.
2001 *Amphipyndax stocki* (CAMPBELL & CLARK) var. C VISHNEVSKAYA. – VISHNEVSKAYA, p. 147, pl. 3, fig. 2 and 4; pl. 14, fig. 1–3.V

NASSELLARIA INCERTAE SEDIS

Genus **AFENS** RIEDEL & SANFILIPPO 1974

Afens liriodes RIEDEL & SANFILIPPO 1974
(Plate 1, Fig. 14)

1974 *Afens liriodes* new genus and new species RIEDEL & SANFILIPPO, p. 775, pl. 11, fig. 11; pl. 13, figs. 14–16.
1985 *Afens liriodes* RIEDEL & SANFILIPPO. – SANFILIPPO & RIEDEL, p. 624, text-figs. 13.3a–c.
1985 *Afens liriodes* RIEDEL & SANFILIPPO. – SCHAAF, text-fig.11, p. 250.
1988 *Afens liriodes* RIEDEL & SANFILIPPO. – THUROW,, pl. 2,fig.1.
1994 *Afens liriodes* RIEDEL & SANFILIPPO. – O'DOGHERTY, p. 246, pl. 42, figs. 23–26.

Planktonic Foraminifera

In the following, we use the ranges proposed by Caron (1985). The first pelagic facies at the top of the Pantokrator Limestone is dated as middle Turonian by the presence of *Helvetoglobotruncana helvetica* (Pl. 3, Fig. 1, early Turonian – middle Turonian) and *Marginoglobotruncana marianosi* (Pl. 3, Fig. 2, middle Turonian – late Turonian).

In the pelagic limestone rich in Globotruncanidae at the top of the radiolarian cherts, the presence of *Dicarinella primitiva* (Pl. 3, Fig. 9, middle Turonian – Coniacian), *Marginotruncana sigali* (Pl. 3, Figs. 5 and 12, middle Turonian – early Santonian), *M. schneegansi* (Pl. 3, Figs. 3 and 7, middle Turonian – early Santonian), *M. coronata* (Pl. 3, Fig. 4, middle Turonian – early Campanian), *M. pseudolinneiana* (Pl. 3, Figs. 6, 10 and 11, middle Turonian – early Campanian) and *M. sinuosa* (Pl. 3, Fig. 8, late Turonian – early Santonian) allow to state a late Turonian to early Santonian age without further precision.

Discussion and Conclusions

The radiolarian chronostratigraphic ranges used for this paper are based on the comparison of taxon ranges established by 10 different authors (see list in introduction). By stacking chronostratigraphic ranges, we obtain a maximum range for the existence of each taxon. Despite possible inaccuracies of calibration, the radiolarian age given by this procedure is consistent with the age based on planktonic foraminifera (late Turonian to Coniacian). In combining the radiolarian and the planktonic foraminifer ages, the samples would be restricted to the late Turonian. However, the major discrepancies of published radiolarian ranges call for an urgent, major revision of the Late Cretaceous radiolarian biochronology, a project that is underway (Jackett et al. 2002 and Diserens et al. 2003). The integration of planktonic foraminifera with radiolarians may greatly enhance biochronologic resolution in sections where both groups occur.

This is the first time that Late Cretaceous radiolarians are described from the Argolis Peninsula.

Acknowledgements

The field work in the Argolis Peninsula was partly financed by the Société Acadèmique Vaudoise and the University of Lausanne, Switzerland. The salaries are paid by the Swiss National Science Foundation (grant n° 2000-063762 and n° 200021-105845). We are thankful for the loan of the Vernez' thin sections by the Musée Cantonal de Géologie, Lausanne, Switzerland. We are grateful for help in the field by France Girault and Thierry Glassey; assistance in the laboratory by Sébastien Bruchez, Marc-Olivier Diserens and Sarah-Jane Jackett; and for discussions with Paulian Dumitrica. The manuscript has benefited from careful reviews by Atsushi Matsuoka and Luis O'-Dogherty.

SPECIES LIST

Acaeniotyle rebellis O'DOGHERTY 1994 (Pl. 1, Figs. 1–2)
Acaeniotyle* sp. *A (Pl. 1, Fig. 3)
Acaeniotyle* sp. *B (Pl. 1, Fig. 4)
Acanthocircus hueyi (PESSAGNO 1976) **sensu** O'DOGHERTY 1994 (Pl. 2, Figs.18–19)
Acanthocircus tympanum O'DOGHERTY 1994 (Pl. 2, Figs. 16–17)
Acanthocircus venetus (SQUINABOL 1914) **sensu** O'DOGHERTY 1994 (Pl. 2, Figs. 14–15)
Afens liriodes RIEDEL & SANFILIPPO 1974 (Pl. 2, Fig. 33)
Archaeocenosphaera* (?) *mellifera O'DOGHERTY 1994 (Pl. 1, Figs. 5–6)
***Archaeocenosphaera* (?) sp.** (Pl. 1, Figs. 7–8)
Crucella cachensis PESSAGNO 1971 (Pl. 1, Figs. 29–30)
Crucella messinae PESSAGNO 1971 (Pl. 1, Figs. 27–28)
***Dactyliodiscus* sp.** (Pl. 2, Figs. 1–2)
Dictyomitra formosa SQUINABOL 1904 (Pl. 2, Figs. 20–21)
Dictyomitra montisserei (SQUINABOL 1903) **sensu** O'DOGHERTY 1994 (Pl. 2, Figs. 23–24)

Dictyomitra* sp. cf. *D. formosa SQUINABOL 1904 (Pl. 2, Fig. 22)
Dictyomitra urakawensis TAKETANI 1982 (Pl. 2, Fig. 25)
***Halesium* sp.** (Pl. 1, Figs. 32–33)
Halesium triacanthum (SQUINABOL 1903) **sensu** O'DOGHERTY 1994 (Pl. 1, Fig. 31)
Patellula ecliptica O'DOGHERTY 1994 (Pl. 2, Figs. 9–10)
Patellula helios (SQUINABOL 1903) **sensu** O'DOGHERTY 1994 (Pl. 2, Figs. 7–8)
Patellula heroica O'DOGHERTY 1994 (Pl. 2, Figs. 11–12)
***Patellula* sp.** (Pl. 2, Fig. 13)
***Pessagnobrachia* sp.** (Pl. 1, Fig. 34)
Praeconocaryomma californiaensis PESSAGNO 1976 (Pl. 1, Figs. 21–22)
Praeconocaryomma lipmanae PESSAGNO 1976 (Pl. 1, Figs. 23–24)
***Praeconocaryomma* sp.** (Pl. 1, Figs. 25–26)
Praeconocaryomma universa PESSAGNO 1976 (Pl. 1, Figs. 19–20)
***Pseudoacanthosphaera* (?) sp.** (Pl. 1, Fig.17)
Pseudoacanthosphaera galeata O'DOGHERTY 1994 (Pl. 1, Fig. 14)
Pseudoacanthosphaera* sp. aff. *P. spinosissima (SQUINABOL 1904) (Pl. 1, Fig. 16)
Pseudoacanthosphaera superba (SQUINABOL 1904) (Pl. 1, Fig. 15)
Pseudoaulophacus putahensis PESSAGNO 1972 (Pl. 2, Figs. 5–6)
Pseudoaulophacus sculptus (SQUINABOL 1904) **sensu** O'-DOGHERTY 1994 (Pl. 2, Figs. 3–4)
Pseudodictyomitra pseudomacrocephala (SQUINABOL 1903) (Pl. 2, Figs. 27–28)
Stichomitra communis SQUINABOL 1903 (Pl. 2, Figs. 29–30)
Stichomitra stocki (CAMPBELL & CLARK 1944) **sensu** O'-DOGHERTY 1994 (Pl. 2, Figs. 31–32)
Tetracanthellipsis euganeus SQUINABOL 1903 (Pl. 1, Fig. 18)
Torculum coronatum (SQUINABOL 1904) (Pl. 2, Fig. 26)
Triactoma cellulosa FOREMAN 1973 (Pl. 1, Figs. 9–10)
Triactoma hexeris O'DOGHERTY 1994 (Pl. 1, Figs. 11–12)
Triactoma* sp. aff. *T. hexeris O'DOGHERTY 1994 (Pl. 1, Fig. 13)

REFERENCES

BANDINI, A. 2004: Sédimentologie et tectonostratigraphie du sud de la péninsule de l'Argolide, Péloponèse (Grèce) & Upper Cretaceous radiolarians of Karnezeika, (Argolis Peninsula, Peloponesus (Greece). Unpublished diploma thesis, University of Lausanne.

BAUMGARTNER, P.O. 1984: Comparison of unitary associations and probabilistic ranking and scaling as applied to Mesozoic radiolarians. Computers and Geosciences, 10 (1), 167–183.

BAUMGARTNER, P.O. 1985: Jurassic sedimentary evolution and nappe emplacement in the Argolis peninsula (Peloponnesus, Greece). Denkschriften der Schweizerischen Naturforschenden Gesellschaft; 99, 111p. Birkhäuser, Basel, Boston, Stuttgart.

BAUMGARTNER, P.O., DE WEVER, P. & KOCHER, R. 1980: Correlation of Tethyan Late Jurassic-Early Cretaceous radiolarian events. Cahiers de Micropaléontologie, 2, 23–86.

BAUMGARTNER, P.O., GUEX, J. & DUMITRICA, P. 1995: Concepts of the systematic and biostratigraphic work. In: BAUMGARTNER, P.O et al. (Eds.): Middle Jurassic to Lower Cretaceous Radiolaria of Tethys: Occurrences, Systematics, Biochronology, 11–15. Mémoires de Géologie (Lausanne), 23. Institut de Géologie et de Paléontologie, Université de Lausanne.

CAMPBELL, A.S. & CLARK, B.L. 1944: Radiolaria from Upper Cretaceous of Middle California. Geological Society of America Special Paper, 57, 1–61.

CARON, M. 1985: Cretaceous planktic foraminifera. In: BOLLI, H.M., SAUDERS, J.B. & PERCH-NIELSEN, K. (Eds.): Plankton stratigraphy, 17–86. Cambridge University Press, Cambridge/ New York.

DE WEVER, P. & ORIGLIA-DEVOS, I. 1982: Datations nouvelles par les Radiolaires de la série des Radiolarites s. l. du Pinde-Olonos, (Grèce). C. R. Acad. Sc. Paris, 294, 399–404.

DE WEVER, P. & THIÉBAULT, F. 1981: Les Radiolaires d'âge jurassique supérieur à crétacé supérieur dans les radiolarites du Pinde-Olonos (Presqu'île de Koroni; Péloponnèse méridional, Grèce). Géobios, 14(5), 577–609.

DE WEVER, P., GEYSSANT, J.R., AZÉMA, J., DEVOS, I., DUÉE, G., MANIVIT, H. & VRIELYNCK, B. 1986: La coupe de Santa Anna (zone de Sciacca, Sicile): une synthèse biostratigraphique des apports des macro-, micro- et nannofosiles du Jurassique supérieur et Crétacé inférieur. Revue de Micropaléontologie, 29 (3), 141–186.

DE WEVER, P., DUMITRICA, P., CAULET, J.P., NIGRINI, C. & CARIDROIT, M. 2001: Radiolarians in the sedimentary record. 533 p, Gordon & Breach Science Publ..

DISERENS, M.-O., BAUMGARTNER, P.O. & DUMITRICA, P. 2003: Age determination of late Cretaceous radiolarites in orogenic environments: an example from accreted teeranes of southern Costa Rica. Interrad X 2003, Abstracts & Programme, University of Lausanne, Switzerland, 49–50.

DUMITRICA, P. 1975: Cenomanian Radiolaria at Podul Dimbovitei, Micropaleontological guide to the Mesozoic and Tertiary of the Romanian Carpathians. In: 14th European Micropaleontological Colloquium, Romania, 87–89. Institute of Geology and Geophysics, Bucarest.

FOREMAN, H.P. 1973: Radiolaria from DSDP Leg 20. In: HEEZEN, B.C. et al. (Eds.): Initial Reports of the Deep Sea Drilling Project, 20, 249–305. U.S. Government Printing Office, Washington, D.C.

FOREMAN, H.P. 1975: Radiolaria from the North Pacific, Deep Sea Drilling Project, Leg 32. In: LARSON, R.L. et al. (Eds.): Initial Reports of the Deep Sea Drilling Project, 32 (Hakodate, Japan to Honolulu, Hawaii, Aug.-Oct., 1973), 579–676. U.S. Government Printing Office, Washington, D.C.

FOREMAN, H.P. 1977: Mesozoic Radiolaria from the Atlantic Basin and its Borderlands. In: F.M. SWAIN (Ed.): Stratigraphic Micropaleontology of Atlantic Basin and Borderlands, 305–313. Elsevier Scientific Publishing Company, Amsterdam, Netherlands.

GIRAULT, F.E. 2005: Sédimentologie et tectonostratigraphie de Deprano et sa région (Argolide – Grèce). Unpublished diploma thesis, University of Lausanne.

GLASSEY, T. 2005: Etude sédimentologique et tectonique d'un secteur de la région de Midhéa, en Argolide septentrionale (Grèce). Unpublished diploma thesis, University of Lausanne.

GUEX, J. 1977: Une nouvelle méthode d'analyse biochronologique. Bulletin de la Société Vaudoise des Sciences Naturelles, 73, 170–216

GUEX, J. 1979: Terminologie et méthodes de la biostratigraphie moderne: commentaires critiques et propositions. Bulletin de la Société Vaudoise des Sciences Naturelles, 74, 170–216.

GUEX, J. 1980: Calcul, caractérisation et identification des associations unitaires en biochronologie. Bulletin de Société Vaudoise des Sciences Naturelles, 75, 111–126.

GUEX, J. 1981: Associations virtuelles et discontinuités dans la distribution des espèces fossiles: un exemple intéressant. Bulletin de la Société Vaudoise des Sciences Naturelles, 75, 179–197.

GUEX, J. 1984: Estimation numériques de la qualité de l'enregistrement fossile des espèces. Bulletin de la Société Vaudoise des Sciences Naturelles, 77, 79–89.

GUEX, J. 1987: Corrélations biochronologiques et associations unitaires. 244p., Presses Polytechniques Romandes, Lausanne.

GUEX, J. 1988: Utilisation des horizons maximaux résiduels en biochronologie. Bulletin de la Société Vaudoise des Sciences Naturelles, 79 (2), 135–142.

GUEX, J. 1991: Biochronological Correlations. Springer-Verlag, Berlin/Heidelberg/New York. 250 p.

GUEX, J. & DAVAUD, E. 1982: Recherche automatique des associations unitaires: option nouvelle et exemple d'application. Bulletin de la Société Vaudoise des Sciences Naturelles, 78 (2), 195–205.

GUEX, J. & DAVAUD, E. 1984: Unitary Associations Method: use of graph theory and computer algorithm. Computers & Geosciences, 10 (1), 69–96.

JACKETT, S.-J., DISERENS, M.-O & BAUMGARTNER, P.O. 2002: Late Cretaceous To Early Cenozoic Radiolarian Biochronology Of Low-Latitude Orogenic Regions. Problems And Solutions. Abstracts and Proceedings of the Norwegian Geological Society, 1, 63.

NEUMANN, P. 2003: Ablagerungsprozesse, Event- und Biostratigraphie kreidezeitlicher Tiefwassersediemente der Tethys in der Olonos-Pindos-Zone Westgriechenlands. Müncher Geowiss. Abh., 40, 1–156.

O'DOGHERTY, L. 1994: Biochronology and Paleontology of Mid-Cretaceous Radiolarians from Northern Apennines (Italy) and Betic Cordillera (Spain). Mémoires de Géologie (Lausanne), 21. 415p. Institut de Géologie et de Paléontologie, Université de Lausanne.

PESSAGNO, E.A. 1971: Jurassic and Cretaceous Hagiastridae from the Blake-Bahama Basin (Site 5A, JOIDES Leg 1) and the Great Valley Sequence, California Coast Ranges. Bulletins of American Paleontology, 60(264), 5–83.

PESSAGNO, E.A. 1976: Radiolarian zonation and stratigraphy of the Upper Cretaceous portion of the Great Valley Sequence, California Coast Ranges. Micropaleontology spec. Publ., 2, 1–95.

RIEDEL, W.R. & SANFILIPPO, A. 1974: Radiolaria from the southern Indian Ocean, DSDP Leg 26. In: DAVIES, T.A., et al. (Eds.): Initial Reports of the Deep Sea Drilling Project, 26 (Durban, South Africa to Fremantle, Australia; Sept.-Oct. 1972), 771–781. U.S. Government Printing Office, Washington, D.C.

SANFILIPPO, A. & RIEDEL, W.R. 1985: Cretaceous Radiolaria. In: BOLLI, H.M., SAUDERS, J.B. & PERCH-NIELSEN, K. (Eds.): Plankton stratigraphy, 631–712. Cambridge University Press, Cambridge/ New York.

SAVARY, J. & GUEX, J. 1991: BioGraph: un nouveau programme de construction des corrélations biochronologiques basées sur les associations unitaires. Bulletin de la Société Vaudoise des Sciences Naturelles, 80 (3), 317–340.

SCHAAF, A. 1985: Un nouveau canevas biochronologique du Crétacé inférieur et moyen: les biozones à radiolaires. Sci. géol. (Strasbourg) Bull., 38(3), 227–269.

SQUINABOL, S. 1903: Radiolarie fossile di Teolo (Euganei). Atti e memorie del'Academia di scienze, lettere ed arti. Padova, new series, 19, 127–130.

SQUINABOL, S. 1904: Radiolarie cretacee degli Euganei. Atti e memorie del'Academia di scienze, lettere ed arti. Padova, new series, 20, 171–244.

SQUINABOL, S. 1914: Contributo alla conoscenza dei Radiolarii fossili del Veneto. Appendice – Di un genera di Radiolari caratteristico del Secondario (Contribution to the knowledge of fossil Radiolaria. Appendix – On a genus of Radiolaria characteristic of the Mesozoic). Memorie dell'Istituto geologico della R. Universita di Padova, 2, 249–306.

TAKETANI, Y. 1982: Cretaceous radiolarian biostratigraphy of the Urakawa and Obira areas, Hokkaido. Sci. Rep. Tohoku Univ. Series 2: Geology = Tohoku Daigaku Rika Hokoku. Dai 2: Shu Chishitsugaku, 52(1–2), 1–76.

THIÉBAULT, F., DE WEVER, P., FLEURY, J.J. & BASSOULET, J.P. 1981: Précisions sur la série stratigraphique de la nappe du Pinde-Olonos de la presqu'île de Koroni (Péloponnèse méridional – Grèce): l'âge des Radiolarites – (Dogger – Crétacé supérieur). Ann. Soc. Géol. Nord, 100, 91–105.

THUROW, J. 1988: Cretaceous radiolarians of the North Atlantic Ocean; ODP Leg 103 (sites 638, 640, and 641) and DSDP legs 93 (Site 603) and 47B (Site 398). In: BOILLOT, G., WINTERER, E.L., (Eds.): Proceedings of the Ocean Drilling Program, Scientific results (Galicia margin; covering Leg 103 of the cruises of the drilling vessel JOIDES Resolution, Ponta Delgada, Azores, to Bremerhaven, Germany, 25 April 1985–19 June 1985), 103, 379–418. Texas A. & M. University, Ocean Drilling Program, College Station, TX, United States.

VERNEZ, G. 1990: Etude géologique et minéralogique de la Vallée de Paliominos entre Karnezeika et Stavropodhi. Unpublished diploma thesis, University of Lausanne.

VISHNEVSKAYA, V. S. 2001: Jurassic to Cretaceous radiolarian biostratigraphy of Russia. GEOS, Moscow, 376 p.

VRIELYNCK, B. 1981: Evolution paléogéographique et structurale de la presqu'île d'Argolide (Grèce). Revue de géologie dynamique et de géographie physique, 23(4), 277–288.

Manuscript received January 2004
Revision accepted February 2005

Plate 1

SEM-illustrations of Upper Cretaceous radiolarians from Karnezeika, Argolis Peninsula (Greece)

Figures 1–2	***Acaeniotyle rebellis*** O'DOGHERTY 1994 Al68_06 (Figs. 1 and 2)
Figure 3	***Acaeniotyle* sp. *A*** Al74_300
Figure 4	***Acaeniotyle* sp. *B*** Al70_090
Figures 5–6	***Archaeocenosphaera* (?) *mellifera*** O'DOGHERTY 1994 Al72_190 (Figs. 5 and 6)
Figures 7–8	***Archaeocenosphaera* (?) sp.** Al73_320 (Figs. 7 and 8)
Figures 9–10	***Triactoma cellulosa*** FOREMAN 1973 Al74_300 (Figs. 9 and 10)
Figures 11–12	***Triactoma hexeris*** O'DOGHERTY 1994 Al73_320 (Figs. 11 and 12)
Figure 13	***Triactoma* sp. aff. *T. hexeris*** O'DOGHERTY 1994 Al72_190
Figure 14	***Pseudoacanthosphaera galeata*** O'DOGHERTY 1994 Al73_320
Figure 15	***Pseudoacanthosphaera superba*** (SQUINABOL 1904) Al74_300
Figure 16	***Pseudoacanthosphaera* sp. aff. *P. spinosissima*** (SQUINABOL 1904) Al73_320
Figure 17	***Pseudoacanthosphaera* (?) sp.** Al73_320
Figure 18	***Tetracanthellipsis euganeus*** SQUINABOL 1903 Al73_320
Figures 19–20	***Praeconocaryomma universa*** PESSAGNO 1976 Al72_190 (Figs. 19 and 20)
Figures 21–22	***Praeconocaryomma californiaensis*** PESSAGNO 1976 Al72_190 (Figs. 21 and 22)
Figures 23–24	***Praeconocaryomma lipmanae*** PESSAGNO 1976 Al74_300 (Figs. 23 and 24)
Figures 25–26	***Praeconocaryomma* sp.** Al74_300 (Fig.25 and 26)
Figures 27–28	***Crucella messinae*** PESSAGNO 1971 Al68_060 (Fig. 27), Al72_190 (Fig. 28)
Figures 29–30	***Crucella cachensis*** PESSAGNO 1971 Al74_300 (Fig. 29), Al73_320 (Fig. 30)
Figure 31	***Halesium triacanthum*** (SQUINABOL 1903) **sensu** O'DOGHERTY 1994 Al68_060
Figures 32–33	***Halesium* sp.** Al68_060 (Fig. 32), Al73_320 (Fig. 33)
Figures 34	***Pessagnobrachia* sp.** Al74_300

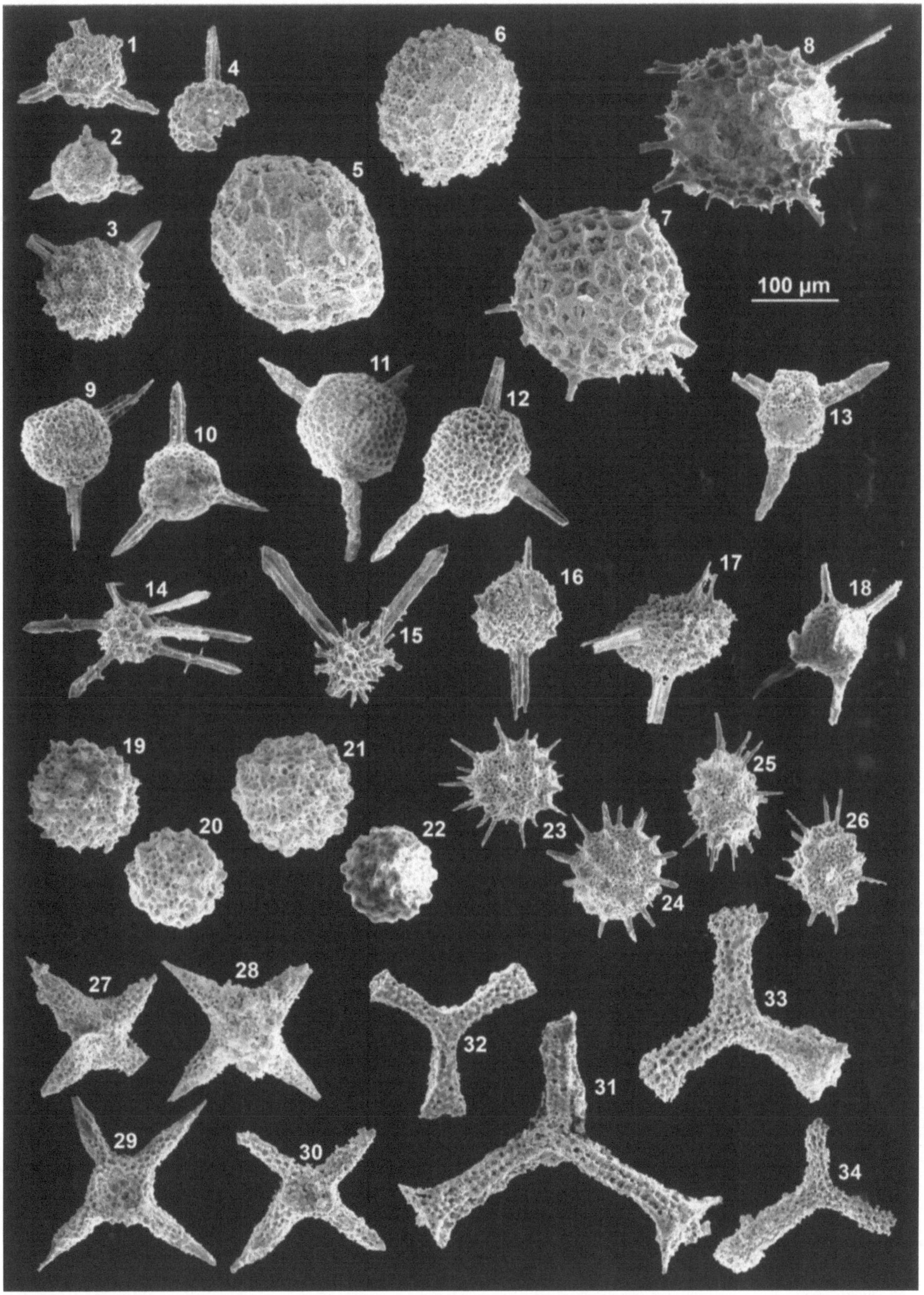

1
2
3
4
5
6
7
8
100 µm
9
10
11
12
13
14
15
16
17
18
19
20
21
22
23
24
25
26
27
28
29
30
31
32
33
34

Plate 2

SEM-illustrations of Upper Cretaceous radiolarians from Karnezeika, Argolis Peninsula (Greece)

Figures 1–2 ***Dactyliodiscus* sp.** Al72_190 (Figs. 1 and 2)
Figures 3–4 ***Pseudoaulophacus sculptus*** (SQUINABOL 1904) **sensu** O'DOGHERTY 1994 Al72_190 (Figs. 3 and 4)
Figures 5–6 ***Pseudoaulophacus putahensis*** PESSAGNO 1972 Al73_320 (Figs. 5 and 6)
Figures 7–8 ***Patellula helios*** (SQUINABOL 1903) **sensu** O'DOGHERTY 1994 Al74_300 (Figs. 7 and 8)
Figures 9–10 ***Patellula ecliptica*** O'DOGHERTY 1994 Al73_320 (Figs. 9 and 10)
Figures 11–12 ***Patellula heroica*** O'DOGHERTY 1994 Al74_300 (Figs. 11 and 12)
Figure 13 ***Patellula* sp.** Al72_190
Figures 14–15 ***Acanthocircus venetus*** (SQUINABOL 1914) **sensu** O'DOGHERTY 1994 Al72_190 (Fig. 14), Al73_320 (Fig. 15)
Figures 16–17 ***Acanthocircus tympanum*** O'DOGHERTY 1994 Al74_300 (Fig. 16), Al73_320 (Fig. 17)
Figures 18–19 ***Acanthocircus hueyi*** (PESSAGNO 1976) **sensu** O'DOGHERTY 1994 Al74_300 (Fig. 18), Al73_320 (Fig. 19)
Figures 20–21 ***Dictyomitra formosa*** SQUINABOL 1904 Al72_190 (Fig. 20), Al73_320 (Fig. 21)
Figure 22 ***Dictyomitra* sp. cf. *D. formosa*** SQUINABOL 1904 Al72_190
Figures 23–24 ***Dictyomitra montisserei*** (SQUINABOL 1903) **sensu** O'DOGHERTY 1994 Al68_060 (Fig. 23), Al72_190 (Fig. 24)
Figure 25 ***Dictyomitra urakawensis*** TAKETANI 1982 Al72_190
Figure 26 ***Torculum coronatum*** (SQUINABOL 1904) Al72_190
Figures 27–28 ***Pseudodictyomitra pseudomacrocephala*** (SQUINABOL 1903) Al73_320 (Figs. 27 and 28)
Figures 29–30 ***Stichomitra communis*** SQUINABOL 1903 Al73_320 (Figs. 29 and 30)
Figures 31–32 ***Stichomitra stocki*** (CAMPBELL & CLARK 1944) **sensu** O'DOGHERTY 1994 Al68_060 (Fig. 31), Al73_320 (Fig. 32)
Figure 33 ***Afens liriodes*** RIEDEL & SANFILIPPO 1974 Al74_300

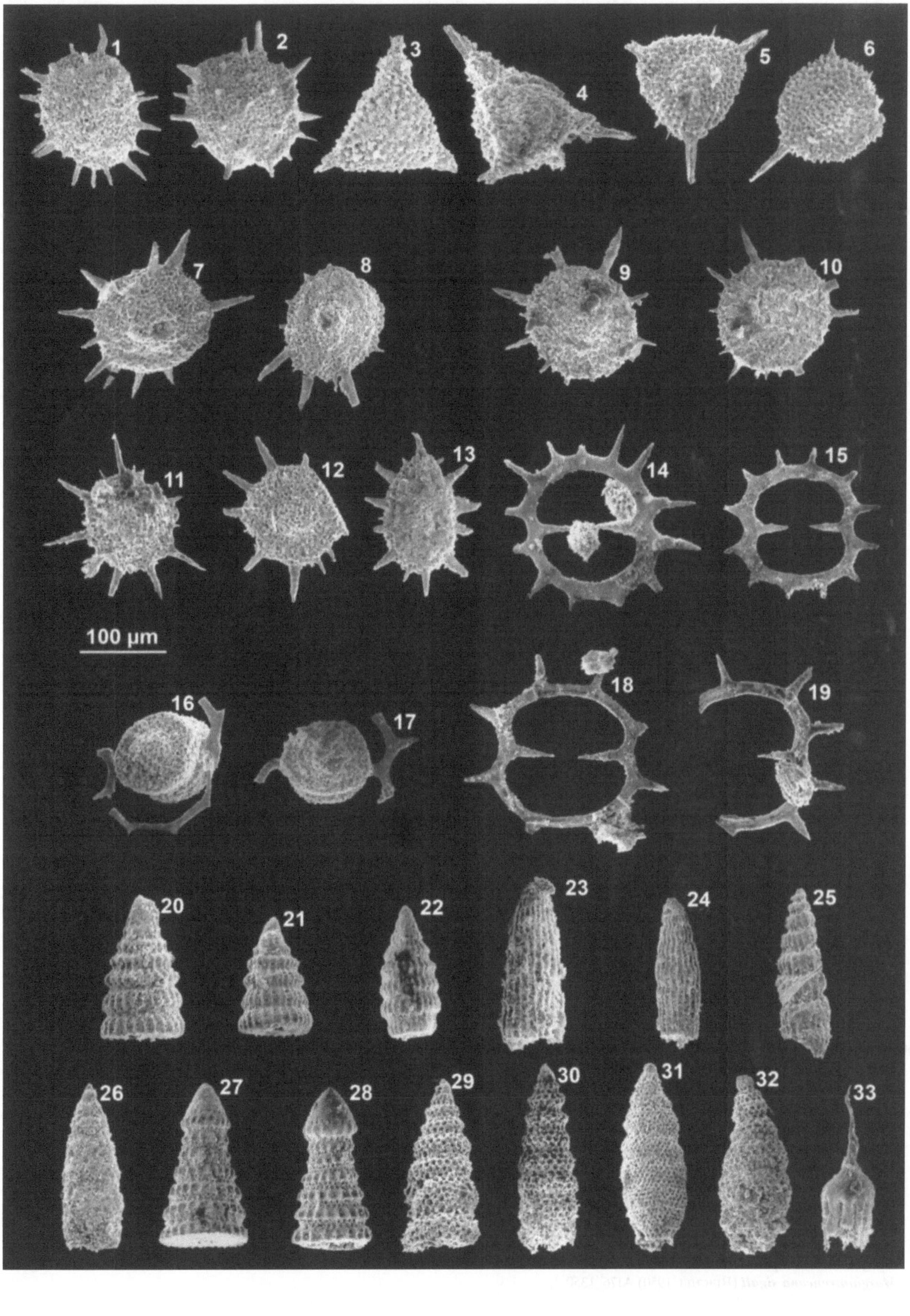
1
2
3
4
5
6
7
8
9
10
11
12
13
14
15
100 µm
16
17
18
19
20
21
22
23
24
25
26
27
28
29
30
31
32
33

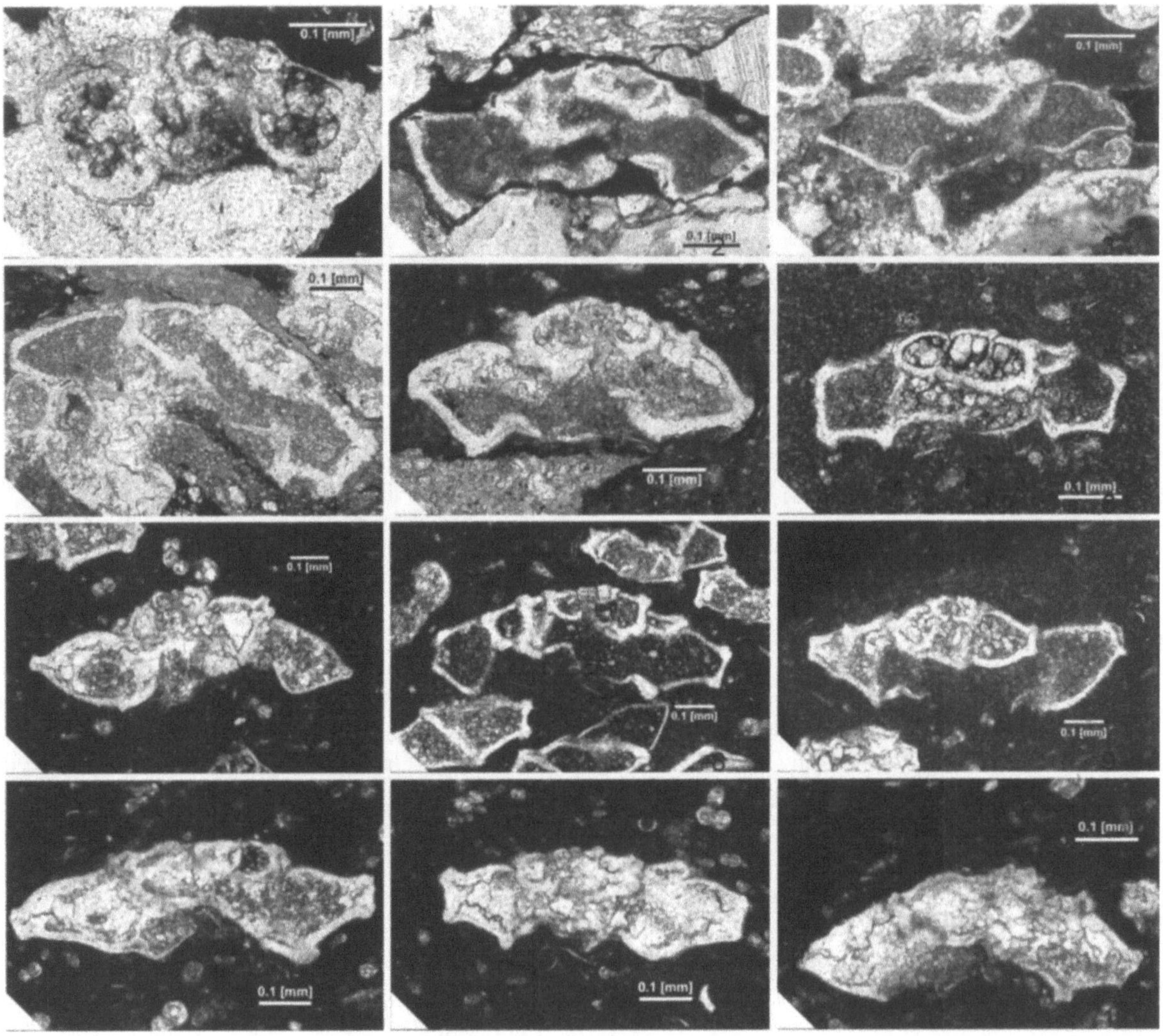

Plate 3

Optical microscope illustrations of Upper Cretaceous planktonic foraminfers from Karnezeika, Argolis Peninsula (Greece)

Figure 1 ***Helvetoglobotruncana helvetica*** (Bolli 1945) CKAR2
Figure 2 ***Marginotruncana marianosi*** (Douglas 1969) CKAR7
Figure 3 ***Marginotruncana schneegansi*** Sigal 1952 CKAR13
Figure 4 ***Marginotruncana coronata*** (Bolli 1945) CKAR13
Figure 5 ***Marginotruncana sigali*** (Reichel 1950) CKAR13
Figure 6 ***Marginotruncana pseudolinneiana*** Pessagno 1967 CKAR14
Figure 7 ***Marginotruncana schneegansi*** Sigal 1952 CKAR14
Figure 8 ***Marginotruncana sinuosa*** Porthault 1970 CKAR14
Figure 9 ***Dicarinella primitiva*** (Dalbiez 1955) CKAR14
Figure 10 ***Marginotruncana pseudolinneiana*** Pessagno 1967 Al76_1350
Figure 11 ***Marginotruncana pseudolinneiana*** Pessagno 1967 Al76_1350
Figure 12 ***Marginotruncana sigali*** (Reichel 1950) Al76_1350

0012-9402/06/01S021-13
DOI 10.1007/s00015-006-0607-0
Birkhäuser Verlag, Basel, 2006

Eclogae geol. Helv. 99 (2006) Supplement 1, S21–S33

Radiolarian correlation of Jurassic siliceous successions of the Rosso Ammonitico Formation in the Southern Alps and Western Sicily (Italy)

PAOLA BECCARO

Key words: Radiolarians, Ammonites, Biostratigraphy, Unitary Associations Zones, Rosso Ammonitico Medio, Middle-Late Jurassic, Italy
Parole chiave: Radiolari, Ammoniti, Biostratigrafia, Zone ad Associazioni Unitarie, Rosso Ammonitico Medio, Giurassico medio-superiore, Italia

ABSTRACT

This paper deals with the radiolarian biostratigraphy of Middle-Upper Jurassic pelagic siliceous successions of the Southern Alps and Western Sicily (Italy). The crucial complement to this research is the occurrence of ammonites in the studied successions (Rosso Ammonitico Medio: the intermediate siliceous member of the Rosso Ammonitico Formation), as well as in the under- and overlying sediments (Rosso Ammonitico Inferiore and Rosso Ammonitico Superiore, respectively). The abundance of radiolarians in all successions allow to analyse them for a twofold purpose: to date directly most of the successions, and to improve the calibration of radiolarian biozones thanks to the occurrence of ammonites. The biostratigraphic analysis has been carried out using the Unitary Associations method, and six new radiolarian biozones have been defined: the combined occurrence of radiolarians and ammonites provided a new Bathonian to late Kimmeridgian radiolarian zonation for the Southern Alps and Western Sicily (Italy). The new Unitary Association Zones show a good reproducibility throughout the investigated successions, and make possible a first direct dating and correlation by radiolarians of the Rosso Ammonitico Medio. Furthermore, the radiolarian biozones reveal a significant diachronism for both the lower and the upper limit of the Jurassic pelagic siliceous facies in the Alpine and Sicilian sections. In the light of new radiolarian biozones, the age assignments of the Ceniga (Southern Alps) and the Sant'Anna (Sicily) sections, and the ranges of some taxa are discussed.

RIASSUNTO

La ricerca riguarda la biostratigrafia a radiolari di successioni pelagiche silicee riferibili al Giurassico Medio-Superiore delle Alpi Meridionali e della Sicilia occidentale (Italia). L'aspetto più importante è la presenza di ammoniti sia nelle successioni studiate (Rosso Ammonitico Medio: porzione intermedia silicea della Formazione del Rosso Ammonitico) sia in quelle sotto- e sovrastanti (rispettivamente Rosso Ammonitico Inferiore e Rosso Ammonitico Superiore). L'abbondanza di radiolari in tutte le sezioni studiate ha permesso di analizzarli sia per datare direttamente la maggior parte delle successioni sia per migliorare la calibrazione delle biozone a radiolari grazie alla presenza delle ammoniti. L'analisi biostratigrafica è stata effettuata con il metodo delle Associazioni Unitarie e sono state definite sei nuove biozone a radiolari: la copresenza di radiolari ed ammoniti ha così fornito una nuova zonazione a radiolari per l'intervallo di tempo Batoniano-Kimmeridgiano delle Alpi Meridionali e della Sicilia occidentale (Italia). Le Zone ad Associazioni Unitarie identificate mostrano una buona riproducibilità tra le sezioni stratigrafiche analizzate ed hanno permesso una prima datazione e correlazione diretta a radiolari del Rosso Ammonitico Medio. Le biozone a radiolari hanno inoltre rivelato un significativo diacronismo sia per il limite inferiore sia per il limite superiore delle facies pelagiche silicee nelle sezioni alpine e siciliane. Alla luce dei nuovi dati biostratigrafici forniti dai radiolari sono infine discusse le età della facies silicea nelle sezioni di Ceniga (Alpi Meridionali) e di Sant'Anna (Sicilia) e la distribuzione stratigrafica di alcuni taxa.

Introduction

The study of siliceous successions in the Southern Alps and in Western Sicily (Italy) benefits from the fact that such successions are approximately coeval and may be referred to analogous paleogeographic settings. These similarities enable to compare radiolarian assemblages of different geographic areas and to correlate different Tethyan paleogeographic regions. The paleogeographic domains are two pelagic plateaux (Trapanese Domain in Western Sicily and Trento Plateau in the Southern Alps) and one basin (Sicanian Domain in Western Sicily) (Figs. 1 and 2). The age of the studied successions ranges from the Bathonian to the Kimmeridgian.

The Sicilian chain and the Southern Alps represent segments of the Alpine collisional belt formed during the "Tertiary" time along the boundary between the European and African plates. The Jurassic paleogeographic settings of Sicily and the Alps result from the breakup of the Triassic carbonate platform and the subsequent extensional movements. The platform was divided in several areas with different subsidence rates, and different paleogeographic domains were formed (pelagic plateaux, basins and limited areas of platforms) (Figs. 1 and 2). During the Middle Jurassic a general deepening took place and the deposition of sediments in the Rosso Ammonitico facies occurred on the pelagic plateaux up to Late Jurassic

[1] Dipartimento di Scienze della Terra, Via Valperga Caluso 35, 10125 Torino, Italy. Email: paola.beccaro@unito.it

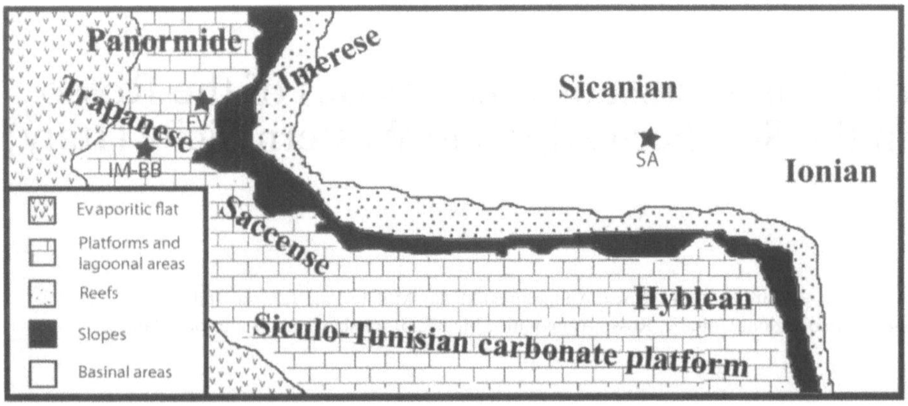

Fig. 1. An attempt of paleogeographic reconstruction of the Sicilian area at the Late Triassic time (Catalano et al. 1996). Favignana, Balata di Baida and Inici Mt. sections belong to the pelagic plateau of the Trapanese Domain; Sant'Anna section belongs to the Sicanian Basin. The paleogeographic location of the stratigraphic sections is only indicative.

(Winterer & Bosellini 1981; Catalano et al. 1996). The Rosso Ammonitico sediments consist of condensed red nodular limestone rich in ammonites, and are subdivided in three members whose ages differ between the Southern Alps and Sicily. The lower part (RAI: Rosso Ammonitico Inferiore) spans the late Bajocian-early Callovian in the Alps and the Bathonian-mid Oxfordian in Sicily. The upper part (RAS: Rosso Ammonitico Superiore) starts in the mid Oxfordian in the Alps and in the early Kimmeridgian in Sicily. RAI and RAS are separated by the intermediate siliceous member (RAM: Rosso Ammonitico Medio), assigned to upper Callovian-mid Oxfordian in the Alps and mid Oxfordian-Kimmeridgian in Sicily. The intermediate siliceous member represents the studied sediments in all the stratigraphic sections except for Sant'Anna (Sicily).

Very few authors have studied the Jurassic radiolarian assemblages of the Southern Alps and Western Sicily. Kocher (1981) was the first author who studied radiolarians from the Southern Alps but mainly in the Lombardian basin. The Ceniga section was first described by Fogelgesang (1975), and then studied for radiolarians by Baumgartner (1984) and Baumgartner et al. (1995b). The Coston delle Vette section was first described by Dal Piaz (1907), then analysed by Bosellini & Dal Cin (1968) and Della Bruna & Martire (1985), and finally studied for radiolarians by Beccaro (1998) and Beccaro et al. (2002). The Cava Vianini was only described by Papa (1994) for sedimentologic purposes. Concerning Western Sicily, several authors studied the geology for different aims (Giunta & Liguori 1972, 1973; Wendt 1964, 1971; Catalano et al. 1989; Cecca et al. 2001; Catalano et al. 2002 among others). Nevertheless, the radiolarian papers regarding the Sicanian Basin have been mainly focused on the Sant'Anna section (Riedel & Sanfilippo 1974; Mascle 1973, 1979; Baumgartner et al. 1980; Origlia-Devos 1983; Baumgartner 1984; De Wever et al. 1986; Aita 1987; De Wever 1995) because of its rich fossil content (ammonites, belemnites, brachiopods, echinoderms, nannofossils). Other Sicanian sections were studied by Kito et al. (1990) and Kito & De Wever (1992, 1994).

Generally, the close association of ammonite-bearing beds with levels containing well-preserved radiolarians is quite rare. In the investigated sections, the ammonites occur both in the under- and overlying sediments and, in some instances, also within the siliceous successions: This fact motivates the analysis of radiolarian assemblages in the selected sections. The abundance of radiolarians enables to study them for a twofold purpose: to date directly most of the siliceous facies (whose age was generally based on the ages of the bracketing formations) and to improve the calibration of radiolarian zones by ammonite zones.

This paper deals with the main results of the author's PhD research (Beccaro 2002) where radiolarian assemblages were studied for the first time in the following sections: Fornazzo Strada, Fornazzo Cava, Castello Inici, Balata di Baida, Favignana (North-western Sicily) (Beccaro 2002, 2004a) and Cava Vianini (Southern Alps) (Beccaro 2002). Concerning the Sant'Anna (South-western Sicily) and the Ceniga (Southern Alps) sections, new radiolarian data improved the former biochronologic assignments. New species of Nassellaria (*Fultacapsa ozvoldovae, Podobursa andreai, Podobursa vannae, Loopus doliolum martae*) and Spumellaria (*Emiluvia peteri, Triactoma enzoi*) from the cited sections have been described in Beccaro 2004b.

Lithologic description of the stratigraphic sections

In Western Sicily five stratigraphic sections of the Rosso Ammonitico Medio (RAM) have been studied in the Trapanese Domain (pelagic plateau) and one section in the Sicanian Basin (Fig. 1). The geographic location of these sections is shown in Figure 3.

Trapanese Domain (North-western Sicily) – Fornazzo sections are located at the Inici Mt. near Castellammare del Golfo. The **Fornazzo Strada** section crops out along the road to Fornazzo quarry. The RAM is 26 m thick and consists of well-stratified red siliceous limestone alternating with calcareous marlstone. Nodules and beds of dark red chert are abundant (Fig. 4). The **Fornazzo Cava** section crops out in the old Fornazzo quarry, and the **Castello Inici** section is located in the southwestern side of the Inici Mt. In both sections the RAM is incomplete and it is 8 m and 13 m thick, respectively. The lithology is the same as at Fornazzo Strada and the main

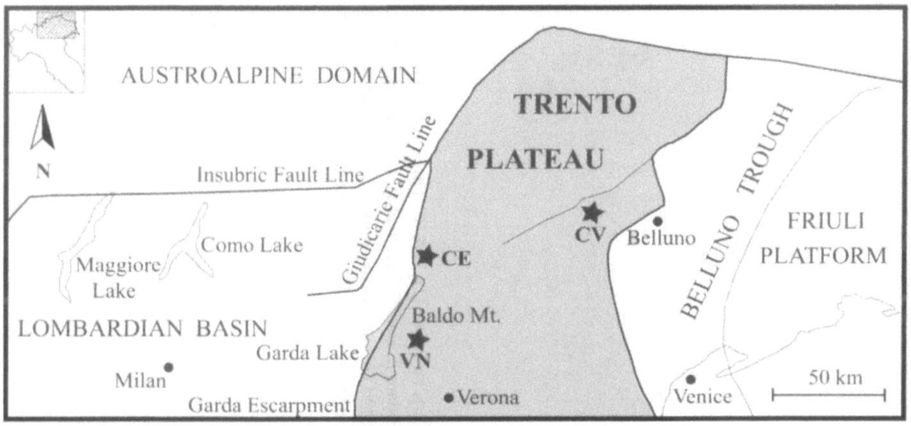

Fig. 2. Geographic location of the stratigraphic sections and present day distribution of Mesozoic paleogeographic domains in the Southern Alps (Italy) (Bosellini et al. 1981; Martire 1992). CE: Ceniga, CV: Coston delle Vette, VN: Cava Vianini.

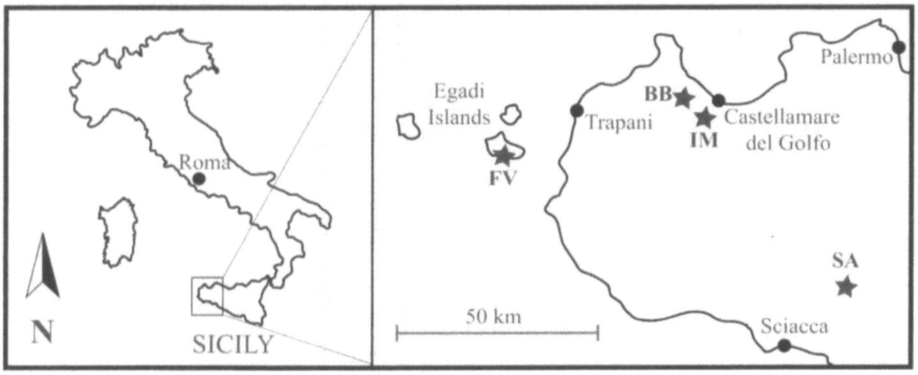

Fig. 3. Geographic location of the stratigraphic sections in Western Sicily (Italy). BB: Balata di Baida, FV: Favignana, IM refers to the three sections of the Inici Mt. (Fornazzo Strada, Fornazzo Cava and Castello Inici), SA: Sant'Anna.

difference is the occurrence of macrofossils (ammonites and belemnites; Fig. 4). The **Balata di Baida** section crops out along the orographic left of the Sarcona River at Balata di Baida village (near Inici Mt.). The RAM is 21 m thick, and it consists of an alternation of nodular red limestone and variously coloured chert beds; thin levels of marlstone are widespread along the section (Fig. 4). The radiolarian preservation in the Inici sections is moderate. The **Favignana** succession is located in the southern coast of the Favignana island (Egadi Archipelago). The section is 2 m thick, and it consists of thin limestone and marlstone alternating with thin chert beds black and red in colour (Fig. 4). Neither the base nor the top of the siliceous succession crop out. The radiolarian preservation is very good.

Sicanian Domain (South-western Sicily) – The basinal succession of **Sant'Anna** crops out at about 1 km NE of Sant'Anna village (near Sciacca). The stratigraphic section is 9 m thick, and consists of a regular alternation of whitish limestone and marlstone; the chert is rare. The base is not exposed (Fig. 4). The radiolarian preservation is good.

In the Southern Alps three stratigraphic sections have been studied in the Trento Plateau (Fig. 2). The **Cava Vianini** section crops out in an active quarry close to Madonna della Corona Sanctuary (eastern side of the Garda Lake). The RAM is 10 m thick, and it consists of an alternation of dark siliceous limestone and beds chert (light brown, red and black in colour). Some levels of bentonites occur at the top of the section (Fig. 4). The radiolarian preservation varies from moderate to very good. The **Ceniga** section crops out near the Sarca River, south of Ceniga village (North of the Garda Lake). In this section it is also possible to observe the Early Jurassic platform deposits of San Vigilio Oolite (oolitic-bioclastic grainstone). The top of San Vigilio Oolite is overlain by a very thin RAI (20 cm) formed by pink pelagic limestone and bearing a hardground at its top. The RAM is 9 m thick, and it consists of an alternation of red siliceous limestone thinly stratified and thin whitish marlstone. Six levels of bentonites occur in the middle part of the section (Fig. 4). The radiolarian preservation is poor. The **Coston delle Vette** section is located in the Feltrine Alps (Dolomiti Bellunesi). Here, the time-equivalent facies of the RAM is the Fonzaso Formation, which reaches a thickness of 100 m. The Fonzaso Fm. consists of a quite regular alternation of packstone and wackestone with subordinate oolitic grainstone and mudstone, and chert ribbons and nodules (Fig. 4) (Della Bruna & Martire 1985; Beccaro 1998; Beccaro et al. 2002). The central part of the formation is characterized by the oolitic grainstone resedimented from the Friuli Platform. The radiolarian preservation varies from moderate to very good.

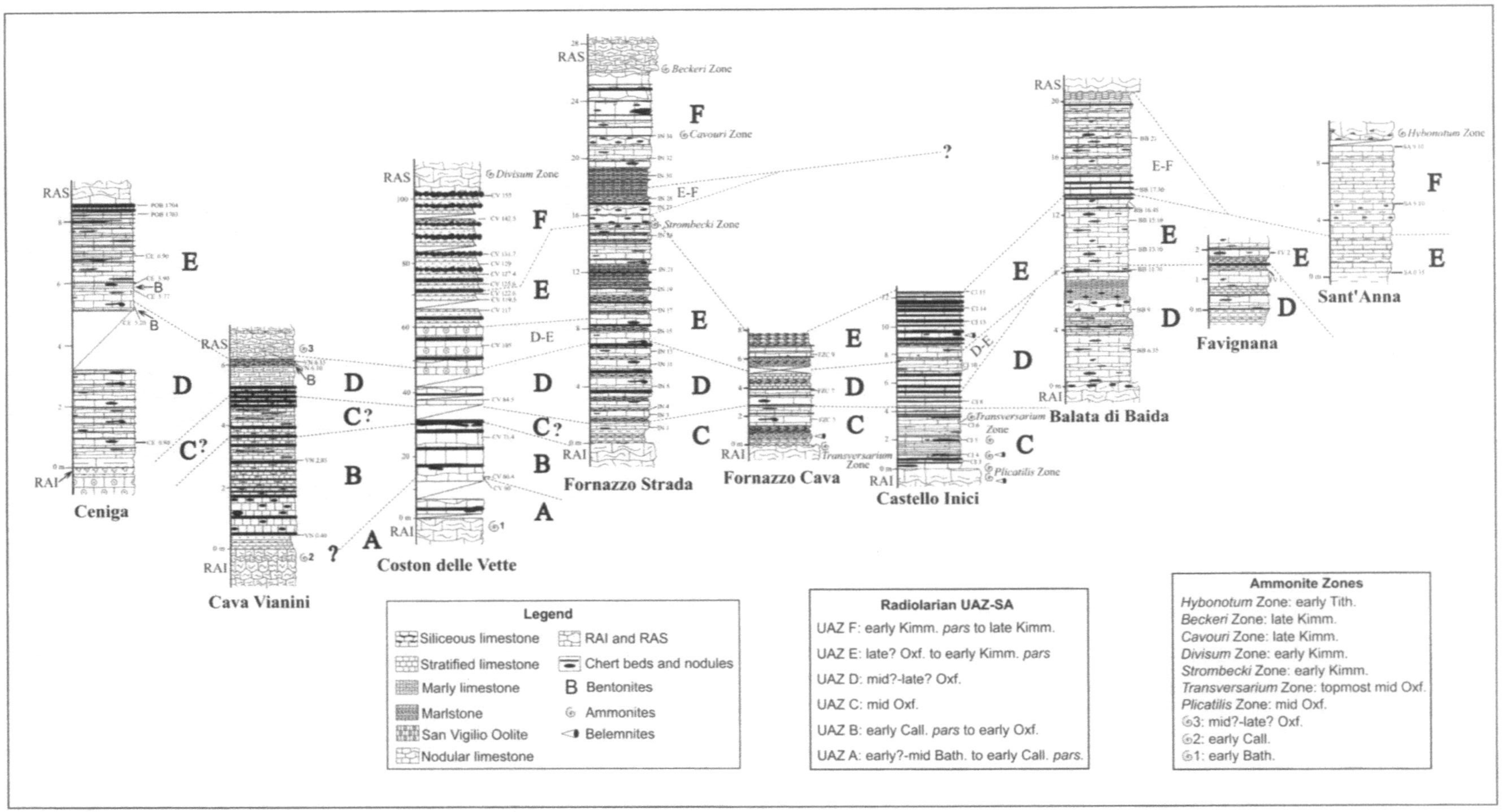

Fig. 4. Radiolarian biostratigraphic correlation of pelagic siliceous facies between Western Sicily and the Southern Alps by means of the six radiolarians biozones UAZ A–F. Three sections from the Southern Alps (Ceniga, Cava Vianini, Coston delle Vette), five sections from North-western Sicily (Fornazzo Strada, Fornazzo Cava, Castello Inici, Balata di Baida, Favignana), and one section from South-western Sicily (Sant'Anna) have been dated and correlated. The investigated successions represent the intermediate pelagic siliceous member (RAM) of the Rosso Ammonitico Fm. in all the sections, except at Coston delle Vette (where the Fonzaso Fm. is the time-equivalent siliceous facies of the RAM) and at Sant'Anna (where a basinal section crops out). The age of the biostratigraphic units UAZ A–F is provided by calibration with ammonite assemblages found both in the studied successions and in the under- and overlying sediments.

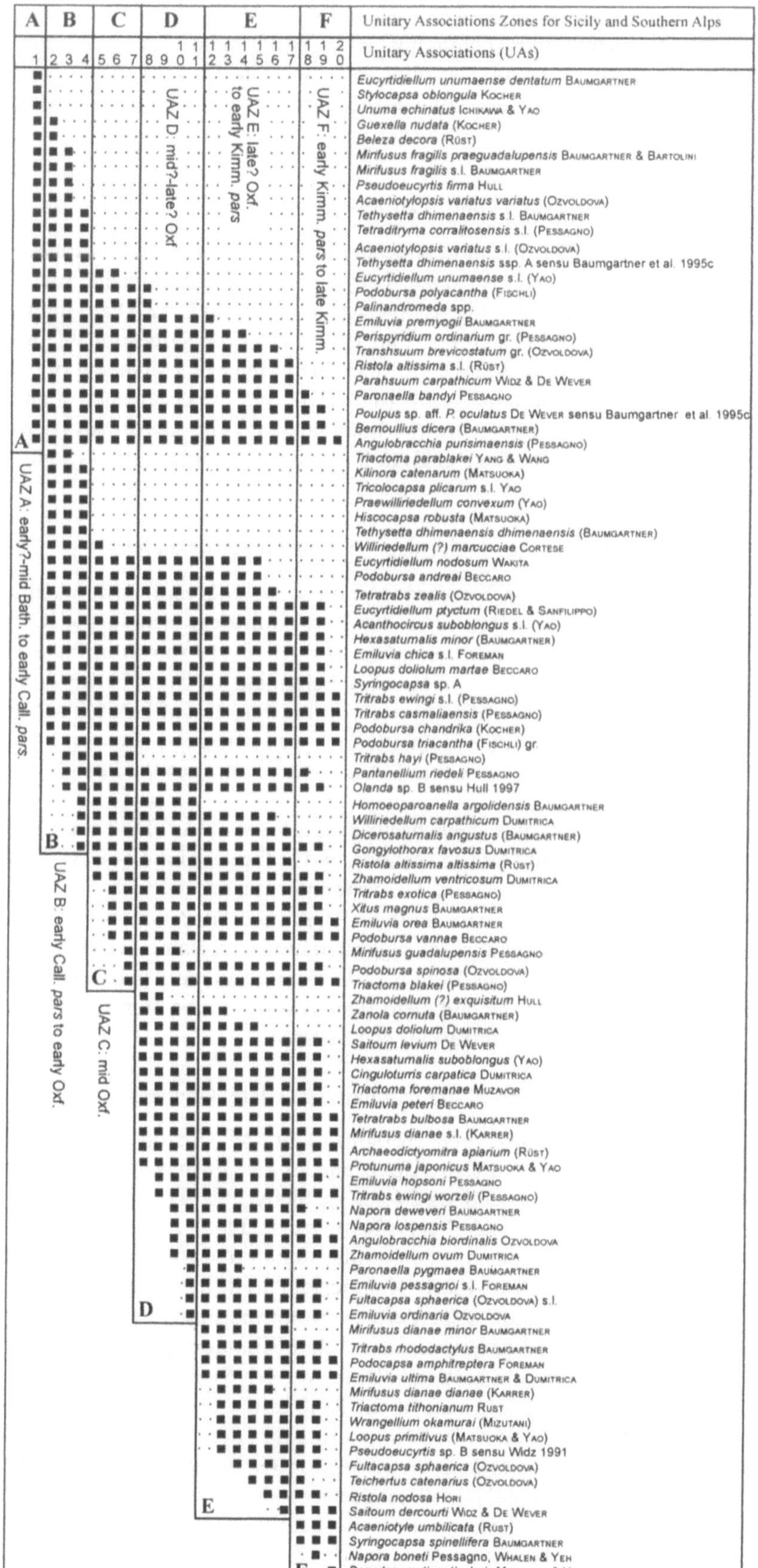

Fig. 5. Range chart: occurrences of radiolarian taxa used for the biozonation. The software BioGraph grouped 100 selected species in 20 Unitary Associations which have been manually assembled in six Unitary Associations Zones: UAZ A–F.

Radiolarian biozones UAZ A–F

The radiolarian biostratigraphy illustrated in this paper has been carried out using the Unitary Associations method (Guex 1977, 1991) and the software BioGraph (Savary & Guex 1991, 1999), and represents a new regional zonation for the Southern Alps and Western Sicily (Italy). About 130 radiolarian species have been identified from 8 stratigraphic sections, and 100 taxa have been retained to construct the zonation (Fig. 5). Also added to this database are the radiolarian occurrences of the Coston delle Vette section (Trento Plateau, Southern Alps) (Beccaro et al. 2002), and the raw data of two Baumgartner's samples (POB 1703 and POB 1704) of the Ceniga section (Trento Plateau, Southern Alps) (Baumgartner 1984; Baumgartner et al. 1995b). The adding of the Baumgartner's samples was necessary in order to get a radiolarian record also from the upper part of the Ceniga section that lacked in well preserved samples from PhD field work. Six Unitary Associations Zones (UAZ A–F) have been defined (Fig. 5), and the chronostratigraphic value of each UAZone is provided by calibration with the ammonites found in the RAM of the studied sections and in the under- and overlying sediments (RAI and RAS).

UAZ A (early?-mid Bathonian – early Callovian *pars*)

UAZ A occurs only at the Coston delle Vette section (Southern Alps; Fig. 4). Some species characteristic of UAZ A are illustrated in Plate 1, Figures 1–3. UAZ A is assigned to early?–mid Bathonian – early Callovian pars thanks to the presence of early Bathonian ammonites at the top of RAI at the Coston delle Vette section (Della Bruna & Martire 1985), and the early Callovian pars age assignment of the following UAZ B.

UAZ B (early Callovian *pars* – early Oxfordian)

UAZ B is present at the Cava Vianini and Coston delle Vette sections (Southern Alps; Fig. 4). Species belonging to UAZ B are shown in Plate 1, Figures 4–11. The base of UAZ B is dated as early Callovian *pars* due to the presence of early Callovian ammonites at the top of RAI (Papa 1994) at the Cava Vianini section. The upper part of UAZ B is constrained to early Oxfordian by the well-dated UAZ C.

UAZ C (mid Oxfordian)

UAZ C has been recognized in the Fornazzo Strada, Fornazzo Cava and Castello Inici sections (North-western Sicily; Fig. 4). Species belonging to UAZ C are illustrated in Plate 1, Figures 12–13 and 15–18. In the Alpine sections UAZ C seems missing but no evidence of stratigraphic gaps has been found. Most likely, UAZ C was not recognized at the Coston delle Vette section due to the cover, and in the Cava Vianini section due to the low frequency sampling. At the Ceniga section the time interval corresponding to UAZ C is probably included in the very condensed RAI (Fig. 4). UAZ C is assigned to mid Oxfordian by the occurrence of ammonite assemblages belonging to *Plicatilis* and *Transversarium* Zones at the Castello Inici section (Savary 2000), and *Transversarium* Zone at the Fornazzo Cava section (Bovero 2000).

UAZ D (mid?-late? Oxfordian)

UAZ D is the most widespread and best recorded biozone in all the sections except Sant'Anna (Sicanian Domain, South-western Sicily; Fig. 4). Species belonging to UAZ D are shown in Plate 1, Figures 14 and 19–24. UAZ D is questionably assigned to mid-late Oxfordian due to the age assignment of the well calibrated UAZ C (mid Oxf.) and the occurrence of ammonites referred to the boundary mid-late Oxfordian (Papa 1994) at the base of RAS in the Cava Vianini section (Southern Alps; Fig. 4).

UAZ E (late? Oxfordian – early Kimmeridgian *pars*)

UAZ E is well recorded in all sections except Cava Vianini (Southern Alps), where the time-equivalent facies is the RAS (Fig. 4). Species belonging to UAZ E are shown in Plate 1, Figures 25–31. Ammonites belonging to *Strombecki* Zone (early Kimmeridgian) occur at the very top of UAZ E at the Fornazzo Strada section (North-western Sicily) (Bovero 2000) so that it is likely that UAZ E also comprises the late Oxfordian.

Bentonite levels occur at the top of UAZ D at the Cava Vianini and Ceniga sections, and at the very base of UAZ E at the Ceniga section. The time intervals expressed by UAZ D (mid?-late? Oxfordian) and UAZ E (late? Oxfordian-early Kimmeridgian *pars*) suggest that the volcanic activity took place also in the late Oxfordian-early Kimmeridgian time, and not only in the mid Oxfordian as stated in the literature (Martire 1989; Baumgartner et al. 1995b).

UAZ F (early Kimmeridgian *pars* – late Kimmeridgian)

UAZ F has been recognized at the Coston delle Vette section (Southern Alps), and at the Fornazzo Strada and Sant'Anna sections (Western Sicily; Fig. 4). Species occurring into UAZ F are illustrated in Plate 1, Figures 32–35. UAZ F spans all the Kimmeridgian due to the presence of *Divisum* Zone (early Kimmeridgian) at the base of the RAS in the Coston delle Vette section (Southern Alps) (Dal Piaz 1907), and of *Cavouri* Zone (late Kimmeridgian) at the Fornazzo Strada section (North-western Sicily) (Bovero 2000). The extremely base of the RAS at the Fornazzo Strada section is dated as late Kimmeridgian (*Beckeri* Zone) (Bovero 2000). In the Sant'Anna section (South-western Sicily) the top of the pelagic siliceous succession is assigned to UAZ F, and is overlain by ammonite-bearing limestone belonging to the *Hybonotum* Zone (early Tithonian) (De Wever et al. 1986).

Diachronism of the Jurassic siliceous facies between the Southern Alps and Western Sicily

The stratigraphic correlation through UAZ A–F reveals a significant diachronism for the lower as well as for the upper limit of the pelagic siliceous facies in the Alpine and Sicilian sections.

In the Southern Alps (Trento Plateau) the siliceous deposition began in the early?-mid Bathonian-early Callovian *pars* (UAZ A) at the Coston delle Vette section, in the early Callovian *pars* – early Oxfordian (UAZ B) at the Cava Vianini, and in the mid?-late? Oxfordian (UAZ D) at the Ceniga section (Fig. 4). In North-western Sicily (Trapanese Domain) the siliceous deposition started in the mid-late Oxfordian: UAZ C (mid Oxf.) at the Fornazzo Strada, Fornazzo Cava and Castello Inici sections; UAZ D (mid?-late? Oxf.) at the Balata di Baida and Favignana sections. In the basinal section of Sant'Anna the onset of the siliceous facies is the youngest: UAZ E (late? Oxf.-early Kimm. *pars*) (Fig. 4). On the other hand, UAZ D (mid?-late? Oxf.) corresponds to the final phase of the siliceous deposition at the Cava Vianini section (Southern Alps), and UAZ E (late? Oxf.-early Kimm. *pars*) corresponds to the final phase of the siliceous deposition in the Ceniga section (Southern Alps). UAZ F (early Kimm. *pars*-late Kimm.) indicates the end of the siliceous facies at the Coston delle Vette (Southern Alps), Fornazzo Strada and Sant'Anna (Western Sicily). The diachronism of the ending of the siliceous facies in the Fornazzo Cava, Castello Inici and Favignana sections is only due to the incompleteness of the successions (Fig. 4).

As can be inferred by the above discussion, the diachronism of the siliceous facies occurs within the same paleogeographic domain as well (Fig. 4). The three Alpine sections were located in different sectors of the Trento Plateau, and the diachronism suggests that local topography and/or tectonic movements primarily controlled the onset of the siliceous deposition. Different bottom morphology and tectonics may have influenced different spatial and temporal distribution of the siliceous facies: the oldest onset and the youngest end took place at Coston delle Vette (located in a transitional area between Trento Plateau and Belluno Basin) while the youngest onset was at Ceniga (located near the strongly tectonic active Garda Escarpment). On the other hand, the Sicilian sections belonging to the Trapanese Domain were paleogeographically very close to each other, and the diachronism of the onset amongst them is minor (only Favignana section shows a rather younger onset but its base does not crop out).

Age assignment discussion on the siliceous facies at Ceniga (Southern Alps) and Sant'Anna (Sicily)

In the light of new radiolarian biozones UAZ A–F, the age assignments of the Ceniga (Trento Plateau, Southern Alps) and the Sant'Anna (Sicanian Domain, South-western Sicily) sections have been discussed.

Concerning the Ceniga section, the time interval stated in the previous papers (Baumgartner 1984; Baumgartner et al. 1995b) spans the mid Callovian to early Tithonian. The radiolarian assemblages of the new samples CE 0.80 and CE 5.20 belong to UAZ D (Fig. 4) and assign most of the section to mid?-late? Oxfordian. The radiolarian content of the new samples CE 5.77, CE 5.90 and CE 6.90 belong to UAZ E (Fig. 4), and refer the upper part of the section to late? Oxfordian-early Kimmeridgian *pars*. The last two samples (POB 1703 and POB 1704) are from Baumgartner's sampling and they were assigned to late Oxfordian-early Tithonian (Baumgartner 1984). The radiolarian assemblages of these samples belong now to UAZ E (late? Oxf.-early Kimm. *pars*). In account of this research, the age of the Rosso Ammonitico Medio at the Ceniga section is restricted to mid? Oxfordian-early Kimmeridgian *pars* (UAZ D–E) (Fig. 4). This age assignment highly differs from those stated for other sections in the Trento Plateau, where the RAM is referred to late Callovian – mid Oxfordian (e.g., Cava Vianini, this paper; Martire 1996).

Concerning the Sant'Anna section, several authors assigned it to a variety of ages: mid Oxfordian to mid Tithonian (Origlia-Devos 1983), mid Oxfordian to early Kimmeridgian (De Wever et al. 1986), mid-late Oxfordian to late Oxfordian-early Kimmeridgian (De Wever 1995). The radiolarian content of the new sample SA 0.35 belongs to UAZ E (Fig. 4) and constrains the first layers of the succession to late? Oxfordian-early Kimmeridgian *pars*. The radiolarian assemblages of the new samples SA 5.10 and SA 9.10 belong to UAZ F (Fig. 4) and date the middle and upper parts of the section to early Kimmeridgian *pars*-late Kimmeridgian. This age assignment is consistent with the early Tithonian age of the ammonite assemblages found at the base of the overlying nodular limestone (De Wever et al. 1986). The present study now restricts the age of the pelagic siliceous succession of the Sant'Anna section to the late? Oxfordian-late Kimmeridgian (UAZ E–F).

Age assignment discussion for some taxa

After the calibration of UAZ A–F through ammonite zones, the stratigraphic distribution of some taxa can be compared with those of Baumgartner et al. (1995a). In the following discussion the lettered UAZones (e.g., UAZ A) refer to the new biozones illustrated in the present paper; the numbered UAZones (e.g., UAZ 8–13) refer to the biozones of Baumgartner et al. (1995a). Some age considerations were illustrated also in Beccaro (2004a) concerning *Eucyrtidiellum unumaense* s.l. (Yao) and *Williriedellum (?) marcucciae* Cortese. The ranges of these species remained unchanged after the new zonation by UAZ A–F: *Eucyrtidiellum unumaense* s.l. (Yao) and *Williriedellum (?) marcucciae* Cortese extended to mid Oxfordian (and not only to early Oxfordian as the previous assignment of Baumgartner et al. 1995a).

Podobursa polyacantha (Fischli) (Pl. 1, Fig. 14)

Its range is assigned to UAZ 5–8 (latest Baj.-early Bath. to mid Call.-early Oxf.) but in the studied sections *P. polyacantha* reaches the very base of UAZ D (mid?-late? Oxf.) (Fig. 5).

Ristola altissima altissima (Rüst) and *Ristola altissima nodosa* Hori (Pl. 1, Figs. 13 and 30, respectively)

The age of *R. altissima altissima* is stated as UAZ 7–12 (late Bath.-early Call. to early-early late Tith.). In the studied sections *R. altissima altissima* first appears in UAZ C (mid Oxf.), and its ancestral form *R. altissima major* Baumgartner & De Wever occurs in samples assigned to UAZ B (early Call. *pars*-early Oxf.) at the Coston delle Vette section (Southern Alps) (Beccaro et al. 2002). Furthermore, *R. altissima altissima* disappears at the top of UAZ E (late? Oxf.-early Kimm. *pars*) when *R. altissima nodosa* first appears. *R. altissima nodosa* was comprised in the synonymy of *R. altissima altissima* in Baumgartner et al. 1995c and this fact explains the long range (UAZ 7–12: late Bath.-early Call. to early-early late Tith.) of the latter species. *R. altissima nodosa* is very probable the transitional form between *R. altissima altissima* and *R. cretacea* (Baumgartner), whose range is UAZ 12–17 (early-early late Tith. to late Val.). It is therefore reasonable that the range of *R. altissima altissima* (UAZ 7–12: late Bath.-early Call. to early-early late Tith.) is restricted to UAZ C–E (mid Oxf. – early Kimm. *pars*).

Syringocapsa spinellifera Baumgartner (Pl. 1, Fig. 34)

Its range is referred to UAZ 9–12 (mid-late Oxf. to early-early late Tith.) but in the studied sections *S. spinellifera* does not appear before UAZ F (early Kimm. *pars*-late Kimm.) (Fig. 5). The age assignment of this taxa should be reconsidered in the light of the present data.

Tetratrabs bulbosa Baumgartner and *Tetratrabs zealis* (Ozvoldova) (Pl. 1, Figs. 21 and 22, respectively).

T. zealis (UAZ 4–13: late Baj. to latest Tith.-earliest Berr.) disappears towards the top of UAZ E (late? Oxf.-early Kimm. *pars*) whereas is stated to reach the earliest Berriasian (Baumgartner et al. 1995c). *T. bulbosa* is assigned to UAZ 7–11 (late Bath.-early Call. to late Kimm.-early Tith.) but it first appears in UAZ D (mid?-late? Oxf.).

Conclusions

The main conclusions of this research concern the definition of six new radiolarian biozones for the Southern Alps and Western Sicily (Italy), the dating of the intermediate pelagic siliceous member (Rosso Ammonitico Medio) of the Rosso Ammonitico Fm. in the studied area, and the correlation of the investigated successions by means of the new radiolarian biozones.

Nine stratigraphic sections of the Rosso Ammonitico Medio or time-equivalent siliceous facies have been analysed in the Southern Alps and Western Sicily. Six new radiolarians biozones (UAZ A–F) have been defined (Figs. 4 and 5): UAZ A (early?-mid Bath.-early Call. *pars*), UAZ B (early Call. *pars*-early Oxf.), UAZ C (mid Oxf.), UAZ D (mid?-late? Oxf.), UAZ E (late? Oxf.-early Kimm. *pars*), UAZ F (early Kimm. *pars*-late Kimm.). The chronostratigraphic value of each UA Zone is provided by calibration with the ammonites found in the same stratigraphic sections. The new UAZones show a good reproducibility throughout the investigated successions, and make possible the first correlation by radiolarians of the intermediate siliceous member of the Rosso Ammonitico Fm. between the Southern Alps and Western Sicily (Fig. 4).

The newly defined radiolarian biozones also confirm the diachronism of the Middle Jurassic siliceous deposition in the studied area. On the Trento Plateau (Southern Alps) the siliceous deposition began in the early?-mid Bathonian – early Callovian *pars* (UAZ A) at the Coston delle Vette section, in the early Callovian *pars* – early Oxfordian (UAZ B) at the Cava Vianini, and in the mid?-late? Oxfordian (UAZ D) at the Ceniga section (Fig. 4). In the Trapanese Domain (Northwestern Sicily) the siliceous deposition started in the mid-late Oxfordian: UAZ C (mid Oxf.) at the Fornazzo Strada, Fornazzo Cava and Castello Inici sections, and UAZ D (mid?-late? Oxf.) at the Balata di Baida and Favignana sections. In the Sicanian Basin (South-western Sicily) the onset of the siliceous facies at the Sant'Anna section was in UAZ E (late? Oxf.-early Kimm. *pars*) (Fig. 4).

In the light of the new radiolarian biozones UAZ A–F, the age assignments of the Ceniga and Sant'Anna sections have been compared with the literature data. The age of the Ceniga section has been restricted to mid? Oxfordian – early Kimmeridgian *pars* (UAZ D–E), and the age of the Sant'Anna section is now referred to the late? Oxfordian – late Kimmeridgian (UAZ E–F) (Fig. 4).

Thanks to the calibration with ammonites, some range considerations have been stated for certain taxa. The age of *Podobursa polyacantha* (Fischli) is extended to mid-late Oxfordian. At least for the studied area, it could be possible that: *Syringocapsa spinellifera* Baumgartner does not appear before early Kimmeridgian, *Tetratrabs bulbosa* Baumgartner does not appear before the mid Oxfordian, the range of *Ristola altissima altissima* (Rüst) is restricted to mid Oxfordian – early Kimmeridgian *pars*.

The radiolarian biostratigraphy by means of new UAZ A–F is a contribution for a better-defined radiolarian zonation of the Jurassic Mediterranean Tethys. All radiolarian data provided by this research will be used both to improve the zonation of the INTERRAD Jurassic-Cretaceous Working Group (Baumgartner et al. 1995d), and to create a database for the definition of new radiolarian biozones for the Jurassic Mediterranean Tethys.

Acknowledgements

I would like to thank the supervisors of my PhD thesis: Prof. Peter O. Baumgartner (Université de Lausanne, Switzerland), Dr. Špela Goričan (Paleontološki Inštitut ZRC-SAZU, Ljubljana, Slovenia), Prof. Marta Marcucci (University of Florence, Italy) and Dr. Luca Martire (University of Turin, Italy). I am also grateful to the reviewer Dr. Luis O'Dogherty (Universidad de Cádiz, Spain) for having improved this paper by helpful comments.
The scanning electron micrographs have been taken at the University of Florence and University of Turin (Italy), at the University of Bristol (Great Britain) and at the Paleontološki Inštitut ZRC-SAZU (Ljubljana, Slovenia).

REFERENCES

AITA, Y. 1987: Middle Jurassic to Lower Cretaceous Radiolarian Biostratigraphy of Shikoku with Reference to Selected Sections in Lombardy Basin and Sicily. Tohokun Univ., Sci. Rep., 2nd ser. (Geol.), 58/1, 1–91.

BAUMGARTNER, P.O. 1984: A Middle Jurassic-Early Cretaceous low-latitude radiolarian zonation based on Unitary associations and age of Tethyan radiolarites. Ecl. geol. Helv., 77/3, 729–837.

BAUMGARTNER, P.O., DE WEVER, P. & KOCHER, R. 1980: Correlation of Tethyan Late Jurassic-Early Cretaceous radiolarian events. Cah. Micropaléontologie, 2, 23–85.

BAUMGARTNER, P.O., BARTOLINI, A., CARTER, E.S., CONTI, M., CORTESE, G., DANELIAN, T., DE WEVER, P., DUMITRICA, P., DUMITRICA-JUD, R., GORIČAN, Š., GUEX, J., HULL, D. M., KITO, N., MARCUCCI, M., MATSUOKA, A., MURCHEY, B., O'DOGHERTY, L., SAVARY, J., VISHNEVSKAYA, V., WIDZ, D. & YAO, A., 1995a: Middle Jurassic to Early Cretaceous radiolarian biochronology of Tethys based on Unitary Associations. In: Middle Jurassic to Lower Cretaceous Radiolaria of Tethys: Occurrences, Systematics, Biochronology. BAUMGARTNER, P.O. et al. (Eds.). Mémoires de Géologie (Lausanne), 23, 1013–1043.

BAUMGARTNER, P.O., MARTIRE, L., GORIČAN, Š., O'DOGHERTY, L., ERBA, E. & PILLEVUIT, A. 1995b: New Middle and Upper Jurassic radiolarian assemblages co-occurring with ammonites and nannofossils from the Southern Alps (Northern Italy). In: BAUMGARTNER, P.O. et al. (Eds.): Middle Jurassic to Lower Cretaceous Radiolaria of Tethys: Occurrences, Systematics, Biochronology. Mémoires de Géologie (Lausanne), 23, 737–749.

BAUMGARTNER, P.O., O'DOGHERTY, L., GORIČAN, Š., DUMITRICA-JUD, R., DUMITRICA, P., PILLEVUIT, A., URQUHART, E., MATSUOKA, A., DANELIAN, T., BARTOLINI, A., CARTER, E.S., DE WEVER, P., KITO, N., MARCUCCI, M. & STEIGER, T. 1995c: Radiolarian catalogue and systematics of Middle Jurassic to Early Cretaceous Tethyan genera and species. In: BAUMGARTNER, P.O. et al. (Eds.): Middle Jurassic to Lower Cretaceous Radiolaria of Tethys: Occurrences, Systematics, Biochronology. Mémoires de Géologie (Lausanne), 23, 37–685.

BAUMGARTNER, P.O., O'DOGHERTY, L., GORIČAN, Š., URQUHART, E., PILLEVUIT, A. & DE WEVER, P. Eds. 1995d: Middle Jurassic to Lower Cretaceous Radiolaria of Tethys: Occurrences, Systematics, Biochronology. Mémoires de Géologie (Lausanne), 23, 1–1172.

BECCARO, P. 1998: Biostratigrafia a Radiolari della Formazione di Fonzaso (Giurassico medio-superiore, Dolomiti Bellunesi). Graduation Thesis, University of Turin, Italy, (unpubl), 1–163.

BECCARO, P. 2002: Radiolarian Biostratigraphy of Middle-Upper Jurassic Pelagic Siliceous Successions of Western Sicily and Southern Alps (Italy). PhD Thesis, XIV ciclo, University of Florence, Italy, (unpubl.), 1–134.

BECCARO, P. 2004a: Upper Jurassic Radiolarians from Inici Mountain area (Northwestern Sicily, Italy): Biochronology and Calibration by Ammonites. Riv. It. Pal. Strat., 110/1, 289–301.

BECCARO, P. 2004b: New Middle and Upper Jurassic radiolaria from western Sicily and Southern Alps (Italy). Razprave IV, Razreda SAZU, XLV-3, 5–27.

BECCARO, P., BAUMGARTNER, P.O. & MARTIRE, L. 2002: Radiolarian biostratigraphy of the Fonzaso Formation, Middle-Upper Jurassic, Southern Alps, Italy. Micropaleontology 48, suppl. 1, 43–60.

BOSELLINI, A. & DAL CIN, R. 1968: Il Giurassico medio-superiore di Fonzaso (Feltrino Occidentale). Istituto di Geologia dell'Università di Ferrara, Sezione IX, Scienze Geologica e Paleontologica, 4, 237–247.

BOSELLINI, A., MASETTI D. & SARTI M. 1981: A Jurassic "Tongue of the ocean" infilled with oolitic sands: the Belluno Trough, Venetian Alps, Italy. Marine Geology, 44, 59–95.

BOVERO, A. 2000: Analisi paleontologica e stratigrafica del Rosso Ammonitico (Giurassico medio-superiore) nella sezione Fornazzo di Monte Inici, Sicilia Occidentale. Graduation Thesis, University of Turin, Italy, (unpubl.), 1–173.

CATALANO, R., DI STEFANO, P. & KOZUR, H. 1989: Lower Permian Albaillellacea (Radiolaria) from Sicily and their stratigraphic and paleogeographic significance. Rend. Accad. Sc. Fis. Mat. Napoli, IV, LVI, CXXVIII, 80–113.

CATALANO, R., DI STEFANO, P., SULLI, A. & VITALE, F.P. 1996: Paleogeography and structure of the Central Mediterranean: Sicily and its offshore area. Tectonophysics 260, 291–323.

CATALANO, R., LO CICERO, G. & SULLI, A. 2002: Geology of Sicily: an introduction. In: SANTANTONIO, M. (Ed.): General Field Trip Guidebook, VI International Symposium on the Jurassic System, 12–22 September 2002, Palermo, Italy, 1–320.

CECCA, F., SAVARY, B., BARTOLINI, A., REMANE, J. & CORDEY, F. 2001: The Middle Jurassic – Lower Cretaceous Rosso Ammonitico succession of Monte Inici (Trapanese domain, western Sicily): sedimentology, biostratigraphy and isotope stratigraphy. Bull. Soc. Géol. France, 172/5, 647–660.

DAL PIAZ, G. 1907: Le Alpi Feltrine. Memorie del Reale Veneto Istituto di Scienze, Lettere ed Arti, 27/9, 1–162.

DELLA BRUNA, G. & MARTIRE, L. 1985: The Jurassic succession (Pliensbachian-Kimmeridgian) in the Feltre Alps, Belluno. Riv. It. Pal. Strat., 91/1, 15–62.

DE WEVER, P. 1995: Radiolarians from the Sciacca Zone, Santa Anna, Sicily (Italy). In: BAUMGARTNER, P.O. et al. (Eds.): Middle Jurassic to Lower Cretaceous Radiolaria of Tethys: Occurrences, Systematics, Biochronology. Mémoires de Géologie (Lausanne), 23, 839–845.

DE WEVER, P., GEYSSANT, J.R., AZÉMA, J., DEVOS, I., DUÉE, G., MANIVIT, H. & VRIELYNCK, B. 1986: La coupe de Santa Anna (Zone de Sciacca, Sicile): une synthèse des apports des macro-, micro- et nannofosilles du Jurassique supérieur et Crétacé inférieur. Revue de Micropaléontologie, 29/3, 141–186.

FOGELGESANG, J.F. 1975: Géologie du Monte Baldo septentrional (Prov. de Trente, Italie) et aspects géochimiques de la sédimentation pélagique tridentine et lombarde au Jurassique. Thèse 3ème cycle, Université Pierre et Marie Curie (unpubl.), Paris, 1–178.

GIUNTA, G. & LIGUORI, V. 1972: Geologia dell'estremità nord-occidentale della Sicilia. Riv. Min. Sic., 136–138: 165–226.

GIUNTA, G. & LIGUORI, V. 1973: Evoluzione palaeotettonica della Sicilia. Boll. Soc. Geol. Ital., 92, 903–924.

GUEX, J. 1977: Une nouvelle méthode d'analyse biochronologique. Bull. Géol. Lausanne, No 224, 309–322.

GUEX, J. 1991: Biochronologic Correlations. Springer-Verlag, 1–252.

KITO, N. & DE WEVER, P. 1992: Nouvelles espèces d'Hagiastridae (Radiolaires) du Jurassique moyen de Sicilie (Italie). Revue de Micropaléontologie, 35/2, 127–141.

KITO, N. & DE WEVER, P. 1994: New species of middle Jurassic Actinommidae (Radiolaria) from Sicily (Italy). Revue de Micropaléontologie, 37/2, 123–134.

KITO, N., DE WEVER, P., DANELIAN, T. & CORDEY, F. 1990: Middle to Late Jurassic radiolarians from Sicily (Italy). Marine Micropaleontology, 15, 329–349.

KOCHER, R.N. 1981: Biochronostratigraphische Untersuchungen oberjurassischer Radiolarienführender Gesteine, insbesondere der Sudalpen. Mitt. Geol. Inst. ETH Univ. Zurich, N.F. 234, 1–184.

MARTIRE, L. 1989: Analisi biostratigrafica e sedimentologica del Rosso Ammonitico Veronese dell'Altopiano di Asiago (VI). PhD thesis, University of Turin, Italy, (unpubl.), 1–166.

MARTIRE, L. 1992: Sequence stratigraphy and condensed pelagic sediments. An example from the Rosso Ammonitico Veronese, northeastern Italy. Paleogeography, Paleoclimatology, Paleoecology, 94, 169–191.

MARTIRE, L. 1996: Stratigraphy, Facies and Synsedimentary Tectonics in the Jurassic Rosso Ammonitico Veronese (Altopiano di Asiago, NE Italy). Facies, 35, 209–236.

MASCLE, G.H. 1973: Etude géologique des Monts Sicani (Sicile). Thèse de Doctorat d'Etat, Université Pierre et Marie Curie, Paris, 1–691.

MASCLE, G.H. 1979: Etude géologique des Monts Sicani (Sicile). Riv. It. Pal. Strat., Memoria, 16, 1–431.

Origlia-Devos, I. 1983: Radiolaires du Jurassique supérieur – Crétacé inferieur: Taxonomie et révision stratigraphique (zone du Pinde-Olonos, Grèce, zone de la Sviacca, Italie, Complexe de Nicoya, Costa Rica et forages du DSDP. Université Pierre et Marie Curie, Paris, (unpubl.), 1–328.

Papa, G. 1994: Analisi stratigrafica del Rosso Ammonitico Veronese nell'area di Spiazzi di Monte Baldo (VR). Graduation Thesis, University of Torino, Italy, (unpubl.), 1–156.

Riedel, W.R. & Sanfilippo, A. 1974: Radiolaria from the southern Indian Ocean, DSDP Leg 26. In: Davies, T.A., Luyendyk, B.P. et al. (Eds.): Initial Rep. Deep Sea Drill. Proj., 26, 771–814.

Savary, B. 2000: L'Ammonitico Rosso du Jurassique moyen et supérieur de la zone Trapanaise (Sicile W, Italie): genèse des structures sédimentaires, discontinuités et implications paléogéographiques. Mémoire du DEA, PalSed, University Claude-Berbard, Lyon 1, France, (unpubl.), 1–50.

Savary, J. & Guex, J. 1991: BioGraph: un nouveau programme de construction des corrélations biochronologiques basées sur les associations unitaires. Bull. Géol. Lausanne, 313, 317–340.

Savary, J. & Guex, J. 1999: Discrete Biochronologic Scales and Unitary Associations: Description of the BioGraph Computer Program. Mémoires de Géologie (Lausanne), 34, 1–281.

Wendt, J. 1964: Stratigraphisch – Palaontologische Untersuchungen im Dogger Westsiziliens. Boll. Soc. Pal. It., 2/1, 57–145.

Wendt, J. 1971: Geologia del Monte Erice (Provincia di Trapani, Sicilia Occidentale). Geol. Rom., 10, 53–76.

Winterer, E.L. & Bosellini, A. 1981: Subsidence and sedimentation on Jurassic Passive Continental Margin, Southern Alps, Italy. Amer. Ass. Petr. Geol. Bull., 65, 394–421.

Manuscript received January 2004
Revision accepted February 2005

Plate 1

Scanning electron micrographs of the most important radiolarians used for the definition of the UAZ A–F are illustrated. For each photo are given the code of the section, the number of the sample, the UAZone of the sample, and the magnification. The codes of the stratigraphic sections are: BB-Balata di Baida, CI-Castello Inici, CV-Coston delle Vette, FZC-Fornazzo Cava, FV-Favignana, IN-Fornazzo Strada, SA-Sant'Anna, VN-Cava Vianini.

1 – *Eucyrtidiellum unumaense dentatum* Baumgartner, CV 60, UAZ A, x250
2 – *Stylocapsa oblongula* Kocher, CV 60, UAZ A, x200
3 – *Unuma echinatus* Ichikawa & Yao, CV 60, UAZ A, x150
4 – *Stylocapsa catenarum* Matsuoka, VN 2.85, UAZ B, x350
5 – *Praewilliriedellum convexum* (Yao), VN 2.85, UAZ B, x200
6 – *Stichocapsa robusta* Matsuoka, VN 2.85, UAZ B, x200
7 – *Triactoma parablakei* Yang & Wang, VN 0.40, UAZ B, x100
8 – *Tricolocapsa plicarum* s.l. Yao, VN 2.85, UAZ B, x250
9 – *Tethysetta dhimenaensis dhimenaensis* (Baumgartner), VN 2.85, UAZ B, x250
10 – *Eucyrtidiellum ptyctum* (Riedel & Sanfilippo), VN 0.40, UAZ B, x250
11 – *Williriedellum carpathicum* Dumitrica, VN 2.85, UAZ B, x200
12 – *Podobursa vannae* Beccaro, CI 4, UAZ C, x100
13 – *Ristola altissima altissima* (Rüst), IN 3, UAZ C, x100
14 – *Podobursa polyacantha* (Fischli), CE 5.20, UAZ D, x100
15 – *Xitus magnus* Baumgartner, CI 5, UAZ C, x120
16 – *Eucyrtidiellum unumaense* s.l. (Yao), FZC 3, UAZ C, x250
17 – *Tritrabs hayi* (Pessagno), CI 5, UAZ C, x75
18 – *Williriedellum (?) marcucciae* Cortese, IN 1, UAZ C, x300
19 – *Zhamoidellum (?) exquisitum* Hull, IN 3, UAZ D, x200
20 – *Angulobracchia biordinalis* Ozvoldova, BB 11.70, UAZ D, x200
21 – *Tetratrabs bulbosa* Baumgartner, CI 8, UAZ D, x60
22 – *Tetratrabs zealis* (Ozvoldova), IN 3, UAZ D, x60
23 – *Napora lospensis* Pessagno, IN 8, UAZ D, x100
24 – *Emiluvia orea* Baumgartner, CI 8, UAZ D, x80
25 – *Mirifusus dianae minor* Baumgartner, CI 15, UAZ E, x100
26 – *Mirifusus dianae dianae* (Karrer), SA 0.35, UAZ E, x100
27 – *Pseudoeucyrtis* sp. B sensu Widz 1991, SA 0.35, UAZ E, x100
28 – *Podocapsa amphitreptera* Foreman, CV 122, UAZ E, x130
29 – *Eucyrtidiellum nodosum* Wakita, IN 19, UAZ E, x300
30 – *Ristola altissima nodosa* Hori, BB 15.10, UAZ E, x150
31 – *Emiluvia ultima* Baumgartner & Dumitrica, SA 0.35, UAZ E, x100
32 – *Pseudoeucyrtis reticularis* Matsuoka & Yao, IN 32, UAZ F, x120
33 – *Acaeniotyle umbilicata* (Rüst), IN 32, UAZ F, x100
34 – *Syringocapsa spinellifera* Baumgartner, CV 125.6, UAZ F, x100
35 – *Napora boneti* Pessagno, Whalen & Yeh, IN 30, UAZ F, x150

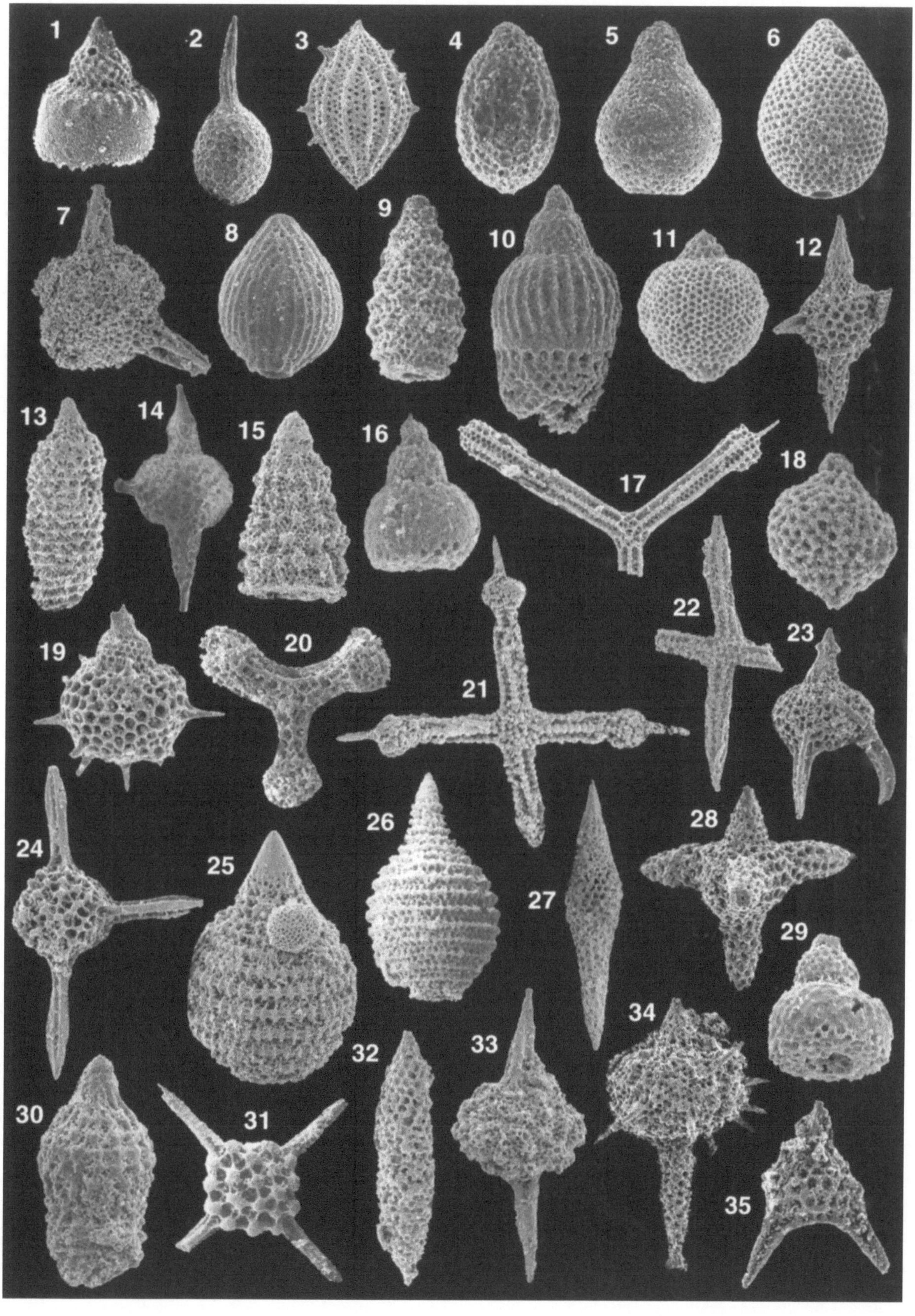
1
2
3
4
5
6
7
8
9
10
11
12
13
14
15
16
17
18
19
20
21
22
23
24
25
26
27
28
29
30
31
32
33
34
35

0012-9402/06/01S035-13
DOI 10.1007/s00015-006-0603-4
Birkhäuser Verlag, Basel, 2006

Eclogae geol. Helv. 99 (2006) Supplement 1, S35–S47

Upper Jurassic Radiolaria from the Vocontian basin of SE France

TANIEL DANELIAN[1], SALIM LAHSINI[1] & MARC DE RAFÉLIS[2]

Key words: Radiolaria, Vocontian basin, France, Upper Jurassic
Mots clés: Radiolaires, Bassin Vocontien, France, Jurassique supérieur

ABSTRACT

Two sections of the Vocontian basin (southeast France) were explored for their radiolarian content. Preservation in calcite impeded their extraction from the Châteauneuf d'Oze section, which is well dated by ammonites. Fortunately, oligospecific assemblages of pyritized radiolaria were yielded by twelve limestone beds of the Méouge section. In general, preservation worsens towards the top of the section. The extracted fauna is dominated by Nassellaria and more particularly by Archaeodictyomitridae, Pseudodictyomitridae and Williriedellidae. It is the first time that some species are reported outside their type area or from Western Tethys. Amongst the published zonations the scheme of Baumgartner et al. (1995a) is found to be the most useful and allows correlation of the studied part of the section with the mid-late Oxfordian to late Kimmeridgian-early Tithonian time interval.

RESUME

Deux coupes du bassin Vocontien (Sud – Est de la France) ont été examinées pour leur contenu en radiolaires. Leur conservation en calcite dans la coupe de Châteauneuf d'Oze (bien datée par des Ammonites) n'a pas permis leur extraction. Par contre, des assemblages oligospécifiques de radiolaires pyritisés ont pu être extraits de douze niveaux calcaires de la coupe de Méouge. En général, la conservation des Radiolaires est moins bonne vers le haut de la coupe. La faune extraite est dominée par des Nassellaires et plus particulièrement des Archaeodictyomitridae, Pseudodictyomitridae et Williriedellidae. Quelques espèces sont signalées pour la première fois en dehors de leur région-type ou dans la Téthys occidentale. Parmi les biozonations publiées, celle de Baumgartner et al. (1995a) s'avère la plus utile et permet de dater la coupe étudiée avec la fourchette d'âge Oxfordien moyen-supérieur à Kimméridgien supérieur-Tithonien inférieur.

Introduction

In spite of the enormous progress made over the last thirty years on our knowledge of Mesozoïc Radiolaria (see De Wever et al. 2001 and references therein) we are still far from unravelling the full biochronological potential of this plankton group. This is due to the large number of species that are still poorly known from a taxonomic and stratigraphic point of view and also to the small number of radiolarian-bearing samples, for which the age is calibrated independently of Radiolaria themselves. The latter point is also important when describing the dynamic of radiolarian macroevolutionary changes through time (Danelian & Johnson 2001).

The Vocontian Basin, situated in the south-eastern part of France (Fig. 1), represents a privileged area for multidisciplinary stratigraphic studies because its pelagic sedimentary sequences are particularly thick and relatively complete (e.g. Jan du Chêne et al. 1993, 2000; Groupe Français d'étude du Jurassique 1997).

Radiolaria were known to be present in the Vocontian Basin (Beaudoin 1977), but observations were based on thin section. Recent reports by Lambert (1999) and Gardin et al. (2000) on Lower Cretaceous (Valanginian-Hauterivian) pyritized Radiolaria constitute the only modern study for the Vocontian basin. We present here radiolarian assemblages extracted from Upper Jurassic strata of the Vocontian Basin and discuss their biostratigraphic implications.

[1] Micropaléontologie, Université Pierre-et-Marie-Curie. – C.N.R.S-UMR 5143 Tour 46–56, 5° étage, Case 104, 4 Place Jussieu, 75252 Paris Cedex 05. Email: danelian@ccr.jussieu.fr
[2] Biominéralisations et Paléoenvironnements, Université Pierre-et-Marie-Curie – FR32. CEPAGE Tour 56–55, 5° étage, Case 116, 4 Place Jussieu, 75252 Paris Cedex 05

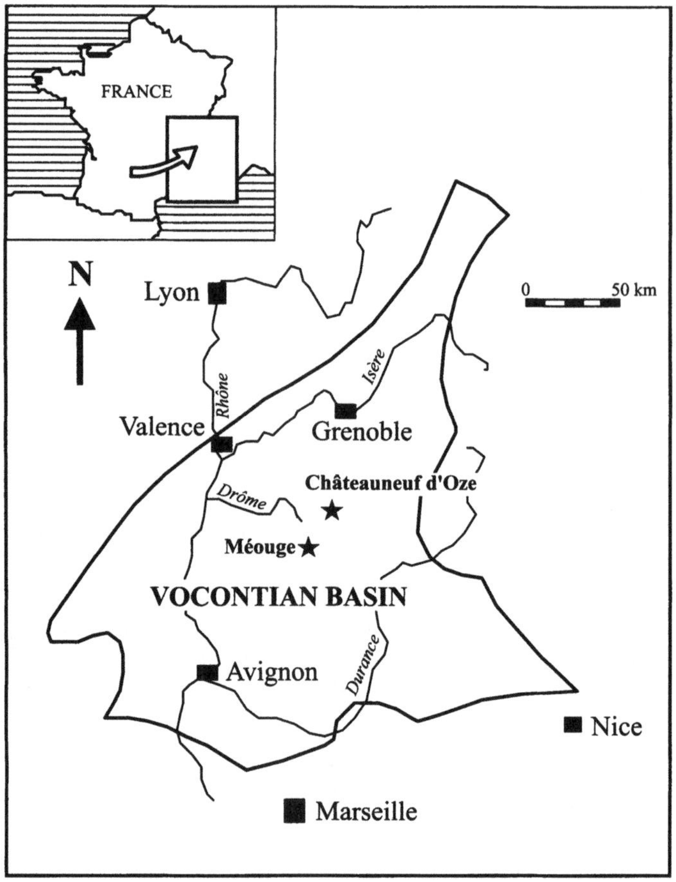

Fig. 1. Outcrops of the Vocontian basin in SE France and location of the two studied sections: Châteauneuf d'Oze and Méouge.

Palaeogeographic and stratigraphic framework

The tectonic evolution of the Vocontian Basin is directly linked to the opening of the Ligurian branch of Tethys and the formation of its European margin (Dercourt et al. 1993). Sedimentary variations reflect the different stages of the opening of Tethys in this area. Break up of the hercynian basement beginning in the Triassic led to the formation of a number of basins (Languedocian, Dauphinois, Provençal and Vocontian), well defined by the presence of submarine ridges (Dubois & Delfaud 1989). The basins were more clearly separated during the Early and Middle Jurassic tectonic subsidence, which was followed by a thermal subsidence during the Late Jurassic and Early Cretaceous. From the middle Oxfordian a generalized subsidence affected the Vocontian passive margin and allowed the development of highly homogeneous sedimentary facies both within the basin and on its edges (Dubois & Delfaud 1989). At this time, marine sedimentary environments reached their maximal extension.

The interest of the Upper Jurassic sequences of the Vocontian Basin lies with the high quality of its outcrops. Pederneiras (1995) established a composite type mega-sequence for its Upper Jurassic sedimentary sequences and highlighted the presence of a number of different lithostratigraphic patterns. These are easily identifiable and correlatable throughout the basin with the bio-chronostratigraphic framework established by Atrops (1982) on the basis of ammonites found in various sections (Fig. 2A). Following an interval dominated by Oxfordian marls, Kimmeridgian sedimentary facies are essentially made of alternating limestones and marls, that become clearly calcareous in the Tithonian (Atrops 1982; Pederneiras 1995; De Rafélis 2000).

Studied material

Two sections of the Vocontian basin are investigated for Radiolaria (Fig. 1). Châteauneuf d'Oze represents one of the reference sections for the Oxfordian – Kimmeridgian boundary interval of the subalpine regions because of its richness in ammonites and quality of outcrops (Atrops 1982; De Rafélis et al. 2001). It is situated a few kilometres to the South-East of Veynes on the left bank of Drouzet, along the Departmental road D20 between Veynes and Châteauneuf d'Oze (44°31'N-5°53'E). Situated on the northern side of the Serres Jurassic high, it represents an area of average sedimentation rates (Moussine-Pouchkine et al. 1998). The section is about 100 metres thick and consists of a regular alternation of marls and limestones interrupted by a 3 metre thick slump. Limestones are fairly homogeneous and consist of slightly bioturbated and organic matter-rich pelagic facies, often rich in ammonites.

The Méouge section is situated along the departmental road D 942 (44°12'N-5°33'E), about twenty kilometers to the south-southwest of Châteauneuf d'Oze. Situated at the southern flank of the Serres Jurassic high, it also represents a depositional environment of average sedimentation rates. The outcrop is about one hundred metres thick and starts close to the point where the road dips downhill beside the river Méouge. Although still poorly studied, the Méouge section has the advantage of being almost continuous and unaffected by the gravitational event present at Châteunauf d'Oze (De Rafélis 2000). It is entirely composed of alternating marls and limestones organised in pluridecimetric beds with an easily workable stratonomy (Fig. 2B). As at Châteauneuf d'Oze, pelagic limestones at Méouge are very homogeneous (micritic). It is only towards the top of the sequence (bed 181 onwards, Fig. 2B) that they display some different facies (nodular limestones, calcarenites with bioclasts or finely laminated calcarenites). Macro- and microfossils are rare. The presence of *Globochaetes*, *Saccocoma*, *Tubiphytes* and a few foraminifera are reported (De Rafélis 2000; Langoisseux 2001). The few ammonites discovered do not provide any precise biostratigraphic information (Atrops, pers. comm.). However, the presence of lithostratigraphic units displaying distinct stratonomic patterns, recognizable throughout the Vocontian basin, are present in the Méouge section and provide some clues as to the age of this section (Pederneiras 1995; Moussine-Pouchkine et

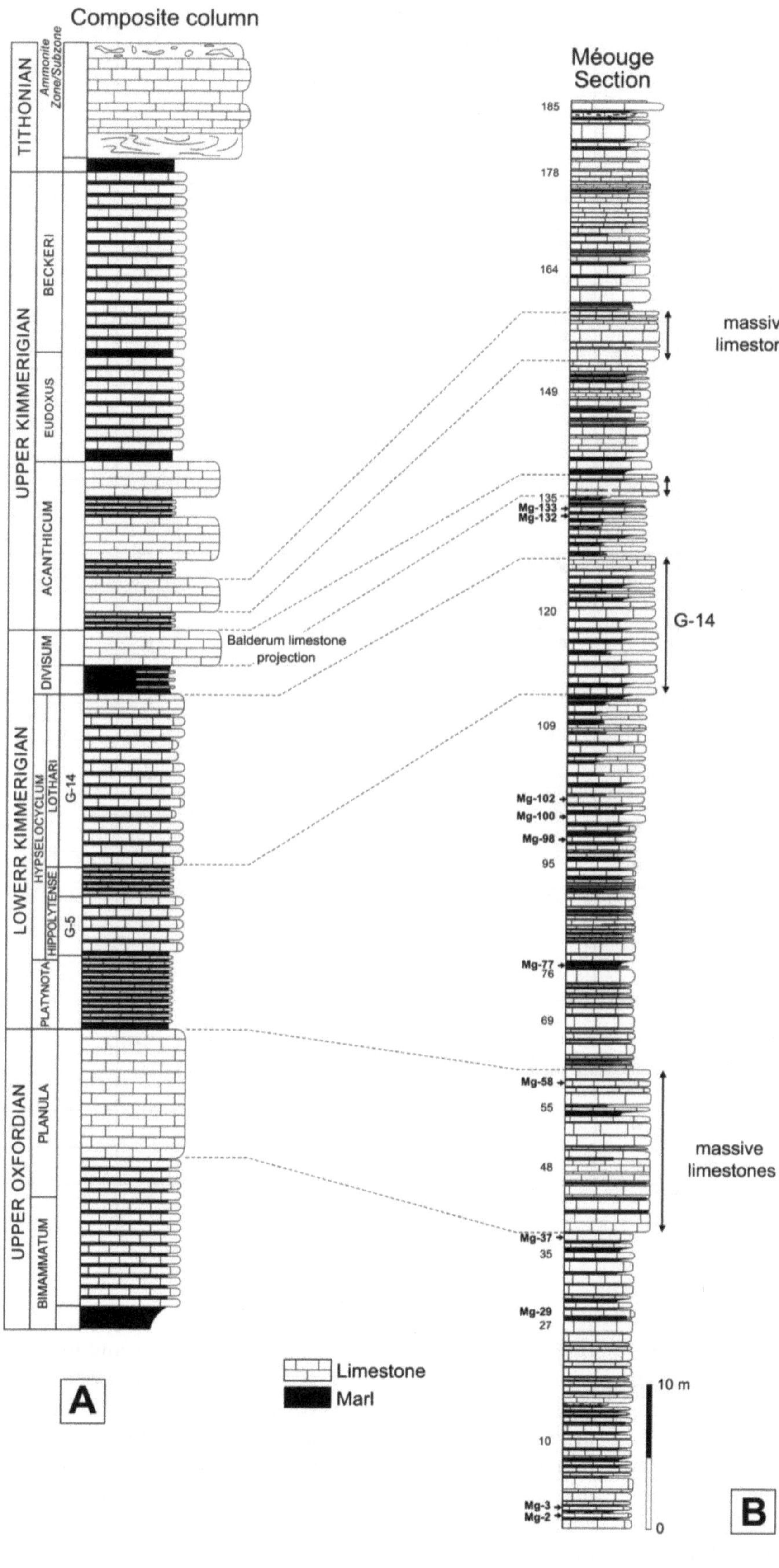

Fig. 2. A) Composite lithostratigraphic column of the Upper Jurassic megasequence of the Vocontian basin, between the "Terres noires" formation at the bottom and the massive Tithonian limestones at the top (modified after Pederneiras 1995). B) Detailed lithostratigraphic column of the Méouge section, including bed numbers and tentative lithostratigraphic correlations with the Upper Jurassic megasequence (after De Rafélis 2000). Arrows indicate limestone beds that yielded radiolaria.

Table 1. Occurrence of radiolarian taxa identified in samples of the Méouge section.

Species/Samples	Mg 2	Mg 3	Mg 21	Mg 29	Mg 37	Mg 58	Mg 77	Mg 98/99	Mg 100	Mg 102	Mg 132	Mg 133
Acastea sp.cf. *A remusa* HULL									■			
Archaeodictyomitra apiarium (RÜST)	■											
Archaeodictyomitra etrusca CHIARI et al.				■								cf.
Archaeodictyomitra patricki KOCHER	■			■								
Archaeodictyomitra sp.aff. *A. patricki* KOCHER	■											
Archaeodictyomitra shengi YANG				■			■					
Archaeodictyomitra spelae CHIARI et al.							■					
Archaeodictyomitra sp.A	■											
Cinguloturris fusiforma HORI							cf.		■			
Emiluvia pentaporata STEIGER & STEIGER									■			
Eucyrtidiellum ptyctum (RIEDEL & SANFILIPPO)	■						■					
Gongylothorax favosus DUMITRICA	■	■		■	■			■	■	■		
Loopus doliolum DUMITRICA				■	■							
Loopus venustus (CHIARI et al.)	■			■	■				cf.	■		■
Pantanellium oligoporum (VINASSA)									■			
Praeconocaryomma scaterba HULL			■									
Protunuma japonicus MATSOUKA & YAO	■				■		■		■			
Ristola altissima ssp.cf. *R. a. altissima* (RÜST)					■							
Saitoum pagei PESSAGNO			■									
Spongocapsula palmerae (PESSAGNO)			■									
Stichocapsa tuscanica CHIARI et al.							■					
Suna sp.							■					
Tethysetta (?) sp.							■					
Triactoma foremanae MUZAVOR							■					
Tripocyclia sp.cf. *T.luciae* JUD									■			
Tritrabs sp.cf. *T. exotica* (PESSAGNO)					■							
Williriedellum carpathicum DUMITRICA			■				■					
Williriedellum crystallinum DUMITRICA		■	■		■							
Zhamoidellum ovum DUMITRICA	■	■	■	■	■	■	■	■	■	■	■	

al. 1998; De Rafélis et al. 1997). Above the "Terres Noires" Formation one can thus recognize from bottom to top (1) an interval of massive limestones (beds 38 to 61) commonly present within the *Planula* Ammonite zone, (2) a bundle of limestone beds known as «the group of 14 (G-14)» (beds 113 to 128), commonly present within the *Lothari* Ammonite subzone, (3) a limestone projection (beds 136 to 140), which is tentatively correlated with the one occurring within the *Balderum* horizon (*Divisum* Ammonite zone), and finally (4) a bundle of massive limestones (beds 153 to 162) which could possibly represent the ones occurring at the base of the Kimmeridgian (*Acanthicum* Ammonite zone). In summary, although no direct biostratigraphic age assignment existed for the Méouge section prior to our study, it could be tentatively considered as covering the upper Oxfordian to upper Kimmeridgian interval.

Radiolarian fauna and age

Samples were first processed with hydrochloric acid, which was followed by a brief (30 min.) bath in diluted hydrofluoric acid (HF 5%). Sample size was in the order of several tens of cm^3.

Although present in abundance in the limestones of Châteauneuf d'Oze section, Radiolaria are entirely calcified and no identifiable forms were extracted. However, twelve limestone and marly limestone samples from Méouge section yielded pyritized identifiable Radiolaria, which display different degrees of preservation and abundance. The occurrence of all radiolarian taxa identified in samples of Méouge section are given on Table 1. The identified fauna is oligospecific, with a dominance of Nassellaria over Spumellaria. Samples Mg-2, Mg-21, Mg-77 and Mg-102 yielded the best preserved assemblages, displaying a diversity of about a dozen species. Amongst Nassellaria, representatives of families Archaeodictyomitridae, Pseudodictyomitridae and Williriedellidae are the most abundant. In general, preservation worsens towards the top of the section, only four taxa being identified in samples Mg-132 and Mg-133.

Amongst the published radiolarian zonations the scheme of Baumgartner et al. (1995a) was found to be the most useful. The co-occurrence of species *Zhamoidellum ovum* and *Gongylothorax favosus* in samples Mg-2 and Mg-100 (table 1) allows their assignment to Unitary Association Zones (UAZ) 9–10, which are correlated with the middle Oxfordian to early Kimmeridgian. This is inferred from the first occurrence (FO) of *Zhamoidellum ovum* and the last occurrence (LO) of *Gongylothorax favosus*. However, following the discovery of an exceptionally well-preserved radiolarian fauna in the Mariana trench (Matsuoka 1998), we now know that *G. favosus* existed until Berriasian time, at least in the Palaeopacific. Consequently, the age of the Méouge section can be only correlated with the mid-late Oxfordian to late Kimmeridgian-early Tithonian, based on the single occurrence of *Z. ovum* (UAZ 9-11) throughout the quasi-totality of the radiolarian-bearing samples. On the other hand, *Emiluvia pentaporata* (= *E. bisellea*) is considered in the biozonation of Baumgartner et al. (1995a) as constrained to UAZ 11. However, it is found to occur in middle Oxfordian to lower Kimmeridgian (UAZ 9–10) samples of the Ionian zone (Danelian 1995) and in Oxfordian strata of Roumania (Dumitrica, pers. commun.).

Conclusion

Radiolaria occur abundantly in the ammonite-rich Chateauneuf d'Oze section, but their preservation in calcite did not allow extraction of any identifiable microfauna. On the contrary, an oligospecific and moderately diverse pyritized Radiolarian fauna was extracted from the Méouge section. Although the biochronologic resolution of Radiolaria does not allow a precise age assignment (mid-late Oxfordian to late Kimmeridgian-early Tithonian), it is currently the only biostratigraphic evidence for the Méouge section. Some species are found for the first time outside their type area (i.e. *Loopus venustus*) or outside the Palaeopacific (*Cinguloturris fusiforma*). Much work is still left to be done in order to refine the biochronological potential of Upper Jurassic radiolaria.

Systematic Palaeontology

Species and genera are discussed in alphabetical order. The suprageneric classification followed in this work is after De Wever et al. 2001 and is not repeated here. The synonymy lists contains, only a few representative studies.

Genus *Acastea* Yang 1993
Type species: *Acaeniotyle diaphorogona* Foreman 1973

Acastea sp.cf. *A. remusa* Hull
(Pl. 1, Fig. 1)
cf. *Acastea remusa* n.sp. Hull 1997, p. 35, pl. 11, figs. 2, 6, 17.

Remarks. – The illustrated morphotype recalls *A. remusa* by its blunt spinal tips and presence of deep distally tapering longitudinal grooves running on the ridges of its massive spines.

Genus *Archaeodictyomitra* Pessagno 1976
Type species: *Archaeodictyomitra squinaboli* Pessagno 1976

Archaedictyomitra etrusca Chiari, Cortese & Marcucci
(Pl. 1, fig. 4)
Archaeodictyomitra etrusca n.sp. Chiari, Cortese & Marcucci *in* Chiari et al. 1997, p. 63, pl. 1, fig. 9–10.

Remarks. – This species can be distinguished from *A. suzukii* Aita by the presence of more numerous costae (13–14 costae visible per half perimeter).

Archaeodictyomitra patricki Kocher
(Pl. 1, Fig. 5–7)
Archaeodictyomitra patricki n.sp. Kocher 1981, p. 57, pl. 12, fig. 14–17.
Archaeodictyomitra sp.cf. *A .patricki* Kocher ; Chiari *et al.* 1997, pl. 1 fig. 16.

Remarks. – This small four-segment species is characterized by its conical proximal part becoming an inverted trapezoid distally. Faint traces of segmental divisions are often visible on the outline. 10–12 longitudinal costae visible per half perimeter run up to the chephalis.

Archaeodictyomitra sp.aff. *A. patricki* Kocher
(Pl. 1, Fig. 8)
aff. *Archaeodictyomitra patricki* Kocher 1981, p. 57, pl. 12, fig. 14–17.
cf. *Archaeodictyomitra* sp.aff. *A.squinaboli* Pessagno, Chiari et al. 1997, pl. 1, fig. 17.

Remarks. – This morphotype differs from *A. patricki* by its more elongated outline and lesser number of costae (about 10 visible per half perimeter).

Archaedictyomitra shengi YANG
(Pl. 1, Fig. 9–10)
Archaeodictyomitra shengi n.sp. YANG 1993, p.64, pl. 1,
Archaeodictyomitra labronica n.sp. CHIARI, CORTESE & MARCUCCI *in* Chiari et al. 1997, p. 64, pl. 1, fig. 11–12.
Not *Archaeodictyomitra shengi* YANG, Hori 1999, fig. 7–13 (= *A. spelae* Chiari et al. 1997).

Remarks. – We do agree with Hori (1999) in considering *A. labronica* as a junior synonym of A. *shengi* YANG 1993 because it is practically impossible to differentiate these elongated, subconical to subcylindrical morphotypes with smooth outline. However, the specimen illustrated by Hori seems to us to be related to *A. spelae* because the last three segments are well individualised on the outline. The known age range of this species is Oxfordian to Tithonian.

Archaeodictyomitra sp. A
(Pl. 1, Fig. 12)
Remarks. – Small (total length : 110 µm) conical form bearing 11 costae visible per half perimeter. It differs from *A. patricki* by its smaller size and absence of an inverted trapezoidal distal part.

Genus *Cinguloturris* DUMITRICA *in* Dumitrica & Mello 1982
Type–species: *Cinguloturris carpatica* DUMITRICA *in* Dumitrica & Mello 1982

Cinguloturris fusiforma HORI
(Pl. 1, Fig. 13)
Cinguloturris fusiforma n.sp. HORI 1999, p. 93, fig. 9.3–9.6, 11.6.

Remarks. – Only two specimens of *Cinguloturris* were found in our material. One of them is incomplete and is questionably assigned to *C. fusiforma.* (pl. 1, fig. 14). This species was previously known only from the Palaeopacific (Japan and Far East Russia). *C. fusiforma* is likely to represent an offshoot lineage of *C. carpatica* from which it differs by its spindle-shaped contour.

Genus *Emiluvia* FOREMAN 1973
Type-species: *Emiluvia chica* FOREMAN 1973

Emiluvia pentaporata STEIGER & STEIGER
(Pl. 1, Fig. 15)
Emiluvia pentaporata n.sp. STEIGER & STEIGER 1994, p. 458, pl. 1, fig. 9–10
Emiluvia bisellea n.sp. DANELIAN in Baumgartner et al. 1995b, p. 196, pl. 4018, fig. 1–4.

Genus *Eucyrtidiellum* BAUMGARTNER 1984
Type-species: *Eucyrtidium (?) unumaensis* YAO 1979

Eucyrtidiellum ptyctum (RIEDEL et SANFILIPPO)
(Pl. 1, Fig. 16–17)
Eucyrtidium ptyctum n.sp.; Riedel & Sanfilippo 1974, p. 778, pl. 5, fig. 7, pl. 12, fig. 14; not fig. 15.
Eucyrtidiellum ptyctum (RIEDEL & SANFILIPPO); Baumgartner et al. 1995b, p. 214, pl. 3017, fig. 1–8; Hull 1997, p. 82, pl. 48, fig. 10.

Remarks. – Two distinct morphotypes were observed in our material. Morphotype A (fig. 16) displays a trapezoidal abdomen, bearing 7 prominent costae visible per half-circumference. Morphotype B (fig. 17) displays a much more rounded abdomen, bearing *c.* 12 faintly developped costae visible per half-circumference. In that respect it resembles *E. bortolotii* which nevertheless displays a more trapezoidal abdomen.

Genus *Loopus* YANG 1993 *sensu* Hull (1997) and Dumitrica et al. (1997)
Type-species: *Pseudodictyomitra primitiva* MATSUOKA & YAO 1985

Loopus doliolum DUMITRICA
(Pl. 1, Fig. 19–20)
Dictyomitra sp. C, YAO 1984, pl. 3, fig. 4.
Pseudodictyomitra sp. C *sensu* Yao, Gorican 1994, p. 84, pl. 22, fig. 15, ? fig. 14.
« Dictyomitra sp. C » *in* Yao, Danelian et al. 1996, fig. 4d–e.
Loopus doliolum n.sp.; DUMITRICA *in* Dumitrica *et al.* 1997, p. 30, pl. 5, fig. 3, 5, 14

Remarks. – Morphotypes that are not clearly constricted at their distal part are included under this species. They are all characterized by a single row of pores and faint costae interrupted by well-marked broad stictures. Its known age range is Oxfordian (Gorican 1994) to Berriasian (Dumitrica et al. 1997).

Loopus venustus (CHIARI, CORTESE & MARCUCCI)
(Pl. 1, Fig. 21–23)
Cinguloturris (?) venusta n.sp. CHIARI, CORTESE & MARCUCCI *in* Chiari et al. 1997, p. 66, pl. 2, fig. 4–5.

Remarks. – This species is characterized by segments displaying a single row of open pores situated in their middle part. Thin costae are placed above and below each pore in such a way that an upper and a lower row of depressions or relict pores are formed on each postabdominal segment. Costae do not loop around the pores. *L. venustus* is closely related to *L. primitivus* and might be its ancestor.

Genus *Pantanellium* PESSAGNO 1977
Type-species: *Pantanellium riedeli* PESSAGNO 1977

Pantanellium oligoporum (VINASSA)
(Pl. 2, Fig. 1)
Ellipsoxiphus oligoporus n.sp. VINASSA 1899, p. 228, pl. 17, fig. 44.
Sphaerostylus oligoporus (VINASSA); Sanfilippo & Riedel 1985, p. 590, fig. 4.5 ; Matsuoka 1992, pl. 3, fig. 4.
Pantanellium oligoporum (VINASSA); Chiari et al. 1997, pl. 3, fig. 8.

Genus *Tethysetta* DUMITRICA in Dumitrica et al. **1997**
Type-species: *Tethysetta pygmaea* DUMITRICA et al. 1997

Tethysetta (?) sp.
(Pl. 2, Fig. 10)
Remarks. – A single specimen found in our material recalls

T. dhimenaensis with its fusiform outline and the presence of spinose ridges on intersegmental divisions. Uncertainties are due to the presence of 4 rows of pores per segment diplayed clearly on the distal part of the shell.

Genus *Tripocyclia* HAECKEL 1881

Type-species: *Tripocyclia trigonum* RÜST 1885

Remarks. – The amended definition of Pessagno et al. (1989, p. 205) is followed in the present study. *Tripocyclia* bears a cortical shell of subcircular to elliptical outline and possesses cortical buttresses. It thus differs from *Triactoma* which bears a nearly spherical cortical shell and lacks cortical butresses. In agreement with Hull (1997) the three secondary spines of species assigned to *Tripocyclia* "may be symmetrically or asymmetrically arranged". This leads us to consider *Neotripocyclia* Pessagno & Yang in Pessagno et al. (1989) as a junior synonym of *Tripocyclia*.

Tripocyclia sp.cf. *T.luciae* JUD

(Pl. 2, Fig. 12)

cf. *Triactoma luciae* n.sp. JUD 1994, p. 115, pl. 23, fig. 8–9.

Triactoma sp.cf. *T.luciae* JUD; Chiari *et al.* 1997, pl. 5, fig. 7.

Triactoma jonesi (PESSAGNO); Beccaro *et al.* 2002, pl. 4, fig. 6.

Remarks. – The single specimen found in our material resembles *T. luciae* based on the rounded triangular outline of its cortical shell and the crown-shaped spinal tips. However, it differs in the presence of deep longitudinal grooves running on the main ridges of its spines, which are thus six-fold in section, rather than tri-fold as in *T. luciae*.

Acknowledgments

This study was funded by the French Ministry of Education (project «coup de pouce» to T. Danelian). C. Abrial et A. Lethiers helped with the drawings. We benefited from discussions with P. Dumitrica and M. Renard on Radiolaria and the Vocontian Basin, respectively. Constructive remarks from P. De Wever and S. Gorican greatly improved the manuscript.

REFERENCES

ATROPS, F. 1982: La sous-famille des *Ataxioceratinae* (Ammonitina) dans le Kimméridgien Inférieur du Sud – Est de la France; systématique, évolution, chronostratigraphie des Genuss *Orthosphincxtes* et *Ataxioceras*. Doc. Lab. Géol. Fac. Sci. Lyon, 83, 300–306.

BAUMGARTNER, P. O. 1984: A Middle Jurassic – Early Cretaceous low latitude radiolarian zonation based on unitary associations and age of Tethyan radiolarites. Eclogae geol. Helv. 77, 729–841.

BAUMGARTNER, P. O., BARTOLINI A., CARTER E. S., CONTI, M., CORTESE, G., DANELIAN T., DE WEVER P., DUMITRICA P., DUMITRICA-JUD R., GORICAN S., GUEX, J., HULL, D.M., KITO, N., MARCUCCI, M., MATSUOKA A., MURCHEY, B., O'DOGHERTY, L., SAVARY, J., VISHNEVSKAYA, V., WIDZ, D. & YAO, A. 1995a: Middle Jurassic to Early Cretaceous radiolarian biochronology of Tethys based on Unitary Associations. In: BAUMGARTNER, P.O., O'DOGHERTY, L., GORICAN, S., URQUHART, E., PILLEVUIT, A. & DE WEVER, P. (Eds.): Middle Jurassic to Lower Cretaceous Radiolaria of Tethys: Occurences, Systematics, Biochronology. Université de Lausanne, Mémoires de Géologie 23, 1013–1048.

BAUMGARTNER, P. O., O'DOGHERTHY, L., GORICAN, S., DUMITRICA-JUD, R., DUMITRICA, P., PILLEVUIT, A., URQUHART, E., MATSUOKA, A., DANELIAN, T., BARTOLINI, A., CARTER, E. S., DE WEVER, P., KITO, N., MARCUCCI, M. & STEIGER, T. 1995b: Radiolarian catalogue and systematics of Middle Jurassic to Early Cretaceous Tethyan genera and species. In: BAUMGARTNER, P.O., O'DOGHERTY, L., GORICAN, S., URQUHART, E., PILLEVUIT, A. & DE WEVER, P. (Eds.): Middle Jurassic to Lower Cretaceous Radiolaria of Tethys: Occurences, Systematics, Biochronology. Université de Lausanne, Mémoires de Géologie 23, 37–685.

BEAUDOIN, B. 1977: Méthodes d'analyse sédimentaire et reconstitution du basin: le Jurassique terminal-Bérriasien des Chaînes Subalpines méridionales. Thèse Sci. Caen, tome 1, 136 p.

BECCARO P., BAUMGARTNER P.O. & MARTIRE L. 2002: Radiolarian biostratigraphy of the Fonzaso Formation, Middle-Upper Jurassic, Southern Alps, Italy. Micropaleontology 48, sup. n° 1, 43–60.

CHIARI, M., CORTESE, G., MARCUCCI, M. & NOZZOLI, N. 1997: Radiolarian biostratigraphy in the sedimentary cover of ophiolites of south-western Tuscany, Central Italy. Eclogae geol. Helv. 90, 55–77.

DANELIAN, T. 1995: Middle to Upper Jurassic Radiolarian biostratigraphy of the Ionian and Maliac zones (Greece). In: BAUMGARTNER, P.O., O'-DOGHERTY, L., GORICAN, S., URQUHART, E., PILLEVUIT, A. & DE WEVER, P. (Eds.): Middle Jurassic to Lower Cretaceous Radiolaria of Tethys: Occurences, Systematics, Biochronology. INTERRAD Jurassic-Cretaceous Working Group. Université de Lausanne, Mémoires de Géologie 23, 865–876.

DANELIAN, T. & JOHNSON, K.G. 2001: Patterns of biotic change in Middle Jurassic to Early Cretaceous Tethyan Radiolaria. Marine Micropaleontology 43: 239–260.

DANELIAN, T., ROBERTSON, A.H.F. & DIMITRIADIS, S. 1996: Age and significance of radiolarian sediments within basic extrusives of the marginal basin Guevgueli Ophiolite (northern Greece). Geological Magazine 133, 127–136.

DE RAFÉLIS, M. 2000: Apport de l'étude de la spéciation du manganèse dans les carbonates pélagiques à la compréhension du contrôle des séquences eustatiques du 3ème ordre. Thèse de doctorat. Univ. Pierre et Marie Curie. 214 p.

DE RAFELIS, M., ROBIN, C. & RENARD, M. 1997: Les alternances marne-calcaire du Kimméridgien du Bassin du Sud-Est (France) : Stratigraphie haute résolution et cycles de Milankovitch – II – Géochimie, 6ème congrès Français de Sédimentologie, Publ. ASF, Montpellier 27, 83.

DE RAFÉLIS, M., EMMANUEL, L., RENARD, M., ATROPS, F. & JAN DU CHÊNE, R. 2001: Geochemical characterization (Mn content) of thrid order eustatic sequences in Upper Jurassic pelagic carbonates of the Vocontian Trough (SE France). Eclogae geol. Helv. 94, 145–152.

DERCOURT J., RICOU L.E. & VRIELYNCK B. (Eds) 1993: Atlas Tethys Palaeoenvironmental Maps. Gauthier-Villars, Paris, 260p.

DE WEVER, P., DUMITRICA, P., CAULET, J.P., NIGRINI, C. & CARIDROIT, M. 2001: Radiolarians in the sedimentary record. 533 p., Gordon & Breach Science Publ.

DUBOIS, P. & DELFAUD, J. 1989: Le bassin du Sud – Est. In : Assoc. Sedim. Fr. Dynamique et méthodes d'études des bassins sédimentaires. 277–297, Editions Technip. Paris.

DUMITRICA, P. & MELLO, J. 1982: On the age of the Meliata Group and the Silica Nappe radiolarites (localities Drzkovce and Bohunovo, Slovak Karst, CSSR). Geologicke prace 77, 17–28.

DUMITRICA, P., IMMENHAUSER, A. & MELLO, J. 1997: Mesozoic Radiolarian biostratigraphy from Masirah Ophiolite, Sultanate of Oman. Part I: Middle Triassic, Uppermost Jurassic and Lower Cretaceous Spumellarians and multisegmented Nassellarians. Bull. National Museum Natural Science 9, 106 p.

FOREMAN, H. 1973: Radiolaria from DSDP Leg 20. In: HEEZEN, B.C. et al. (Eds.): Initial Reports of the Deep Sea Drilling Project. U.S. Goverment Printing Office, Washington D.C. 20, 249–305.

GARDIN, S., BULOT, L.G., COCCIONI, R., DE WEVER, P., ISHIDA, K. & LAMBERT, E. 2000: The Valanginian to Hauterivian hemipelagic successions of the Vocontian Basin (SE France): new high resolution integrated biostratigraphical data. 6th Intern. Cretaceous Symposium, Vienna, p. 34.

GORICAN, S. 1994: Jurassic and Cretaceous radiolarian biostratigraphy and sedimentary evolution of the Budva Zone (Dinarides, Montenegro). Mémoires de Géologie (Lausanne) 18, 120 pp., 28 pls.

GROUPE FRANÇAIS D'ÉTUDE DU JURASSIQUE 1997: Biostratigraphie du Jurassique Ouest-européen et Méditéranéen: zonations parallèles et distribution des invertébrés et microfossiles In: CARIOU, E. & HANTZPERGUE, P. (Eds.). Bull. Centre Rech. Elf Explor. Prod. Mém. 17, 440 p.

HAECKEL, E. 1881: Entwurf eines Radiolarien-Systems auf Grund von Studien der Challenger – Radiolarien. Jenaische Zeitschrift für Naturwissenschaft 15 (new series vo. 8, pt. 3), 418–472.

HORI, N. 1999: Latest Jurassic radiolarians from the northeastern part of the Torinoku Block, Yamizo Mountains, central Japan. Science Reports of the Institute of Geoscience, University of Tsukuba, Section B = Geological Sciences 20, 47–114.

HULL, D. 1997: Upper Jurassic Tethyan and southern Boreal radiolarians from western North America. Micropaleontology. 43, supplement 2. 202 p.

JAN DU CHÊNE, R., BUSNARDO, R., CHAROLLAIS, J., CLAVEL, B., DECONINCK, J.-F., EMMANUEL, L., GARDIN, S., GORIN, G., MANIVIT, H., MONTEIL, E., RAYNAUD, J.-F., RENARD, M., STEFFEN, D., STEINHAUSER, N., STRASSER, A., STROHMENGER, C. & VAIL, P. 1993: Sequence-stratigraphic interpretation of Upper Tithonian-Berriasian reference sections in South-East France: a multidisciplinary approach. – Bull. centres Rech. Explor.-Prod. Elf Aquitaine 17, 151–181.

JAN DU CHÊNE, R., ATROPS, F., DE RAFÉLIS, M., EMMANUEL, L. & RENARD, M. 2000: Palynology and sequence stratigraphy in the tethyan upper oxfordian-lower kimmeridgian, S-E France, Bull. centre Rech. Explor. Prod. Elf-Aquitaine 22, 273–321.

JUD, R. 1994: Biochronology and Systematics of Early Cretaceous Radiolaria of the Western Tethys. Mémoires de Géologie (Lausanne) 19, 147 p., 24 pls.

KOCHER, R.N. 1981: Biochronostratigraphische Untersuchungen oberjurassischer radiolarien führender Gesteine, insbesondere der Südalpen. Mitteilungen aus dem Geologischen Institut der Eidgenossischen Technischen Hochschule und der Universität Zürich, Neue Folge 234, 1–184.

LAMBERT, E. 1999 : Les Radiolaires crétacés du Bassin Vocontien : Stratigraphie, Paléoenvironnement. Thèse MNHN, 209 p., 5 pls.

LANGOISSEUX, O. 2001: Stratigraphie séquentielle basée sur une approche multidisciplinaire: le Kimméridgien inférieur de la fosse vocontienne (S – E de la France). Mémoire. Diplôme d'études supérieures d'ingénieur géologue. Univ. de Genève. 89 p.

MATSUOKA, A. 1992: Jurassic and Early Cretaceous radiolarians from Leg 129, Sites 800 and 801, western Pacific Ocean. In: LARSON, R.L. et al. (Eds.): Proceedings of the Ocean Drilling Program, Scientific Results, College Station, TX (Ocean Drilling Program) 129, 203–220.

MATSUOKA, A. 1998: Faunal composition of earliest Cretaceous (Berriasian) radiolaria from the Mariana Trench in the western Pacific. News of Osaka Micropaleontologists, Spec. Vol. 11: 165–187.

MATSUOKA , A. & YAO, A. 1985: Latest Jurassic Radiolarians from the Torinosu Group in Southwest Japan. Osaka City Univ., J. Geosci. 28, 125–145.

MOUSSINE-POUCHKINE, A., AMARD, B., ATROPS, F. & AMARD, B. 1998: Lithochronostratigraphie à haute résolution dans le Kimméridgien du Bassin du Sud-Est: Implications sur les taux de sédimentation et la durée des unités biostratigraphiques. 2 ème congrès français de Stratigraphie, Paris, 125.

PEDERNEIRAS RAJA GABAGLIA, G. 1995: Stratigraphie et faciès de tempête de la rampe carbonatée du Jurassique supérieur du centre du Bassin du Sud - Est (France): calcarénites, brèches, corps glissés. Thèse Univ. Montpellier II. 239 p.

PESSAGNO, E. A. Jr. 1976: Radiolarian zonation and stratigraphy of the Upper Cretaceous portion of the Great Valley Sequence, California Coast Ranges. Micropaleontology, Special Publication 2, 1–95.

PESSAGNO, E. A. Jr. 1977: Upper Jurassic Radiolaria and radiolarian biostratigraphy of the California Coast Ranges. Micropaleontology. 23 (1): 56–113, pls. 1–12.

PESSAGNO, E. A. Jr., SIX, W. M. & YANG, Q. 1989: The Xiphostylidae Haeckel and Parvivaccidae, n. fam., (Radiolaria) from the North American Jurassic. Micropaleontology. 35 (3), 193–255.

RIEDEL, W. R. & SANFILIPPO, A. 1974: Radiolaria from the southern Indian Ocean, DSDP leg 26. In: DAVIES, T. A.et al. (Eds.): Initial reports of the Deep Sea Drilling Project. U. S. Government Pinting Office, Washington, D. C., 26, 771–814.

RÜST, D. 1885: Beiträge zur Kenntnis der fossilen Radiolarien aus Gesteinen der Kreide. Paleontographica, 34, 181–213.

SANFILIPPO, A. & RIEDEL, W.R. 1985: Cretaceous Radiolaria. In: BOLLI, H.M., SAUNDERS, J.B. & PERCH-NIELSEN, K. (Eds.): Plankton Stratigraphy. Cambridge Univ. Press, 573–630.

STEIGER E. & STEIGER T. 1994: New Radiolaria from the "Ruhpoldiger Marmor" of Urschlau (Late Jurassic, Chiemgau Alps, Bavaria). Abhandl. Geol. Bundesanstalt-A. 50, 453–466.

VINASSA DE REGNY, P.E. 1899: I radiolari delle ftaniti titoniane di Carpena (Spezia). Palaeontographia italica 4, 217–238.

YAO, A. 1979: Radiolarian fauna from the Mino Belt in the northern part of the Inuyama Area, Central Japan, Part II: Nassellaria I. Journal of Geosciences, Osaka City University 22, 21–72.

YAO, A. 1984: Subdivision of the Mesozoic complex in Kii-Yura area, southwest Japan and its bearing on the Mesozoic basin development in the southern Chichibu terrane. Journal of Geosciences, Osaka City University 27, 41–103.

YANG, Q. 1993: Taxonomic studies of Upper Jurassic (Tithonian) Radiolaria from the Taman Formation, east-central Mexico. Palaeoworld, Special Issue 3, Nanjing University Press, 164 p., 26 pls.

Manuscript received January 2004
Revision accepted February 2005

Plate 1

Scanning Electron Micrographs of Radiolaria extracted from samples of the Méouge section. Bar scale (upper right) is equal to 100 μm for all figures.

1) *Acastea* sp.cf. *A. remusa* HULL, Mg-100; 2-3) *Archaeodictyomitra apiarium* (RÜST), Mg-2; 4) *Archaeodictyomitra etrusca* CHIARI et al. Mg-29; 5-6) *Archaeodictyomitra patricki* KOCHER, Mg-29; 7) *A. patricki*, Mg-2; 8) *Archaeodictyomitra* sp.aff. *A. patricki* KOCHER, Mg-2 ; 9) *Archaeodictyomitra shengi* YANG, Mg-29; 10) *A. shengi*, Mg-77; 11) *Archaeodictyomitra spelae* CHIARI et al., Mg-77; 12) *Archaeodictyomitra* sp. A, Mg-2; 13) *Cinguloturris fusiforma* HORI, Mg-100; 14) *Cinguloturris* sp.cf. *C. fusiforma* HORI, Mg-77; 15) *Emiluvia pentaporata* STEIGER & STEIGER, Mg-100; 16) *Eucyrtidiellum ptyctum* (SANFILIPPO & RIEDEL), Morphotype A, Mg-77 ; 17) *E. ptyctum*, Morphotype B, Mg-2; 18) *Gongylothorax favosus* DUMITRICA, Mg-37; 19) *Loopus doliolum* DUMITRICA, Mg-37; 20) *L. doliomum*, Mg-29; 21) *Loopus venustus* (CHIARI et al.), Mg-2; 22) *L. venustus*, Mg-2; 23) *L. venustus*, Mg-133.

PLATE 1

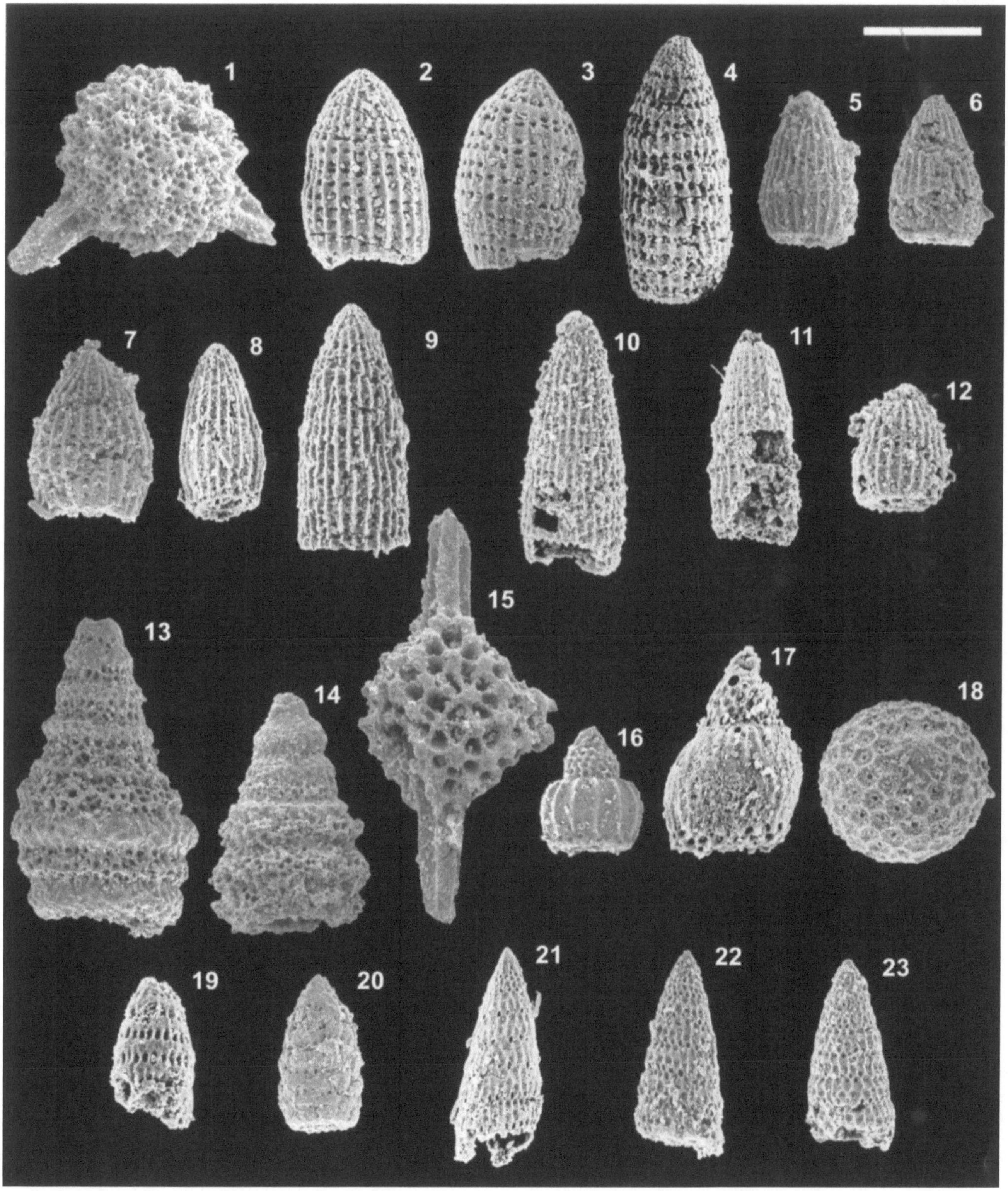

Plate 2

Scanning Electron Micrographs of Radiolaria extracted from samples of the Méouge section. Bar scale (upper right) is equal to 100 µm for all figures except of fig. 7 (=200 µm).

1) *Pantanellium oligoporum* (VINASSA), Mg-2; 2) *Praeconocaryomma scatebra* HULL, Mg-21; 3) *Protunuma japonicus* MATSUOKA & YAO, Mg-37; 4) *P. japonicus*, Mg-37; 5) *Ristola altissima* cf.ssp. *R. a. altissima* (RÜST) *sensu* Baumgartner et al. 1995b, Mg-37; 6) *Saitoum pagei* PESSAGNO, Mg-21; 7) *Spongocapsula palmerae* (PESSAGNO), Mg-21; 8) *Stichocapsa tuscanica* CHIARI, CORTESE & MARCUCCI, Mg-77; 9) *Suna* sp., Mg-77; 10) *Tethysetta* (?) sp., Mg-77; 11) *Triactoma foremanae* MUZAVOR; Mg-77; 12) *Tripocyclia* sp.cf. *T.luciae* JUD, Mg-100; 13) *Tritrabs* sp.cf. *T. exotica* (PESSAGNO), Mg-37; 14) *Williriedellum carpathicum* DUMITRICA, Mg-77 ; 15) *W. carpathicum*, Mg-21; 16) *Williriedellum crystallinum* DUMITRICA, Mg-37; 17) *Zhamoidellum ovum* DUMITRICA, Mg-37 ; 18) *Z. ovum*, Mg-21; 19) *Z. ovum*, Mg-2.

PLATE 2

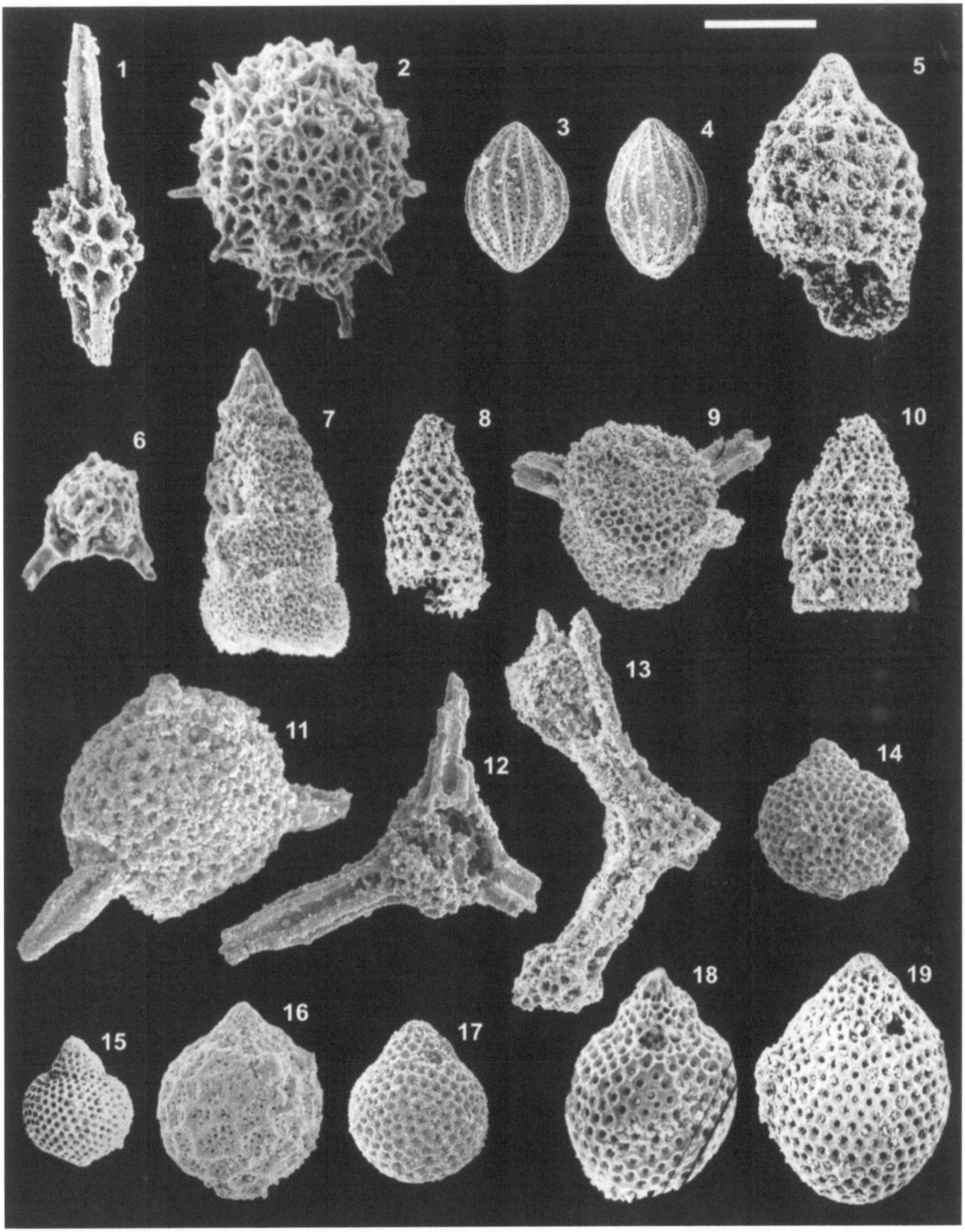

0012-9402/06/01S049-14
DOI 10.1007/s00015-006-0609-y
Birkhäuser Verlag, Basel, 2006

Eclogae geol. Helv. 99 (2006) Supplement 1, S49–S62

The plankton turnover at the Permo-Triassic boundary, emphasis on radiolarians

PATRICK DE WEVER[1], LUIS O'DOGHERTY[2] & ŠPELA GORIČAN[3]

Key words: Permian, Triassic, extinction, radiation, protists, Radiolaria
Mots clés: Permien, Trias, extinction, radiation, protoctistes, radiolaires

ABSTRACT

The examination of plankton biodiversity through Permian-Triassic period seems to display different patterns of evolution depending of the scale of study (taxonomy stratigraphy or biogeography). In this paper we present the state of the art of the plankton turnover at the Permo-Triassic and we review more precisely the pattern of extinction and recovery of radiolarians during such period, because at that time, plankton was essentially represented by radiolarians. At a global scale the end-Permian to early Triassic period is marked by several strong extinctions in the marine realm, and in the radiolarians they occur progressively as exemplified by two orders of radiolarians: Albaillelaria and Latentifistularia. Nevertheless, this period is marked by a tremendous post-crisis diversification in both at family and generic level, more than extinction; an actualized revision of the diversity at the family level is also offered in our review. At a moderate scale (i.e. genera and in a region) the modifications appear impressive while at specific and regional domain the message is not so clear, one can wonder if some crisis manifestation results from a taxonomic accident or from preservative conditions. In fact, strangely enough more the number of studies is, more the diversity is high, and oppositely less they are, more the provincialism is evoked.

RESUME

L'examen de la biodiversité du plancton au passage Permien-Trias semble montrer différents types de manifestation selon l'échelle d'observation aussi bien en terme de stratigraphie que de taxonomie ou de géographie. A cette époque, le plancton était surtout représenté par les radiolaires. A une échelle globale cette période est certes marquée par des extinctions mais elles sont progressives comme le prouvent les deux ordres de radiolaires: les *Albaillelaria* et des *Latentifistularia*. Néanmoins cette période est surtout marquée par une énorme diversification post-crise, plus encore que par l'extinction. A une échelle moyenne (ex. niveau générique et celui d'un pays) les renouvellements de faune semblent impressionnants, alors qu'à l'échelle des espèces et de bassins particuliers, on est conduit à se demander si des modifications résultent de réelles crises biologiques ou de problèmes de conservation. Il est amusant de constater que plus le nombre d'études est élevé, plus la diversité apparaît grande et à l'opposé, plus elles sont rares, plus le provincialisme est évoqué.

1. Introduction

Mesozoic deep-sea sedimentary rocks are widespread, the successions are relatively complete and little altered, and they have been sampled extensively. Undoubtedly, the appearance of evolutionary innovation among the plankton is enhanced by these circumstances, furthermore, many of the best preserved Paleozoic sedimentary rocks were deposited in shallow, epicontinental seaways where open-ocean plankton might not be present or abundant. The minute calcite plates of calcareous nannoplankton, delicate opaline silica tests (radiolarians) and frustules (diatoms) might well be destroyed in more highly altered sediments, and if early dinoflagellates lacked archeopyles, their affinities would not be recognizable. Nonetheless, the spectacular diversity changes recorded after the Permo-Triassic crisis within the sedimentary successions containing well-preserved fossils indicate that during the Early Mesozoic a genuine radiation of microplankton took place, one that included heterotrophic as well as photoautotrophic protists. Undoubtedly, the radiolarians are the most important and dominant group among the microplankton after the Permo-Triassic crisis. This is the reason why we analyse in this paper the changes that occurred in this group.

1.1 Bioevents concept and faunal turnover

The idea of global bioevents is under discussion since the very beginning of the science of Earth history. Georges Cuvier (1769–1832) already recognized that most intervals occurred in Earth history from time to time, each of them with a strong

Correspondance: Patrick De Wever
[1] Muséum National d'Histoire Naturelle, 43 Rue Buffon, F-75005 Paris, France. Email: pdewever@mnhn.fr
[2] Facultad de Ciencias del Mar, Universidad de Cádiz, 11510 Puerto Real, Spain. Email: lodogher@uca.es
[3] Paleontoloski Institut Ivana Rakovca ZRC SAZU, Novi trg 2, SI-1000 Ljubljana, Slovenia. Email: spela@zrc-sazu.si

Table 1. Effects of the main crisis of biodiversity on families and genera with indication of calculated species loss (simplified from Jablonski 1996)

	Families		Genera	
	Observed extinction (%)	**Calculated species loss (%)**	**Observed extinction (%)**	**Calculated species loss (%)**
End Ordovician	26	84	60	85
Late Devonian	22	79	57	83
End Permian	51	95	82	95
End Triassic	22	79	53	80
End Cretaceous	16	70	47	76

faunal turnover, later called catastrophe, extinction event, faunal change, bio-event, and so on. The concept of catastrophe seems somewhat exaggerate as described and emphasized later by d'Orbigny, but one should set this idea in its proper context: at that time the Earth was supposed to be approximately 6000 years old only.

Bioevents, often in connection with a significant change in lithology, were used by d'Orbigny and most scientists concerned in the first three quarters of the 19th century in order to subdivide the Phanerozoic. The result of these works is the stratigraphic scale. Nevertheless, three decades ago the fact of bio-event was disputed and even if these have now been accepted, the global synchronicity of the events has often been doubted, and the question as to causation has been discussed very controversially. If we still use the term catastrophism, we are not using it in its original meaning (Cuvier's definition), but accepting his valid observation of the existence of bio-events. According to the main phases these bioevents are named extinction-radiation or faunal turnover sequence.

1.2 Scale of time

Some supposed catastrophes were in fact only the result of a lack of record and they are no more considered as catastrophic events. One should also be aware of global *vs.* regional catastrophes. If the distinction is quite easy in theory, practically it is not so obvious, as it will be discussed later herein.

Biotic crises and events are sometimes distinguished. Differentiating from an event, a crisis spans a relatively long time interval. This is the case where over a relatively long period the extinction rate exceeds the origination rate. Crises do not only occur in times of long-term environmental changes or fluctuations, but also as a terminal phase of a taxon, when the diversity of a group of organisms decreases, sometimes towards to zero.

1.3 Crisis: at which taxonomic level?

Biodiversity changes in the paleontological record show that mass extinctions and recoveries occurred during some relatively short time intervals. These events are named and analysed as "evolutionary crises". When considering crisis one might discuss the taxonomic level involved, the message can vary according to the category. Plots of generic turnover should provide a better indication of morphological innovation than those of species, because genera are erected according to major features of morphology.

The problem of estimating the importance of extinctions and recovery depends upon the involved taxonomic level and the way the counting is conducted. Sepkoski's family and genus level (1989) summaries have allowed the species-level extinction intensities of the five major extinctions to be calculated (Table 1).

The fossil record of planktonic life has the advantage of abundant specimens, good stratigraphic control, closely spaced samples, and broad geographic distribution. The quality, however, varies substantially from group to group on the one hand, and is largely dependent on the desired stratigraphic level of precision on the other hand.

At a global scale, the Permian-Triassic boundary was characterized by the extinction of 50% of invertebrate families and 90% of invertebrate species; extinction of fusulinids, rugose and tabulate corals (orders), 2 orders of bryozoans, productids (order) and several orders of articulate brachiopods, trilobites (class), euryperids (orders), blastoids (sub-phylum) and several subclasses of attached echinoderms (mostly crinoids).

1.4 Estimation and counting: how variations can be illustrated

For several years, numerous researchers have been greatly interested in the study of biodiversity and its evolution through time. These studies pointed out several crisis periods. Nevertheless, it must be mentioned that the results depend on the methodology used. Some researchers simply counted the number of taxa presented in a publication without any taxonomic analysis. That is to say that the number of taxa in a particular work often depends on the taxonomic point of view of the author(s) of the used publications. Hence simple counting might be a source of mistakes. It is, therefore, advantageous if the counting of taxa is done by one or several specialists with practice in the respective group and after a thorough taxonomic review.

1.5 Recovery

A study by David Jablonski (1996) shows that recovery from mass extinctions differs from one geographical region to another. Regions differ greatly from each other not only in terms of which and how quickly species diversified, but also in the ratio of surviving local species to foreign invaders. In general,

Phanerozoic patterns of phytoplankton radiation and extinction parallel those documented for skeletonized marine invertebrates, both augmenting and constraining thought about evolution in the oceans. Plants and animals may be the most conspicuous fossils in Phanerozoic rocks, but planktonic protists are the most abundant.

Much of our ability to discriminate among various hypotheses explaining the evolutionary changes within geologic times not only depends on careful and detailed comparisons of evolutionary patterns, recognised in trophically, ecologically, reproductively, and developmentally different groups of organisms, but also is essential to know how the earth system climate drives the evolutionary process during critic periods of the past.

The time for recovery in the earliest Triassic was also highly variable depending on marine oxygen levels, like it has been recently shown (Twitchett et al. 2004). These authors observed a faster recovery in well-oxygenated environments (seamounts) after the P-T extinction in the middle Griesbachian. An apparent delay in recovery after the end-Permian extinction event was observed in deeper-marine settings due to widespread and prolonged benthic oxygen restriction.

2. The Permo-Triassic Setting: Scenario for a crisis

2.1 Environmental constraints

According to the classic Permian paleogeography, only one continent (Pangea) was present at that time and surrounded by only one super-ocean (Panthalassa). In this land-sea distribution, the concomitant drastic decrease of shelf surface added by the Carboniferous-Permian glaciation (Scheffler et al. 2003) resulted in a drastic dynamic change of the global ocean circulation (Fluteau et al. 2001).

An important event, occurring over a very brief span of geological time (initiated by 251.7 ± 0.4 Ma and terminated by 250.2 ± 0.3 Ma; Kamo et al. 2003) took place in this scenario: the Siberian flood volcanism. A comparatively extensive flood volcanism occurred somewhat earlier in China (Emeishan traps, about 259 ± 3 Ma old; Courtillot & Renne 2003). The large amount of volcanic gases played a primary role in the Permo-Triassic extinction (Wignall 2001) but the rapid and strong negative $\delta^{13}C$ excursion (c.a. 4–6‰), (Benton & Twitchett 2003) is too great to be explained solely by the catastrophic collapse of ocean primary productivity in the late Changxingian (Wang et al 1994; Hallam & Wignall 1997). In this scenario and an additional source of C^{12} to the ocean-atmosphere system is required, but the causes of the isotopic excursion seem to be complex and multiple (Korte et al. 2004). In this sense, the Siberian volcanism triggered an initial global warming at the P-Tr boundary, which in turn melted clathrate bodies, releasing a large input of methane into the atmosphere causing more warming (Heydary & Hassanzadeh 2003). The process continued in a positive feedback spiral that has been termed the 'runaway greenhouse' phenomenon (Benton & Twitchett 2003) leading to the most dramatic extinction known.

In this palaeogeographic reconstruction, the Late Permian was characterized by falling sea levels (Ross & Ross 1988) exposing large areas in the northern high latitudes. But superimposed on this longer trend a large-scale transgressions is record in China, Iran, western Tethys, and in central Europe. This transgression became even more pronounced in the latest Permian (Wignall et al. 1998), just before the Permo-Triassic transition, and reached its maximum in the Early Triassic, with flooding of many areas of the Tethys and North America. In fact the Permo-Triassic boundary interval was a time of second order highstand of sea-level. This reinforces the fact that the end-Permian mass extinction cannot, in any regard, be considered as the product of sea-level fall (Hallam & Wignall 1999). Increasing values of $^{87}Sr/^{86}Sr$ since the latest Permian-Early Triassic time (Korte et al. 2003) are related to enhanced continental weathering under humid climatic conditions in the latest Permian and the lack of a dense land vegetation in the Early Triassic, prior to the Spathian, which are in agreement with the observed sea-level trend. A consequence of this flooding was the spread of anoxic bottom waters into the shallow marine habitat during the early stages of transgression, which in turn may have played an important role during the aftermath of the P-Tr extinction event (Hallam 1989).

In order to establish the duration of the Permo-Triassic extinction several attempts were made and estimates for the duration of the extinction vary according to different inclinations in fixing relevance to particular fossil groups or sampling locations. Holser & Magaritz (1992) suggest a duration of 5–10 Ma for the crisis, whereas Hallam & Wignall (1997) prefer a rapid but not instantaneous mass extinction, although unfavourable climatic and environmental conditions prevailed during most of Early Triassic (Wignall & Hallam 1992; Bottjer et al. 1996). Radiometric ages point to a duration of 0.7–0.3 Ma for the extinction interval, or at least for the interval between the $\delta^{13}C$ decline and the P-Tr boundary (Bowring et al. 1998). Other studies using estimates of sedimentation rates suggest more rapid rates of ecosystem collapse, about 10–30 kyr for marine extinctions (Twitchett et al. 2001; Rampino & Alder 1998). However, analysis of terrestrial flora (Looy et al. 2001) gives a different picture. The time lag between terrestrial ecosystem collapse and plant extinction had a duration of 0.5–0.6 million years, showing also a temporary diversity increase after the negative $\delta^{13}C$ excursion.

2.2 For various groups: real extinctions or incompleteness of the fossil record?

Early Mesozoic radiating protists include taxa characterized by both siliceous and calcareous skeletons, as well as unmineralized groups. Possibly the break up of continents and the partitioning of the world's oceans, including strengthened circulation and upwelling of nutrients, contributed to the ecologic opportunities for diversification of planktonic organisms. After the Permian crisis during the Mesozoic, planktonic protists underwent remarkable evolutionary innovations and radiations.

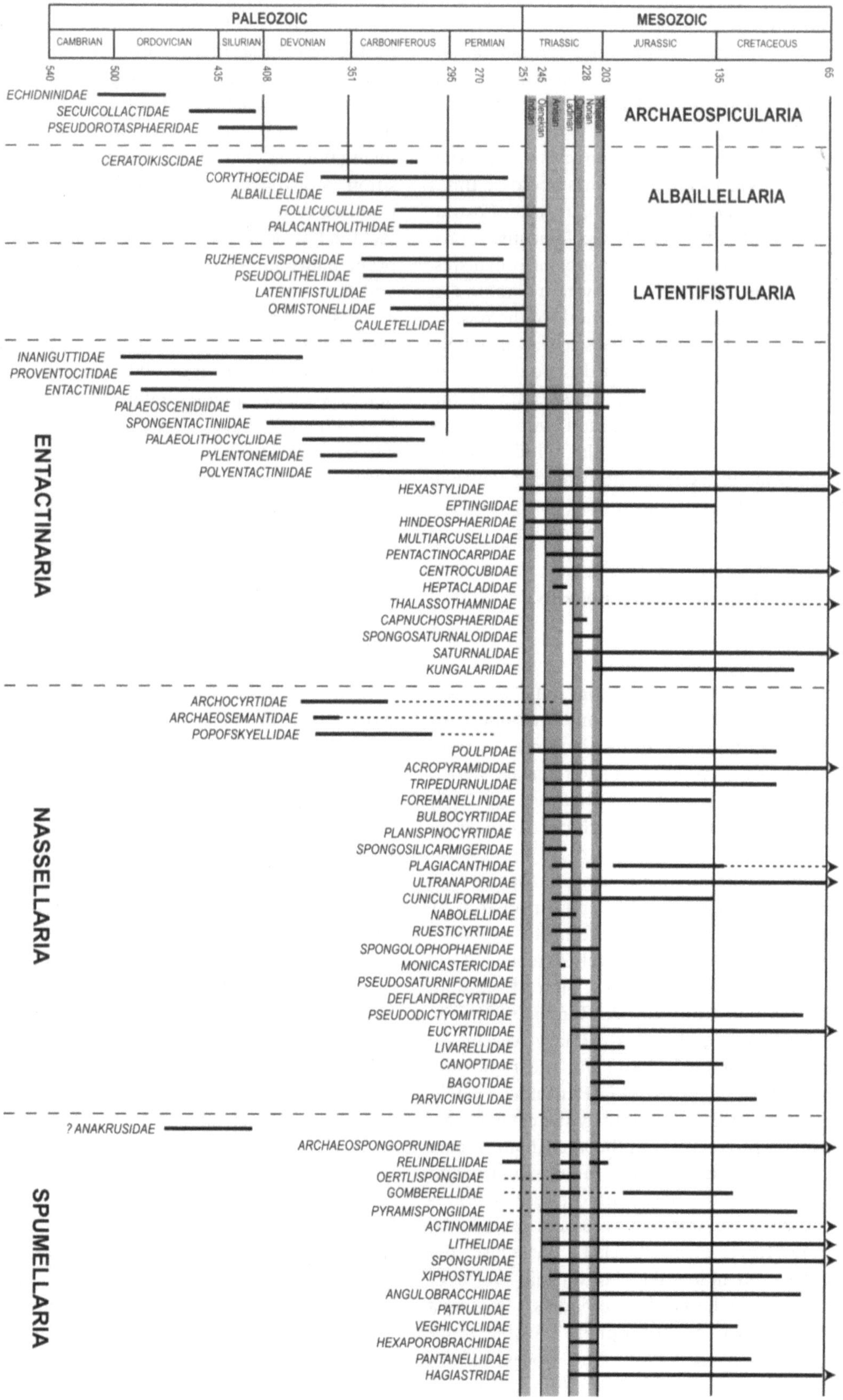

Fig. 1. Stratigraphic ranges of radiolarian families through time (Redrawn from De Wever et al. 2001).

Radiolarians, dinoflagellates, calcareous nannoplankton, diatoms, silicoflagellates, and planktonic foraminifers either appeared and radiated for the first time, or diversified from preexisting forms into a greater variety of new types. The causes of the early Mesozoic plankton revolution remain uncertain; probably they are as complex as the factors that promoted the radiation of skeleton metazoans and protists at the beginning of the Paleozoic. The fluctuations of diversity in planktonic protists have been suggested many times to be a function of variations in primary productivity and heterogeneity of water masses. During intervals of strong heterogeneity of water masses and high productivity, the oceans can support many specialized species that live effectively, at least for a part of their life-cycles, within a narrow range of environments. When productivity drops and oceanic water masses become less structured and more homogeneous, narrowly adapted species disappear. Obviously, as it has been shown by paleontology, petrography and geochemistry, the oceans' physical structure provides opportunities for plankton speciation and the biologic inter-actions of constituent populations. Unfortunately, such biologic interactions are not understood, even for living plankton, and so explanations concentrate on types of oceanographic changes that are known to correlate with stratigraphic changes in diversity.

The fossil record indicates that recovery of ecosystems affected by the crisis needs about 5 Ma (Bottjer et al. 1996; Schubert & Bottjer 1995), which seems to be achieved during Anisian time (Gall et al. 1998). The apparently low recovery together with the poor quality of the fossil record characterizing the interval aftermath the P-Tr extinction modifies the perception of the magnitude of the end-Permian extinction, as well as, it masks the paths undergone by the recovery. These particular patterns have led to explain the recovery in terms of recolonization of damaged Early Triassic environments by the so-called "Lazarus taxa", that is, super-specialized taxa that migrate to "unknown" refuges during extinction periods and that reappear "unchanged" after millions of years when adequate environmental conditions are re-established. One may wonder if it is just the simple variation in the quality of fossil record which masks our perception of the extinction event magnitude and hides devlopments during the recovery phase.

According to Twittchett (2001) the incompleteness of the fossil record in the Early Triassic was caused by a sharp decrease in productivity at the end of the Permian. The drop of productivity during the crisis in turn lead to a reduction of the biomass, which must be readjusted by decrease in abundance (but retaining body size) or decrease in body size (but with similar population size). Those forms that reduce their population sizes play a "Lazarus role" giving an apparent extinction, because they reduce in number so much that they are below paleontological detection levels (Wignall & Benton 1999). Only when the primary productivity returns to pre-crisis normal levels, Lazarus taxa seem to "reappear". Obviously, those taxa that reduce their body size, but retain or even increase their population size would be the best candidates to become fossilized. In this sense the picture displayed by the fossil record is biased. We will show later how this mechanism is illustrated in the radiolarian record at the Permo-Triassic crisis.

Among Paleozoic plankton, remains of radiolarians (together with those of acritarchs) are particularly abundant, although representatives of other planktonic groups such as tintinnids and chrysophytes are found sporadically. We therefore focus hereafter on radiolarians, undoubtedly the dominant plankton fossils in the Paleozoic and Triassic oceans. They first occurred in the Cambrian (Won & Below 1999) and became morphologically complex through time (like Foraminifera) with the number of species increasing significantly in the Ordovician and Devonian.

3. The radiolarian record

3.1 Diversity of radiolarian families at a global scale

Some evaluation of biodiversity for the Phanerozoic was attempted several years ago for all fossil groups at the family level and these included radiolarians (Benton 1993). This tentative analysis, however, does not present a correct image of the radiolarian world: only 35 families are referred to the whole geologic period whereas more than 130 are known (De Wever et al. 2001, 2003). Some analyses were also done at the Permo-Triassic (Erwin 1993; Yao & Kuwahara 1997; Kozur 1998) but these curves also show contradictory results. In order to give a realistic picture we present herein curves of radiolarian diversity following the recent work of De Wever et al. (2001) who have homogenised the taxonomic criteria used for the Paleozoic, Mesozoic and Cenozoic, and updated the stratigraphic ranges for all families known. At a global scale (Figs. 1, 2), rapid changes in the biodiversity of radiolarian assemblages appear to be fairly rare throughout the geological record except during the Early to Middle Triassic, after the major mass extinction event which occurred at the Permian/Triassic boundary. During this time interval of nearly 20 Ma a large number of families occurred for the first time (Fig. 2).

Among the six radiolarian orders occurring in Paleozoic-Mesozoic deposits (Fig. 1), three were more or less extinct at the Late Paleozoic crisis (Albaillellaria, Archaeospicularia and Latentefistularia). Only one order clearly crossed the event (Entactinaria), and two (Nassellaria, Spumellaria) sporadicaly occured before, but clearly diversified after the crisis.

Representatives of the **Archaeospicularia** are the oldest radiolarian order and the only one that clearly became extinct before the Permo-Triassic crisis. The **Entactinaria** had a continuous and regular diversification during the Paleozoic, and crossed the Permian/Triassic boundary to be rather common and diversified in the Mesozoic.

Evolution within the **Albaillellaria** seems to have been continuous and regular during the Paleozoic. During the Carboniferous and the Permian, the Albaillellidae underwent a great and biostratigraphically important diversification. In the Late Carboniferous, simplifications of the internal spicule gave

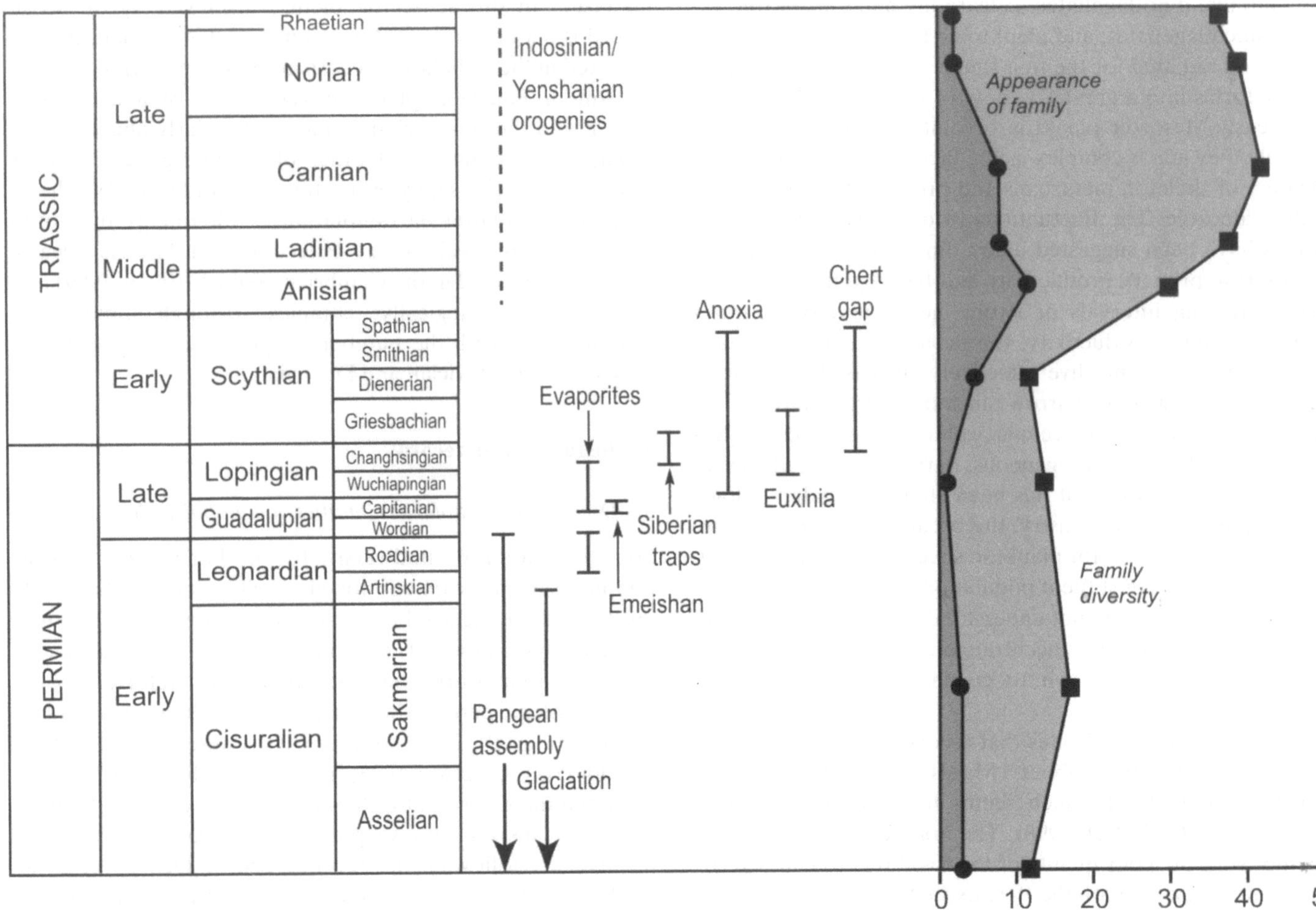

Fig. 2. Radiolarian diversity at the family level facing the main events of the Permo-Triassic boundary. The number of appearance of new families clearly corresponds to the periods of main stresses; modified after Kidder & Worsley (2004).

rise to the last Albaillellaria (Follicucullidae and Palacantholitidae). During the Late Permian the evolution of some albaillellarians underwent two different directions: one was to build a complex shell (*Neoalbaillella*), the other was to simplify this shell (*Follicucullus*); only the latter crossed the boundary (Bragin 1991; Sugiyama 1992, 1997; Yao & Kuwahara 1997). It is known that under environmental stresses, some radiolarian species seem to be able to adopt a regressive/primitive character. Often this process happens along with a reduction or even loss of skeleton (Guex 1981, 2001; De Wever et al. 2003).

The evolution of **Nassellaria** was very slow during the Late Paleozoic and they practically disappeared during the Permian so that one can say that the main nassellarian groups appeared during the Triassic. The oldest Triassic nassellarians are morphologically similar and possibly related to Paleozoic Archaeosemantidae.

The Permian is not marked by the appearance of new orders of radiolarians, although some representatives of probable spumellarians occur together with the last Latentifistularia (Cauletellidae). Some representatives of Latentifistularia (*Ishigaum* and *Cauletella*, *cf.* Takemura et al. 2003) seem to cross the P/T crisis, and then the order became extinct in the Early Triassic. The presence of Spumellaria in the Paleozoic is still uncertain; in fact, many of the Paleozoic spherical radiolarians, previously considered to be Spumellaria contain the remains of an inner spicule, indicative of Entactinaria (De Wever et al. 2001). Since their identification depends on the study of the innermost skeleton, and in most cases this is absent or not well preserved, they could be confused with some entactinarians (in which the initial spicule is dissolved), with which they share their general morphology. One could consider, for example, that they are represented by some many-layered spongy skeletons occurring in the Permian assigned to the Relindellidae and Archaeospongoprunidae. Nevertheless, studies in Oregon and South China (Blome & Reed 1992; Shang et al. 2001) show that true spumellarians (Pantanelliidae-like forms) seem to be already present before the P/T boundary. More recently some representative of several families were reported from uppermost Permian from China (Feng et al. in progress); they are respectively Pyramispongidae

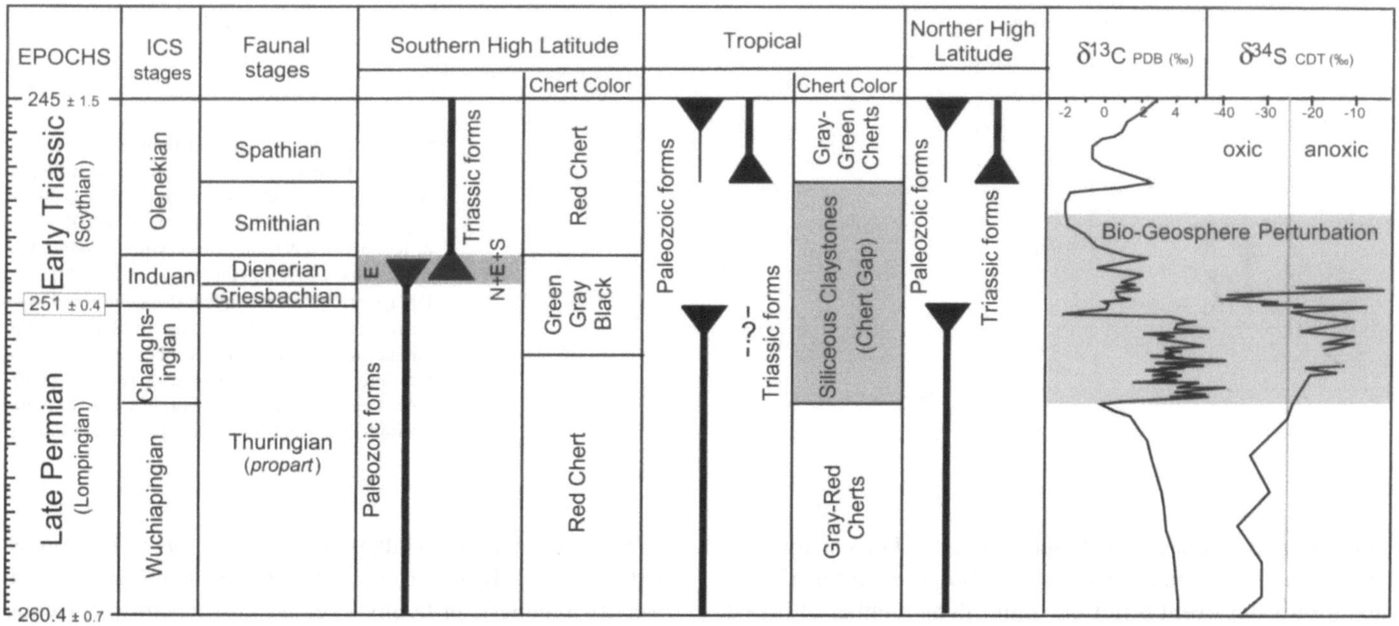

Fig. 3. Stratigraphy of events according to their geographic (latitudinal) position. It points out the lack of information for the critical levels; Carbon isotopes from Benton & Twitchett (2003) and Atudorei (1998), Sulfur isotope from Kato et al. (2002).

(*Tetrapaurinella, Paurinella*), Gomberellidae (*Tamonella*) and Oertlispongidae (*Paraoertlispongus*). Since these genera extent upward, but have not been yet mentioned in the Lower Triassic, they are considered as Lazarus taxa by these authors.

According to the data reviewed above, no new orders of radiolarians appeared after the Permian-Triassic crisis. The only exception might be Nassellaria, if we consider that Mesozoic nassellarians are not direct descendants of Paleozoic representatives, all of which disappeared by the end of the Carboniferous. The existing lineages of radiolarians, nevertheless, produced an impressive number of new morphotypes grouped into new families. Of the 14 families known in the Permian, 9 continued to exist in the Mesozoic and among the 55 families known in the Triassic, 25 appeared during the Early and Middle Triassic. The Middle Triassic was undoubtedly the main epoch for radiolarian radiation; more than a third of the total number of radiolarian families, recognized from the Cambrian to the Present originated during this important period of plankton radiation (Figs. 1, 2).

What could have caused these tremendous radiations and evolutionary innovations? For the most part the causes remain uncertain, but one of them could be the break-up of continents and the partitioning of the world's oceans that produced strengthened circulation, as suggested by ocean modelling, and induced up-welling of nutrients (De Wever et al. 1994). It is known that during the Early Mesozoic, modern oceanic basins began to form (Neotethys spreading) and new oceanic current systems appeared. Numerous microplates began their separation in the Tethys Ocean, creating new seaways in the western Tethys and modifying oceanic circulation. (De Wever & Baudin 1996).

In the aftermath of the mass extinction, when the last representatives of the orders Albaillellaria and Latentifistularia completely disappeared, with the exception of some genera: *Follicucullus, Ishigaum, Cauletella* which are mentioned till the Early Triassic, (Sugiyama 1997; Takemura et al. 2002), the Early Triassic faunas were characterized by extremely low diversity and poor preservation (Hori et al. 2003; Sashida 1983, 1991; Sugiyama 1992, 1997; Suzuki et al. 2002). These faunas mostly contain entactinarians (*Cryptostephanidium, Entactinia, Hegleria, Parentactinia, Pseudostylosphaera, Tiborella*), some sparse mono- and dicyrtid nassellarians (*Archaeosemantis, Hozmadia, Poulpus, Spongosilicarmiger?*), and rare spumellarians with inner spicule (*Glomeropyle, Plafkerium?*). In addition to this extremely low diversity, the poor preservation and the scarcity of good outcrops, limit the information that is now available about Early Triassic faunas (Fig. 3). Furthermore, we stress the importance played by the apparent abrupt disappearance of radiolarian cherts that were common in the deep-marine sections of South China, Japan, and Western Canada in the Late Permian (Fig. 4). They reappeared only in the Spathian after a "chert gap" of 7 to 8 Ma (Isozaki 1994; Kakuwa 1996). This is, in part, the reason why the abrupt change in radiolarians is amplified after this long period of instability. For the reasons mentioned above the current information available for the radiolarians during the Early Triassic should be considered quite incomplete and the evolutionary pattern undergone by the group at this interval remains poorly known.

Following the "chert gap", the rapid diversification of Triassic radiolarian assemblages during the Early to Late Triassic is the most prolific that has ever been recorded in the Phanerozoic. For example, the extinction of radiolarians between the

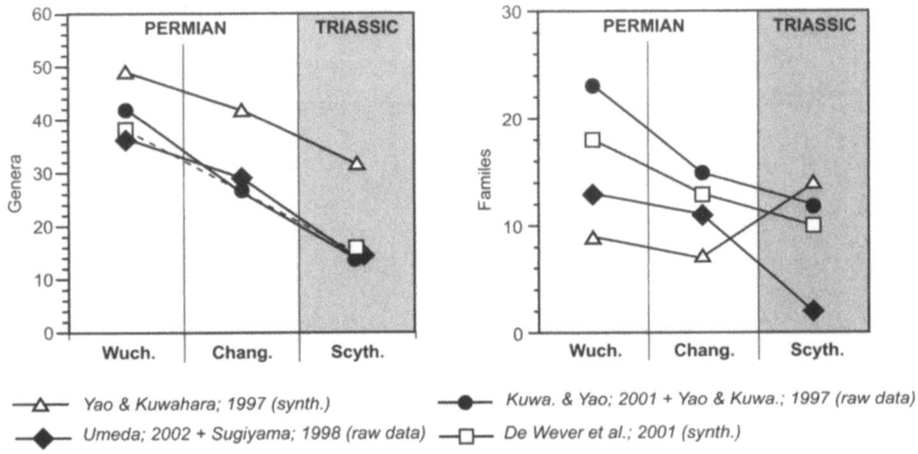

Fig. 4. Numbers of quoted families and genera near the Permian-Triassic Boundary according to several authors, and from raw data (Umeda 2002 + Sugiyama 1997 and Kuwahara & Yao 2001 + Yao & Kuwahara 1997) and from synthetic data (Yao & Kuwahara 1997 and De Wever et al. 2003). For genera all curves follow the same trend, they decrease from Wuchiapingian to Scythian. For families appears a discrepancy between curves. In fact information from Yao & Kuwahara (1997) have been complied but not previously homogenised. Data treatment influences the conclusions.

Permian and Triassic (Fig. 1) seems to have been a slow progressive and important phenomenon. But, the most distinctive feature for this event is not the diminishing number of taxa at both family and generic levels near the P/T boundary, but rather an important diversification that followed it, during the Triassic: nine families cross the P/T boundary (six of these Paleozoic families became extinct by the end of the Triassic) and 45 new families originated during the Triassic; in other words the fauna is almost completely renewed. The Figure 4 displays a similar tendency between family and generic curves, but a reverse tendency is observed in the data from Yao & Kuwahara (1997). The differences can be explained because the supra-generic assignments into families followed by these authors is different from that we have used for our curves (De Wever et al. 2001). Like it was mentioned in the foreword of this article, Figure 4 well-displays how the biodiversity (especially at family level) is strongly biased by the taxonomy employed in the analysis of faunal turnovers. Irrespective to this correlation, the Middle Triassic is a period of unprecedented blooming and accelerated evolution of Entactinaria, Nassellaria and Spumellaria (Fig. 1). This figure clearly shows that Entactinaria represent the most numerous radiolarian group that survived both, the P/T boundary and the Early Triassic crisis, however, the nassellarians represent the group with the most rapid evolution and diversification during the Triassic and later. This fact well illustrates how the nassellarians display a powerful recovery during post-extinction periods as has been signalled for the Mesozoic time (O'Dogherty & Guex 2002) Since the Anisian the entactinarians, nassellarians and spumellarians diversified simultaneously resulting in new families that will become the base for the spectacular radiation at generic and species level.

Usually under anoxic environments radiolarian diversity minima occur simultaneously with the $\delta^{13}C$ positive excursion (O'Dogherty & Guex 2002). During the Changxingian a minor positive peak is concomitant with the end Permian extinction (cf. Holser et al. 1991), but just before the Permo-Triassic boundary a conspicuous negative carbon isotopes excursion (5–6‰) is recorded (Hallam & Wignall 1997; Benton & Twitchett 2003; Korte et al. 2004). The available data for the diversity at generic or family level are insufficient to correlate to the carbon isotope curve. We can only observe that extinction (and the drop on diversity) at family or generic level during the end-Permian (cf. Fig. 3) seems to be linked with decreasing values of $\delta^{13}C$. But, it seems obvious that any interpretation of the $\delta^{13}C$ signal without taking into account other geoindicators and a detailed correlation has actually not much paleobiological significance.

3.2 P-T radiolarian diversity at a detailed scale

Available data

Only 21 years of investigation covers all our information on Early Triassic radiolarians; a rather moderate period of research, and with still few data. Table 2 reports the available taxonomic data for the Early Triassic: 19 papers from the circum-Pacific region and only 12 papers from the Tethyan area (South China, Thailand and Turkey) (Fig. 5). It is also interesting to note that the majority of publications deal with Spathian radiolarians. The earliest Triassic (Griesbachian) radiolarians, on the other hand, are known from New Zealand only and were first described in 2002. The review clearly shows the weakness of our knowledge about the radiolarian stratigraphy during this phase of recovery, and the number of taxa identified at genera or species level is extremely low. It is now known how poor preservation can influence the apparent composition of the fauna (O'Dogherty et al. 2003). The small number of species illustrated or described by publication (Table 2) also strongly contributes to an unrealistic vision about the taxonomic re-organization and the radiation of radiolarians during the Early Triassic. This insufficiency of data, evidently, can be, in some cases, provisionally translated into a wrong interpretation about the paleobiogeography and phylogeny at both order and family levels.

All these papers report the presence of highly abundant

Table 2. Synthesis of published papers concerning Early Triassic radiolarians with respectively: Age, author(s), number of recorded genera/number of species and nov. sp. This table underlines how new study implies new taxa, as a result of the very poor knowledge of these fauna of that precise period.

Faunal stage	Author	Gen./Sp.	n.sp
Griesbachian	Takemura et al. 2002, 2003 (Waipapa, New Zealand)	5/?	
Dienerian	Kozur et al. 1996 (Karakaya, Turkey)	2/2	2
	Sashida et al. 1998 (Phatthalung, Thailand)	5/5	
	Feng et al. 2001 (Hunan, China)	2/2	
	Kamata et al. 2003 (Waipapa, New Zealand)	7/?	
Smithian	Blome & Reed 1992 (Oregon, USA)	4/4	
	Yao & Kuwahara 1997 (Inuyama, Japan)	4/7	
	Yao & Kuwahara 1999 (Ziyun,China)	2/2	
Spathian	Sashida 1983, 1991 (Kanto, Japan)	7/13	8
	Zhang et al 1999 (Nadanhada, China)		
	Bragin 1991 (Shikote Alin, North Eastern Russia)	4/4	
	Sashida & Igo 1992, Sashida et al. 1993 (Phatthalung, Thailand)	6/7	1/4
	Feng 1992 (Southwest China)		
	Feng & Liu 1993 (Yunnan, China)		
	Sugiyama 1992, 1997 (Gifu, Japan)	22/26	19
	Nagai & Mizutani 1993 (Gifu, Japan)	11/14	
	Kusunoki & Imoto 1996 (Kyoto, Japan)		
	Yao & Kuwahara 1997 (Inuyama, Japan)	7/17	
	Isogawa et al. 1998 (Ashio, Japan)	12(17)	60
	Kamata 1995, 1999 (Tochigi, Japan)	16/30	5
	Feng et al. 2000 (Yunnan, China)	12/27	
	Sashida et al. 2000 (Northern Thailand)	38/50	4
	Yao & Kuwahara 2000 (Guizhou,China)	11/14	
	Suzuki et al. 2002 (Ashio, Japan)	15/30	
	Hori et al. 2003 (Kaka Point, New Zealand)	7/15	3

Entactinaria along with rare Nassellaria and undeterminable Spumellaria. The Late Permian extinction undoubtedly played an important role in the evolutionary history of Radiolaria. A quasi-complete renovation took place during the aftermath of the Permian crisis at generic level (only a few genera belonging to 10 families), although at order level all the groups seems to cross the critical boundary (Fig. 1). It is interesting to note the presence of some radiolarians with Triassic affinities (i.e.: pantanelliids) in the uppermost Paleozoic strata (Shang et al. 2001), as well as some Paleozoic forms (Fig. 3) which have been recorded in the Early Triassic (Sugiyama 1992, 1997; Yao & Kuwahara 1997, 1999).

Discussion

Evoking a crisis requires precise definition of the involved level (taxonomic, geographic or stratigraphic) because a phenomenon may appear sharp at one scale but not at a different one. That is why at the family level and global scale radiolarians did not seem to be strongly affected by the Permo-Triassic crisis, but seems very sharp at specific level (i.e. 96% of species disappeared, *cf.* Kozur & Mostler 1982). Other authors (Suzuki et al. 2002) from a database of published data propose at least three distinct turnovers during the Early Triassic time, the most conspicuous turnover being that of the Late Olenekian-Anisian boundary (143 species disappearing and 118 new species appearing).

The small quantity of data is undoubtedly related to the scarcity of localities with a complete record for the Permo-Triassic boundary. As was mentioned earlier, the preservation of radiolarian faunas were so strongly affected by the end Permian environmental collapse that a conspicuous "radiolarite gap" has been described (Isozaki 1994, 1997; Kakuwa 1996) from the uppermost Permian (*Clarkina iranica* Zone of conodont) until the end of the lower Olenekian (top of the Smithian) lasting about 10 Ma. But recent investigations show that this gap, spread over the world, does not exist in high southern latitudes (New Zealand) where radiolarites have been recently described (Kamata et al. 2003; Takemura et al. 2002, 2003).

At the family level and global scale, we have seen that the number of radiolarian families was not affected (Figs. 1, 4). But, oppositely, radiolarian species (and genera) were strongly affected at low latitudes but not in high southern latitudes (Kozur 1998). This discrepancy has been interpreted by Kozur (op. cit) as a differential effect of extinction in southern high latitude with respect to low or northern latitudes due to a huge volcanic activity in northern hemisphere (Siberia and China). The long phase of recovery would be due to immigration from the southern high latitude cold water (less affected than those of high northern and low latitudes) rather than from tropical waters, which would be a faster process (Kozur 1998, Kozur et al. 1996). For this author, this process would be slow (c.a. 5 Ma; late Scythian to early Anisian) because it would require at first the adaptation of some elements of the cold water fauna to warm tropical conditions. In addition, the recovery of the radiolarians, for this author, started at the base of the Spathian where, however, not yet a distinct radiation of Triassic elements occurred but mainly Paleozoic (Late Devonian-Missis-

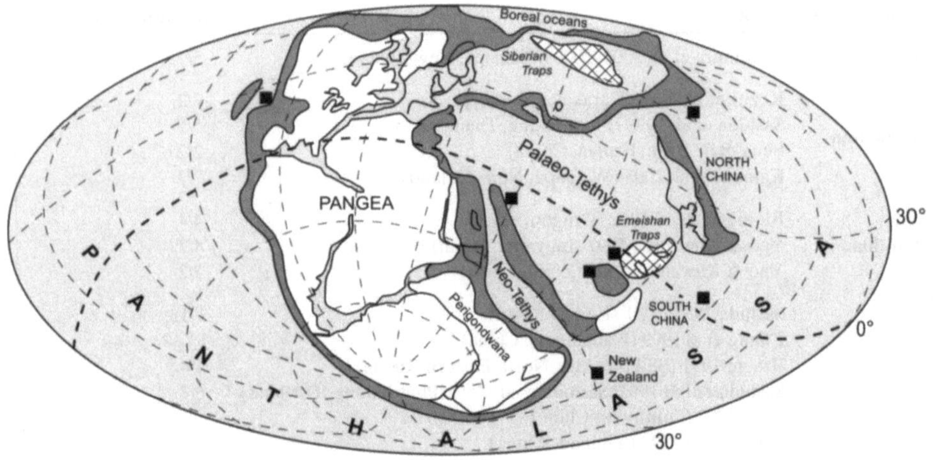

Fig. 5. Paleogeographic map for the Early Triassic (adapted from Hallam & Wignall (1997), displaying the heterogeneous distribution of studied sites for radiolarians. The number of publication for each region is as follow: Oregon: 1; Turkey: 1, Thailand: 4; S.China: 7; Japan: 13; NE Russia:1; New Zealand: 4.
It clearly reveal that besides tropical western Panthalassa almost nothing is known (only 1 locality for Oregon, Turkey, New Zealand and NE Russia). It is therefore difficult to have a representative scope of events which happens at that time. Note the type of projection (with a sigmoid equator) which seems to locate New Zealand close to south pole while it is at 30°S, as mentioned on figure.

sippian) "Lazarus taxa" reappeared from "unknown refuges", which are the origin for some phylogenetic connection at family or even at order level. The proposed "Lazarus recovery" is very big, about 50% of the genera that disappeared at the P/Tr boundary re-appeared only in the Olenekian-Middle Triassic interval. As shown here above, real data neither support nor contradict this interpretation, they are just insufficient at present time.

From a paleobiogeographic point of view, some paleolatitudinal differences were mentioned for Anisian radiolarians, where stratigraphic and taxonomic data are quite abundant and consistent (Bragin 1991; Sugiyama 1997; Kozur & Mostler 1994). In this sense, Aita & Bragin (1999) established the genus *Glomeropyle* for a large, thick-walled form occurring exclusively in high (south and north) latitudes, where other non-Tethyan species co-occur with well known Tethyan representatives (Hori et al. 2003). The same type of reconstruction using a compilation of the published data for the late Early Triassic (Olenekian) has been presented (Suzuki et al. 2002). These authors find a strong provinciality between the radiolarian faunas of the western Panthalassa and the eastern Tethys. Nevertheless, the non-homogeneous data used in the comparison (4 studies for the eastern Tethys against 14 papers for the western Panthalassa) and different stratigraphic levels led them to consider that radiolarian provinces were significantly distinct during late Olenekian because there is only a 5% of coincidence at specific level. Comparing the same data at a generic level rises up the coincidence to 50%. Such differences, in our opinion, probably derived from an inhomogeneous taxonomy (mostly open nomenclature assignations) in the utilised synthesis, partly at least due to the bad preservation recorded in this interval worldwide.

Alternative explanation

It is clear that the plankton response to the adverse environment of the Late Permian is complex and unfortunately badly documented. Forms with thin skeletons have not been preserved, being more fragile, so some may have appeared and evolved during this time of environmental crisis without any fossil record. The strong reduction of biomass at time of the crisis is undoubtedly related to the collapse of the primary productivity. In order to equilibrate the global biomass reduction, some strategies can be drawn (Fig. 6). Like in several groups (e.g. Foraminifera, Leven & Korgachin 2001) radiolarians may strongly decrease their population but maintain their body size, as it is the case of some Paleozoic relict forms (some albaillellids belonging to the genus *Follicucullus*) that could be considered to be "Lazarus forms" (in the sense of scarce record due to a poor preservation rate). Also some forms that reduce the complexity of their skeleton by partial loss of the outer components (=proteromorphosis *sensu* Guex 2001), but moreover having a more delicate skeleton, are more susceptible to be dissolved, and therefore reinforce the Lazarus effect (i.e. Latentifistulidae). On the other hand, other forms reduce their body size under environmental stress by, i) reducing the size of the entire skeleton (as is the case of large spherical spumellarian-type forms of the late Permian (Shimizu et al. 2004), or ii) by losing the external skeleton (proteromorphosis, i.e. Entactinaria forms); the population size being not necessarily reduced. It seems likely that many spicular forms belonging to entactinids in the earliest Triassic in the aftermath of the crisis are the result of a severe loss of the outer skeleton. This simple and economic architecture is a potential source for new phylogenetic developments and would explain also the sudden appearance of so many new groups (families, genera and species) since the beginning of the Middle Triassic when the preservation of radiolarian skeletons improved dramatically. The strong diversification also coincides with the rising values of strontium isotope ratios (Korte et al 2003), interpreted as the consequence of Early Triassic sea level rise and the return to normal geo-biosphere conditions.

It is interesting to note that in spite of the low diversity assemblages displayed in the earliest Triassic, all the information available shows a clear domination by Entactinaria, with many spicular forms, some Spumellaria and mono- or dicyrtid Nas-

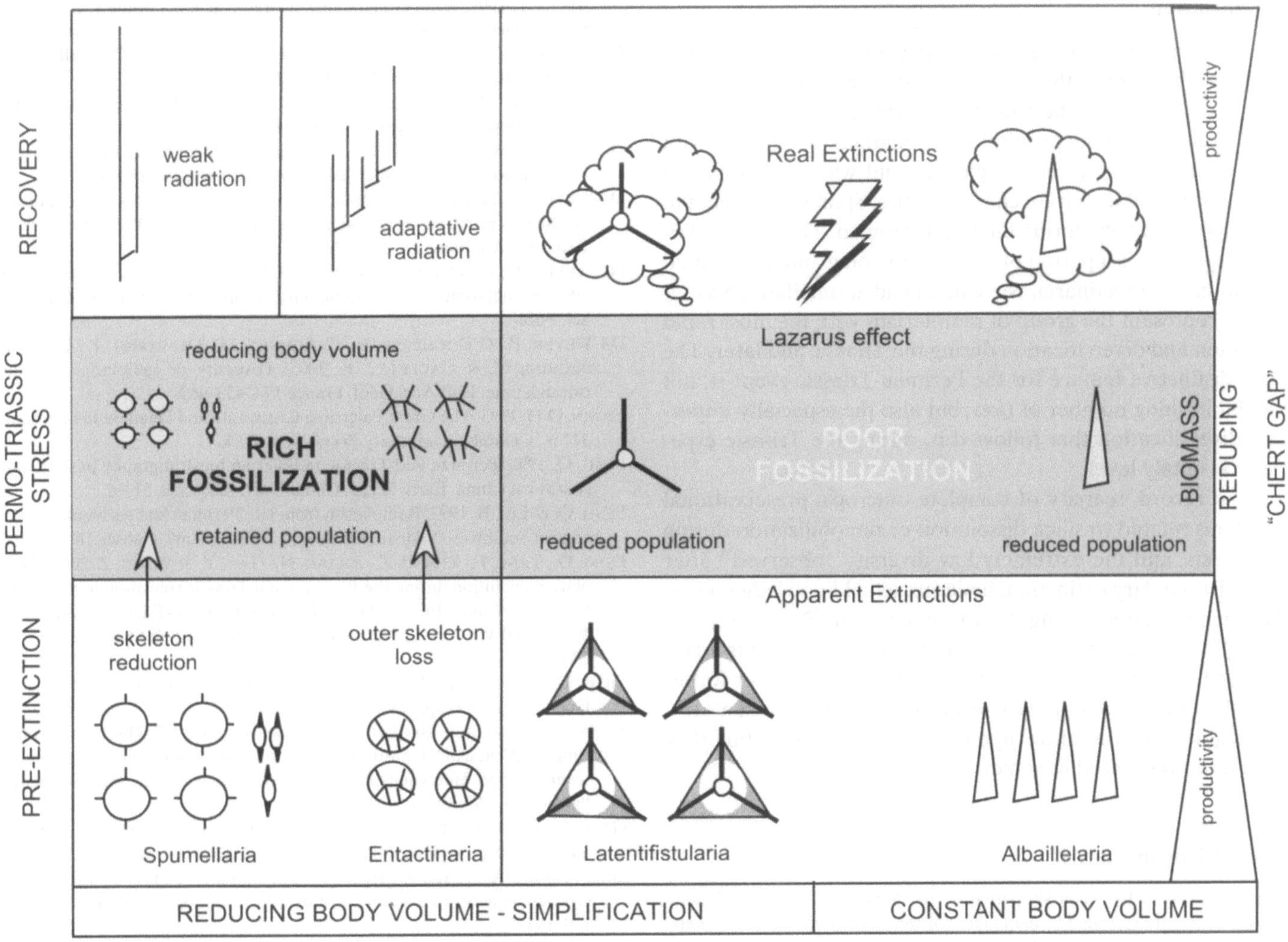

Fig. 6. Stress effects on radiolarian populations (see explanation in text)

sellaria. Multicycrtid Nassellaria do not occur before the Anisian. These shapes fit perfectly with the ability of some species to adopt a regressive/primitive character such as spicular forms and monocyrtids (De Wever 1982, De Wever et al. 2003).

All the data available today show that the picture we have for Early Triassic radiolarian fauna is very far from being complete. We agree with Wignall & Benton (1999) who emphasize that the abundance of Lazarus taxa (largely favoured by Kozur 1998, in the aftermath of the Late Permian extinctions is most probably the reflection of the extreme rarity of these organisms at this time, as well as, the incompleteness of the stratigraphic record (radiolaritic gap) and its poor preservation.

Final consideration

The low diversity displayed by the currently available data reflects the poor quantity of the pelagic fossil record at the Early Triassic, and can be due to the extremely low abundances of planktonic fauna (basically radiolarians) but it could also be a result of exceptionally poor preservation. Both phenomena are undoubtedly connected. For this reason, the apparent faunal dissimilarity displayed between some local studies, actually can not be considered as a solid argument for discussing an early and high provincialism because it would lead to an unrealistic paleobiogeography model for the group at Early Triassic time. A supplementary problem is that larger the interval lacking data is, smaller the probability is to illustrate the origin of the phylogenetic lineages. Indeed, the main period of diversification (Spathian) corresponds to the period for which the number of publications is the highest (2 for Griensbachian, 4 for the Dienerian, 2 for Smithian and 17 for Spathian). This coincidence reminds that of Sheehan (1977) who showed that the highest specific diversity of Phanerozoic marine invertebrate per Ma corresponds to the largest areal exposure of rocks of a given age.

4. Conclusion

We can summarize that the biodiversity of radiolarian families increased constantly throughout the Paleozoic and decreased only slightly towards the end of the Permian. While during the Late Paleozoic time primitive nassellarians appeared, with their characteristic internal spicules, and with skeletons still having some entactinarian characteristics, Spumellaria seem to have appeared only during the latest Permian. The Middle Triassic especially is a period of unique blooming and accelerated evolution of Entactinaria, Nassellaria and Spumellaria. Nassellarians represent the group of radiolarians with the most rapid evolution and diversification during the Triassic and later. The most distinctive feature for the Permian-Triassic event is, not the diminishing number of taxa, but also the especially important diversification that followed it, during the Triassic especially at family level.

Bad record, scarcity of complete outcrops, preservational problems related to silica dissolution or remobilization during diagenesis, and the extremely low diversity "observed" after the P-T boundary extinction, limit our actual knowledge on radiolarian radiation during the aftermath of the Permo-Triassic extinction. Even if some information is available (strong stress leading to diminishing diversity, more fragile or scarcer forms, etc;) the number of data remains too few to be representative and significant, therefore hypotheses are possible, but they should be considered as somewhat speculative guess.

Acknowledgements

The present work has been supported by the Muséum National d'Histoire Naturelle. We are indebted to Jean Guex for thorough comments on an early version of the manuscript and Paulian Dumitrica for his valuable information. We express sincere thanks to Agathe Cambreleng for her efficiency in the drawing operation. Martial Caridroit and an two anonymous reviewer are acknowledged for their critical reading of the manuscript.

REFERENCES

AITA, Y. & BRAGIN, N.Y. 1999: Non-Tethyan Triassic Radiolaria from New Zealand and northeastern Siberia. In: DE WEVER, P. & CAULET, J.-P. (Eds.): InterRad VIII. Geodiversitas 21, 503–526.

ATUDOREI, N.C. 1998: Constraints on the Upper Permian to Upper Triassic marine carbon isotope curve: Case studies from the Tethys, PhD Uni. Lausanne, 160 p.

BENTON, M.J. 1993: The fossil record 2. 845 p., Chapman & Hall.

BENTON, M.J & TWITCHETT, R.J. 2003: How to kill (almost) all life: the end-Permian extinction event. Trends Ecol. & Evol. 18, 358–365.

BLOME, C.D. & REED, K.M. 1992: Permian and Early (?) Triassic radiolarian faunas from the Grindstone Terrane, central Oregon. J. Paleont. 66, 351–383.

BOTTJER, D.J. SCHUBERT, J.K & DROSER, M.L. 1996: Comparative evolutionary palaeoecology: Assessing the changing ecology of the past. In: HART, M. (Ed.): Biotic Recovery from Mass Extinction Events. Geol. Soc. London Spec. Publ. 102, 1–13.

BOWRING, S.A., ERWIN, D.H., JIN, Y.G., MARTIN, M.W., DAVIDEK, K. & WANG, W. 1998: U/Pb Zircon Geochronology and Tempo of the End-Permian Mass Extinction. Science 280, 1039–1045.

BRAGIN, N.Y. 1991: Radiolaria and lower Mesozoic units of the U.S.S.R. east regions. Geol. Inst. Trans. 469, 1–126.

COURTILLOT, V.E. & RENNE, P.R. 2003: On the ages of flood basalt events. Comptes Rendus Geosci. 335, 113–140.

DE WEVER, P. 1982: Radiolaires du Trias et du Lias de la Téthys (Systématique, Stratigraphie). Public. Soc. Géol. Nord 7, 1–599.

DE WEVER, P. & BAUDIN, F. 1996: Palaeogeography of radiolarite and organic-rich deposits in Mesozoic Tethys. Geol. Rundschau 85, 310–326.

DE WEVER, P., AZEMA, J. & FOURCADE, E. 1994: Radiolaires et radiolarites: production primaire, diagenèse et paléogéographie. Bull. Centres rech. et Expl.- Elf Aquitaine 18, 315–379.

DE WEVER, P., DUMITRICÁ, P., CAULET, J.-P., NIGRINI, C. & CARIDROIT, M. 2001: Radiolarians in the sedimentary record. 533 p., Gordon & Breach Sci. Publ.

DE WEVER, P., O'DOGHERTY, L., CARIDROIT, M., DUMITRICÁ, P., GUEX, J., NIGRINI, C. & CAULET, J.-P. 2003: Diversity of radiolarian families through time. Bull. Soc. Géol. France 174, 453–469.

ERWIN, D.H. 1993: The Great Paleozoic Crisis: Life and Death in the Permian. 327 p., Columbia University Press, New York.

FENG, Q. 1992: Permian and Triassic radiolarian biostratigraphy in south and southwest China. Earth Sci., J. China Univ. Geosci. 3, 51–62.

FENG, Q. & LIU, B. 1993: Radiolarian from late Permian and early-middle Triassic in Southwest Yunnan. Earth Sci., J. China Univ. Geosci. 18, 552–563.

FENG, Q., YANG, F., ZHANG, Z., ZHANG, N., GAO, Y. & WANG, Z. 2000: Radiolarian evolution during the Permian and Triassic transition in South and Southwest China. In: YIN, H. et al. (Eds.): Permian-Triassic Evolution of Tethys and Western Circum-Pacific. Dev. Palaeont. & Strat. 18, 309–326. Elsevier.

FENG, Q., GU, S. & LI, M. 2001: Early Triassic Radiolarians from Sangzhi, Hunan. Acta micropalaeont. sinica 18, 249–253.

FLUTEAU, F., BESSE, J., BROUTIN, J. & RAMSTEIN, G. 2001: The Late Permian climate. What can be inferred from climate modelling concerning Pangea scenarios and Hercynian range altitude? Palaeogeogr. Palaeoclimatol. Palaeoecol. 167, 39–71.

GALL, J.C., GRAUVOGEL-STAMM, L., NEL, A. & PAPIER, F. 1998: The Permian mass extinction and the Triassic recovery. C.R. Acad. Sci. Paris 326, 1–12.

GUEX, J. 1981: Associations virtuelles et discontinuités dans la distribution des espèces fossiles: un exemple intéressant. Bull. Soc. Vaud. Sci. Nat. 359, 179–197.

GUEX, J. 2001: Environmental stress and atavism in ammonoid evolution. Eclog. Geol. Helv. 94, 321–328.

HALLAM, A. 1989: The case for sea-level change as a dominant causal factor in mass extinctions of marine invertebrates. Phil. Trans. Roy. Soc. London, B325, 437–455.

HALLAM, A. & WIGNALL, P.B. 1997: Mass extinctions and their aftermath. 370 p., Oxford Univ. Press.

HALLAM, A. & WIGNALL, P.B. 1999: Mass extinctions and sea-level changes. Earth-Sci. Rev. 48, 217–250.

HEYDARI, E. & HASSANZADEH, J. 2003: Deev Jahi Model of the Permian-Triassic boundary mass extinction: a case for gas hydrates as the main cause of biological crisis on Earth. Sed. Geol. 163, 147–163.

HOLSER, W.T. & MAGARITZ, M. 1992: Cretaceous/Tertiary and Permian/Triassic boundary events compared. Geochim. Cosmochim. Acta 56, 3297–3309.

HOLSER, W.T., SCHÖNLAUB, H.P., BOECKELMANN, K., MAGARITZ, M. & ORTH, C.J. 1991: The Permian-Triassic of the Gartnerkofel-1 core (Carnic Alps, Austria): synthesis and conclusion. Abh. geol. Bundesanst. (Wien) 45, 213–232.

HORI, R.S., CAMPBELL, J.D. & GRANT-MACKIE, J.A. 2003: Triassic Radiolaria from Kaka Point Structural Belt, Otago, New Zealand. J. r. Soc. N. Zealand 33, 39–56.

ISOGAWA, J., AITA, Y. & SAKAI, T. 1998: Early Triassic radiolarians from the bedded chert in the Minowa quarry, Kuzuu Town, Tochigi Prefecture. News Osaka Micropaleont. Spec. 11, 81–93.

ISOZAKI, Y. 1994: Superanoxia across the Permo-Triassic boundary; record in accreted deep-sea pelagic chert in Japan. In: EMBRY A.F. (Ed.): Pangea: global environments and resources. 805–812, Canad. Soc. Petrol. Geol..

ISOZAKI, Y. 1997: Permo-Triassic boundary superanoxia and stratified superocean: records from lost deep sea. Science 276, 235–238.

Jablonski, D. 1995: Extinction in the fossil record. In: May, R.M. & Lawton, J.H. (Eds.): Extinction rates. Oxford University Press, 25–44.

Jablonski,. D. 1996: Mass Extinctions: Persistent problems and new directions. In: Ryder, G. et al. (Eds.): The Cretaceous-Tertiary Event and Other Catastrophes in Earth History (Snowbird III). Geol. Soc. Am. Spec. Paper 307, 1–11.

Kakuwa, Y. 1996: Permian-Triassic mass extinction event recorded in bedded chert sequence in southwest Japan. Palaeogeogr. Palaeoclimatol. Palaeoecol. 121, 35–51.

Kamata, Y. 1995: Early Triassic radiolarians from black siliceous shale and black chert in the Kuzu area of the Ashio Terrane, central Japan. Fossils 59, 23–31.

Kamata, Y. 1999: Lower Triassic (Spathian) radiolarians from the Kuzu area (Tochigi Prefecture, central Japan). In: De Wever, P. & Caulet, J.-P. (Eds.): InterRad VIII. Geodiversita 21, 657–673.

Kamata, Y., Matsuo, A., Takemura, A., Yamakita, S., Aita, Y., Sakai, T., Suzuki, N., Hori, R., Sakakibara, M., Fujiki, T., Ogane, K., Takemura, S., Sakamoto, S., Kodama, K., Nakamura, Y., Campbell, H.J. & Spörli, K.B. 2003: Late Induan (Dienerian) Radiolarians from Arrow rocks in the Waipapa terrane, North Island, New Zealand. In: Diserens, M.O. and Jackett, S-J. (Eds.) Interrad X Univ. Lausanne, Abstracts & Programme, 70.

Kamo, S.L., Czamanske, G.K., Amelin, Y., Fedorenko, V.A., Davis, D.W. & Trofimov, V.R. 2003: Rapid eruption of Siberian flood-volcanic rocks and evidence for coincidence with the Permian–Triassic boundary and mass extinction at 251 Ma. Earth Planet Sci. Lett. 214, 75–91.

Kato, Y., Nakao, K. & Isozaki, Y 2002: Geochemistry of Late Permian to Early Triassic pelagic cherts from southwest Japan: implications for an oceanic redox change. Chemical Geology 182, 15–34.

Kidder, D.L. & Worsley, T.R. 2004: Causes and consequences of extreme Permo-Triassic warming to globally equable climate and relation to the Permo-Triassic extinction and recovery. Palaeogeogr. Palaeoclimatol. Palaeoecol. 203, 207–237.

Korte, C., Kozur, H.W., Bruckschen, P. & Veizer, J. 2003: Strontium isotope evolution of Late Permian and Triassic seawater. Geoch. Cosmoch. Acta 67, 47–62.

Korte, C., Kozur, H.W., Joachimski, M.M., Strauss, H., Veizer,·& J. Schwark, L. 2004: Carbon, sulfur, oxygen and strontium isotope records, organic geochemistry and biostratigraphy across the Permian/Triassic boundary in Abadeh, Iran. Int. J. Earth Sci. 93, 565–581.

Kozur, H.W. 1998: Some aspects of the Permian-Triassic boundary (PTB) and of the possible causes for the biotic crisis around this boundary. Palaeogeogr., Palaeoclimatol., Palaeoecol., 143, 227–272.

Kozur, H.W. Extinction and recovery patterns of the radiolarians during and after the Permian-Triassic biotic crisis and its importance for the radiolarian biostratigraphy of the Lower and Middle Triassic, (this volume)

Kozur, H.W. & Mostler, H. 1982: Entactinaria subordo Nov., a new radiolarian suborder. Geol. Paläont. Mitt. Innsbruck 11, 399–414.

Kozur, H.W. & Mostler, H. 1994: Anisian to middle Carnian radiolarian zonation and description of some stratigraphically important radiolarians. Geol. Paläont. Mitt. Innsbruck, Sond. 3, 39–255.

Kozur, H.W., Kaya, O. & Mostler, H. 1996: First evidence of Lower to Middle Scythian (Dienerian-Lower Olenekian) radiolarians from the Karakaya Zone of Northwestern Turkey. Geol. Paläont. Mitt. Innsbruck, Sond. 4, 271–285.

Kusunoki, T. & Imoto, N. 1996: Early Triasssic (Spathian) radiolarians in the chert from southern Kameoka City, Kyoto Prefecture. Earth Sci., J. Assoc, geol. Coll. Japan 50, 183–187.

Kuwahara, K. & Yao. A. 2001: Late Permian radiolarian faunal change in bedded chert of the Mino Belt, Japan. In: Takemura A. & Furutani H. (Eds.): Proceedings of the Seventh Japanese Radiolarian Symposium News Osaka Micropaleont. spec. vol. 12, 33–49.

Leven, E.J. & Korchagin, O.A. 2001: Permian-Triassic Biotic Crisis and Foraminifers. Stratigr. geol. Correl. 9, 364–372.

Looy, C.V. Twitchett, R.J. Dilcher, D.L. Van Konijnenburg-Van Cittert, J.H.A. Visscher, H. 2001: Life in the end-Permian dead zone. Proc. Natl. Acad. Sci. 98, 7879–7883.

Nagai, H. & Mizutani, H. 1993: Early Triassic Radiolarians from Tsuzuya, Minokamo city, Gifu Prefecture, Central Japan. Bull. Nagoya Univ. Furukawa Mus. 9, 1–23.

O'Dogherty, L. & Guex, J. 2002: Rates and pattern of evolution among Cretaceous radiolarians: relations with global paleoceanographic events. In: Carter et al. (Eds.): Micropaleontology of radiolarians. Proceedings of Interrad IX. Micropaleont. 48, 1–22.

O'Dogherty, L., De Wever, P. & Guex, J. 2003: Is the nassellarian/spumellarian diversity ratio a paleonvironmental proxy indicator in the geological record? In: Diserens, M.O. and Jackett, S-J. (Eds.) Interrad X Interrad 2003, Univ. Lausanne, Abstracts & Programme, 90–91.

Rampino, M.R. & Adler, A.C. 1998: Evidence for abrupt latest Permian mass extinction of foraminifera: results of tests for the Signor-Lipps effect. Geology 26, 415–418.

Ross, W.C. & Ross, J.R.P. 1988: Late Paleozoic Transgressive -Regressive Deposition. In: Wilgus, C.K. et al. (Eds.): Sea-Level changes – An integrated approach. SEPM Special Pub. 42, 227–248.

Sashida, K. 1983: Lower Triassic Radiolaria from the Kanto Mountains, Central Japan. Part 1: Palaeoscenidiidae. Trans. Proc. Palaeont. Soc. Japan, new ser. 131, 168–176.

Sashida, K. 1991: Early Triassic radiolarians from the Ogamata Formation, Kanto Mountains, central Japan, Part 2. Trans. Proc. Palaeont. Soc. Japan, new ser. 161, 681–696.

Sashida, K. & Igo, H. 1992: Triassic radiolarians from a limestone exposed at Khao Chiak near Phatthalung, Southern Thailand. Trans. Proc. Palaeont. Soc. Japan, New Ser. 168, 1296–1310.

Sashida, K., Igo, H., Adachi, S., Koike, T., Hisada, K.I., & Nakornsri, N. 1993: Occurrences of Paleozoic and Mesozoic radiolarians from Thailand and Malaysia and its geologic significance (preliminary report). News Osaka Micropaleont., spec. vol. 9, 1–17.

Sashida, K., Igo, H., Adachi, S., Ueno, K., Nakornsri, N. & Sardsud, A. 1998: Late Paleozoic radiolarian faunas from northern and northeastern Thailand. Sci. Rep. Inst. Geosci. Univ.Tsukuba, Sec. B: Geol. Sci. 19, 1–27.

Sashida, K., Igo, H., Adachi, S., Ueno, K., Kajiwara, Y., Nakornsri, N. & Sardsud, A. 2000: Late Permian to Middle Triassic radiolarian faunas from Northern Thailand. J. Paleont. 74, 789–811.

Scheffler, K., Hoernes S. & Schwark, L. 2003: Global changes during Carboniferous-Permian glaciation of Gondwana: Linking polar and equatorial climate evolution by geochemical proxies. Geology 31, 605–608.

Schubert, J.K. & Bottjer, D.J. 1995: Aftermath of the Permian-Triassic mass extinction event: Paleoecology of Lower Triassic carbonates in western USA. Palaeogeogr. Palaeoclimatol. Palaeoecol. 116, 1–39.

Sepkoski, J.J. Jr. 1989: Periodicity in extinction and the problem of catastrophism in the history of life. J. Geol. Soc. London 146, 7–19.

Shang, Q., Caridroit M. & Wang, Y. 2001: Radiolarians from the Uppermost Permian Changhsingian of Southern China. Acta Micropal. Sinica 18, 229–240.

Sheehan, P.M. 1977: Species diversity in the Phanerozoic, a reflection of labor by systematists? Paleobiology, 3, 325–329

Shimizu, N., Isozaki, Y., Matsuda, T., Yao, J. & Ji, Z. 2004: Detailed stratigraphy across the Permo-Triassic boundary at Chaotian in Northern Sichuan, China. J. Geograph. (Tokyo) 113, 87–106.

Sugiyama, K. 1992: Lower and Middle Triassic radiolarians from Mt. Kinkazan, Gifu Prefecture, Central Japan. Palaeont. Soc. Japan, New Ser. 167, 1180–1223.

Sugiyama, K. 1997: Triassic and Lower Jurassic radiolarian biostratigraphy in the siliceous claystone and bedded cherts units of the Southeastern Mino Terrane, Central Japan, Bull. Mizunami Fossil Mus. 24, 79–193.

Suzuki, N., Akiba N. & Kanoh, H. 2002: Late Olenekian radiolarians from bedded chert of Ashio Terrane, northeast Japan, and faunal turnovers in western Panthalassa during Early Triassic. J. China Univ. Geosci. 13, 124–140.

Takemura, A., Aita, Y., Hori, R.S., Higuchi, Y., Sporli, K.B., Campbell, H.J., Kodama, K. & Sakai, T. 2002: Triassic radiolarians from the ocean-floor sequence of the Waipapa Terrane at Arrow Rocks, Northland, New Zealand. N. Zealand J. Geol. & Geophys. 45, 289–296.

TAKEMURA, A., SAKAI, M., YAMAKITA, S., KAMATA, Y., AITA, Y., SAKAI, T., SUZUKI, N., HORI, S.R., SAKAKIBARA, M., KODAMA, K., TAKEMURA, S., SAKAMOTO, S., OGANE, K., KOYANO, T., SATAKE, A., NAKAMURA, Y., CAMPBELL, H.J. & SPÖRLI, K.B. 2003: Early Triassic radiolarians from arrow rocks in the Waipapa Terrane, Northern Island, New Zealand. In: Diserens, M.O. and Jackett, S-J. (Eds.) Interrad X, Univ. Lausanne, Abstracts & Programme 106–107.

TWITCHETT, R.J., LOOY, C.V., MORANTE, R., VISSCHER H. & WIGNALL, P.B. 2001: Rapid and synchronous collapse of marine and terrestrial ecosystems during the end-Permian biotic crisis. Geology 29, 351–354.

TWITCHETT, R.J., KRYSTYN, L., BAUD, A., WHEELEY, J.R. & RICHOZ, S. 2004: Rapid marine recovery after the end-Permian mass-extinction event in the absence of marine anoxia. Geology 32, 805–808.

UMEDA, M. 2002: Taxonomy and Diversity History of Paleozoic Radiolarians: Seven Extinction Events. Journal of Geography (Tokyo) 11, 33–54.

WANG, K., GELDSETZER, H.H.J. & KROUSE, H.R. 1994: Permian-Triassic extinctions: organic $\delta^{13}C$ Evidence From British Columbia, Canada. Geology 22, 580–584.

WIGNALL, P.B. 2001: Large igneous provinces and mass extinctions. Earth-Sci. Rev. 53, 1–33.

WIGNALL, P.B. & BENTON, M.J. 1999: Lazarus taxa and fossil abundances at times of biotic crisis. J. Geol. Soc. London 156, 453–456.

WIGNALL, P.B. & HALLAM, A. 1992: Anoxia as a cause of the Permian/Triassic mass extinction: facies evidence from northern Italy and the western United States. Palaeogeogr. Palaeoclimatol. Palaeoecol. 93, 21–46.

WIGNALL, P.B., MORANTE R. & NEWTON, R. 1998: The Permo-Triassic transition in Spitsbergen: $\delta^{13}C_{org}$ chemostratigraphy, Fe and S geochemistry, facies, fauna and trace fossils. Geol. Mag. 135, 47–62.

WON, M. & BELOW, R. 1999: Cambrian Radiolaria from the Georgina Basin, Queensland, Australia. Micropaleont. 45, 325–363.

YAO, A. & KUWAHARA, K 1997: Radiolarian faunal change from Late Permian to Middle Triassic times. News Osaka Micropaleont. spec. vol. 10, 87–96.

YAO, A. & KUWAHARA. K. 1999: Permian and Triassic radiolarian assemblages from the Yangzi Platform. In: YAO, A. et al. (Eds.): Biotic and Geological Development of the Paleo-Tethys in China. 1-16, University Press.

YAO, A. & KUWAHARA, K. 2000: Permian and Triassic radiolarians from the southern Guizhou Province, China. J. Geosci., Osaka City Univ. 43, 1–19.

ZHANG, Z., FENG, Q. & SHI,W. 1999: Preliminary studies of Triassic Radiolarians from Muyinhe Formation in Southwest Yunnan, China. In: HONGFU, Y. & JINNAN T. (Eds.): Pangea and the Paleozoic-Mesozoic Transition. Proceedings of the International Conference, 74–78.

Manuscript received July 2005
Revision accepted July 2005

0012-9402/06/01S063-4
DOI 10.1007/s00015-006-0602-5
Birkhäuser Verlag, Basel, 2006

Eclogae geol. Helv. 99 (2006) Supplement 1, S63–S66

An online micropaleontology database: Radiolaria.org

JANE K. DOLVEN[1] & HANS A. SKJERPEN[2]

Key words: Online databases, paleoinformatics, www.radiolaria.org

ABSTRACT

Growing amounts of available radiolarian data require better information-management systems. Managing, preserving and exchanging information can be done by using online databases with a web interface. Radiolaria.org (www.radiolaria.org), an example of such a web-database, contains species information and other resources, and has been designed by and for the radiolarian community. In Radiolaria.org researchers can contribute with information and this body of data is then made freely available. All submitted information is reviewed by appointed web-editors to ensure quality and correctness. Species names, descriptions, synonyms and references are based on published material, and this is linked to a picture-archive showing representatives of the species from different geographical areas and/or periods. In addition, Radiolaria.org also serves as a communal website with e.g. news, events and discussion forums.

1. Introduction

Paleoinformatics, the area of paleontology concerned with management of information, including preservation of systematic information and expertise, was first introduced in the paleontological literature by MacLeod & Guralnick (2000) and MacLeod et al. (2000). These authors pointed out that, due to the increasing pace of micropaleontological data gathering, there is a need for more efficient ways of storing, accessing, exchanging, analyzing and presenting data. These demands are met by the adoption of more sophisticated information-management technology, paired with training in the development and use of such technologies. There is also a need to coordinate initiatives in these areas to ensure maximum benefits for the research community as a whole.

Online databases (defined herein as databases with a web interface) are one effective way of managing micropaleontological data. The information is entered or accessed easily, at any time, from anywhere with the use of a web browser. These web-databases enable easy searching and comparison of data, and make it easier to extract needed data in a suitable format, which is important for data exchange.

Building online databases requires large amounts of time and work: 1) to develop a robust and user-friendly paleoinformatic system; 2) to gather and ensure that any data added into the system is of a high quality; 3) to transform the data into a suitable format for storage; 4) to enter the data into the system; and 5) to have the data quality controlled (by web-editors). However, once such a system is operational, it has the potential to make micropaleontological work more efficient by making better use of the information and resources that already exist. Eventually, similar initiatives may be interconnected creating networks of knowledge bases.

In the following sections, we will present the website Radiolaria.org (www.radiolaria.org). The idea of making Radiolaria.org came up during late spring 2000, due to our need of having a database with information (images, descriptions, references, synonyms, geographic and stratigraphic distribution etc.) about radiolarian species. Previously we had used books, reprints, library-copies, and various computer documents to keep information at hand, but a database would make the information more easily available. Instead of making a local system, we decided to put the database online so other radiolarian researchers also could access and add data. The first version of Radiolaria.org was presented at the Interrad-meeting in 2000 in California, where it received positive feedback from fellow researchers. Since then, there has been both functionality and content added to the system. Radiolaria.org is now an established source for radiolarian information online, with several thousand requests every week from all over the world.

[1] Natural History Museum, University of Oslo, Pb. 1172 Blindern, 0318 Oslo, Norway. Email: jane@radiolaria.org
[2] Virtual One as, Pb. 229, 2021 Skedsmokorset, Norway. Email: hans@one.no

2. Radiolaria.org

Radiolaria.org includes general pages explaining what radiolarians are, as well as more detailed pages with taxonomic data on different species. It can be accessed free of charge, through a standard web browser, and uses hot-links for navigation. Most of the information is contained in a relational database, and the webpages are generated from this content.

The most important feature of Radiolaria.org is probably the "species information", and the link between what researchers name different radiolarians (or used to name them, i.e. synonyms) and pictures of how the species look like in different areas/periods. Having this information online, where researchers can contribute and discuss species concepts, may be one step in the right direction in the process of bringing taxonomy into conformity, which is very important for both local and regional biostratigraphic and paleoclimatic work.

2.1. Home page

The web-interface of Radiolaria.org consists of several sections that divide content and functionality into more manageable reports and tools. On the home page (Fig. 1) you will find links to most pages and sections included in the site. A detailed description and screenshot of every page is beyond the scope of this paper, we will here rather give an overview and some examples of the database features and functionality.

Under "About radiolaria.org" you will find general information about the website, the idea behind it, how the system works, and a user manual (.pdf) on how to contribute information to the system. "Species list" contains names (listed alphabetically), pictures, and links to all radiolarian species in the database. The "Archive" section includes (scanned) plates and captions of important taxonomic publications, which we have got the necessary permissions to reproduce. As many books and journals with original plates and species descriptions are getting old, out of print, or somehow unavailable, it is of great importance to digitize and make these publications available online. "Web links" contains links to other online radiolarian resources available on the Internet. "InterRad" informs about the International Association of Radiolarian Paleontologist (InterRad) meetings, membership, key people and the "Radfolks" mailing list. The official "Newsletter" can be downloaded from a password protected page. "News" keeps researchers updated on what is happening within the website and the radiolarian community in general. Under "Mystery Rad" it is possible to post a picture and information about an unknown species and let the community share their opinions and suggestions as to which species it may be. "Radiolarian art" is a page that links science and art. At present (October 2005) it contains pictures of eleven radiolarian models made in white clay by the Swedish artist Eva Bjerke, several pictures of a large quilt made by the Canadian artist Barbara J. West, and pictures of three plexiglass models of living radiolarians made by the German artist Dagmar Borgwart. "Contributors" lists the people that are (or have been) contributing to Radiolaria.org. It also gives the names and contact information for the web-editors responsible for the different geographical areas/time periods. "Radpeople" is a username- and password-protected section providing addresses and e-mails of radiolarian researchers. (If interested, contact the webmaster on Radiolaria.org). To locate specific information it is possible to type in one or more words on the "Search"-page. The user selects whether it is a search on species, synonym, reference, description or archive. The result is displayed with hot-links to the relevant pages. Introduction pages like "What are Radiolarians", "Quaternary and Tertiary radiolarians" written by various authors, contain general information about radiolarians for non/novice researchers.

2.2. Species information

Information about a radiolarian species can be found in several ways:

1) By scrolling the species list and looking for names (listed alphabetically) or pictures;
2) By navigating via the stratigraphic time scale (adopted from the International Commission on Stratigraphy) and geographic links listed on the right-hand side of the home page (e.g. by clicking on "Holocene (Recent)" and selecting "Nordic Seas"). When selecting the area of interest, the different species are listed under either Nassellaria, Spumellaria, or Phaeodaria. The number of species found in each category is also indicated;
3) By using the search-engine.

Under each species it is possible to find the original and/or other descriptions, images, synonyms, references and web links (if any) to other resources on the internet. There is also a discussion forum where radiolarian researchers can exchange ideas and opinions, add comments and suggestions, and post related non-published material. Each species is linked to a geographic/stratigraphic distribution. The user can navigate between different pages by clicking on the links in the submenu:

The "Images"-link displays a list of photographs for the selected species. Information about location, sample, latitude, longitude and photographer is shown next to each picture. Availability of various specimen pictures from different oceanic regions and time intervals may help point out regional/stratigraphic variations within the species. This is of importance, e.g. in comparative morphological analysis.

The "Synonyms" displays a list of registered synonyms for the species, sorted in ascending chronological order. The full reference to each synonym is found when selecting "References". This online facility may give the user a more updated and complete overview of available information relating to a species compared to the consultation of single publications.

All areas/periods where the species has been registered are found under "Distribution". (As this is work in progress, not

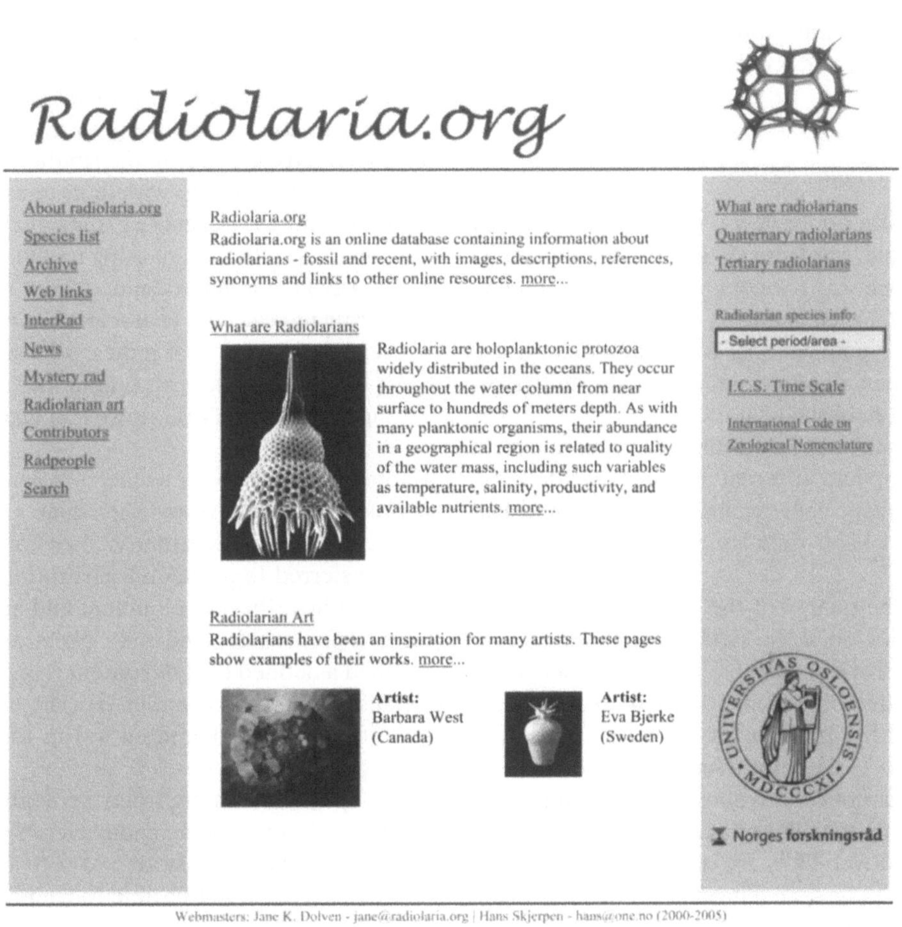

Fig. 1. The home page of Radiolaria.org (October 2005).

all areas/periods may have yet been included.) The presence of a species should be documented by a picture and preferably a reference to a book or paper. The name of the person who contributed the information is indicated in parenthesis.

The display of species names, descriptions, synonyms and pictures together has proven very important in making the species concept more homogenous between users, as scientists use different taxonomies and will therefore give different names to the same specimen. Unfortunately, Radiolaria.org does not prevent the possibility of the same species being registered under two different names in two geographical areas/ periods.

2.3. Contributing information

One important aspect of Radiolaria.org is that any radiolarian researcher can add information (provided it meets certain criteria, see section 2.4.) into the database. This can be done in several ways:

1) A contributor can add a species by selecting the stratigraphic/geographic location where the species is known to be present and click on the "Add species" link on top of the page. This will take the contributor through several steps (pages), where different forms are filled out. In form one (Species and original description) and two (Other description), the genus and species name along with at least one description (including author and year) is entered. Synonym(s) and Reference(s) are inserted in form three, and Image(s) with related information is added in form four. All fields marked with an asterisk (*) are required fields that need to be filled in order to submit the data into the system. Once the data entry is completed, the user can preview the contributed information. After approval, the data is submitted along with the contributor's name and e-mail address. The contributor is also suggested to notify the web-editors by e-mail about the new additions.
2) A user may also add single pieces of information to a species that is already registered in the system. This is done by selecting the species of interest, locating the "Add"-link on top of each page in the submenu (e.g. Add description, Add image, Add synonym/reference), filling out form(s), and submitting the information.

When data are submitted into the system (either way described above), it is not instantly made visible on the site, but placed in an administration area and flagged for review. The review process is described in more detail under the following section.

2.4. *Administration and Quality control*

The goal of the Radiolaria.org-project is to gather high-quality information about radiolarians from all time periods and make this available online. One way to acquire data of good quality is to use information published mainly in peer-reviewed books, journals and electronic publications. In Radiolaria.org, the content of species names, descriptions, synonyms and references is based solely on published data.

Copyright laws affects online information, as this is a form of publishing. Ownership and different use-rights for data can be a source of conflicts, especially regarding already published material. We are aware of copyright issues and are working on solving them.

To further ensure the quality of the submitted data, web-editors are allowed to log on to the system and check submitted material flagged for review. A web-editor is assigned a certain geographic or stratigraphic area, according to his/her expertise, and reviews all information related to this area. The web-editors ensure that: 1) the information submitted is (with regards to species descriptions, synonyms and references) based on published material with correct citations; and 2) the images, distribution and web-links are correctly assigned. If a web-editor finds errors, judges the data entry to be insufficient or wrong, or needs to contact and discuss with the contributor, he/she can use the name and email that was submitted with the content. The review process (checking of submitted material before it is made available online on the website) also prevents intentional sabotage by online pranksters.

3. Future and goals

A complete online catalogue with high quality information requires different people with different expertise to donate many hours of work. While some work in Radiolaria.org is sponsored by grants, most is based on voluntary work without monetary compensation. An important part of reaching the goals of Radiolaria.org is to recruit area/period-experts willing to contribute with their time and expertise. Although most radiolarian researcher already have a fully booked schedule, we hope that many will see the importance of this project and will be able to provide help to build up the database in the future.

Information systems evolve over time, with new functionality ever adapting to new users and needs. What new features will be requested by the radiolarian community is hard to predict, but some of the currently planned features for Radiolaria.org include:

- Support for a wider variety of data, particularly quantitative radiolarian data (including sample information like location, geographic coordinates, age etc.). This data could then be used further, e.g. in a Geographical Information System, to plot species distribution maps.
- A Software Archive with links to useful applications for radiolarian researchers (e.g. data analysis packages including common statistical, plotting and modeling functions).
- A Taxonomic Identification Tool that displays species according to selected characteristics (e.g. number of spheres/segments, number of spines, pore size/shape). Each species is registered with a certain number of characteristics, allowing searches and drill-downs on taxonomic keys.

Radiolaria.org belongs to the radiolarian community and the people that have invested their time and expertise building it. While the system is currently hosted on a private server, it will be transferred to a research institution (doing radiolarian research) for further development and hosting – preferably long term. This transfer will take place after Radiolaria.org has reached a defined goal of content (e.g. when the Recent radiolarians catalogue is completed). The new hosting research institution must be committed to run Radiolaria.org on the following conditions:

- Keep Radiolaria.org freely accessible for all.
- Maintain the system, and preferably develop it further.
- Agree to hand the system over to a new institution, should the institution not be able to comply with the above conditions.

Radiolaria.org was built for the radiolarian community, and will exist as long as this community shares the work and the benefits in our quest for better science.

Acknowledgement

Many thanks to Dr. Giuseppe Cortese for reviewing the paper and providing useful comments and suggestions. We also like to acknowledge Dr. Norman MacLeod and Prof. Kjell Bjørklund for reading and correcting an earlier version of this manuscript. Radiolaria.org is from 2003-2005 funded by the Norwegian Research Council (Project number: 157834/432).

REFERENCES

MacLeod, N. & Guralnick, R., 2000: Paleoinformatics. In: Lane, R.H., Steininger, F., Kaesler, R.L., Ziegler, W., Lipps, J.H. (Eds.): Fossils and the future; paleontology in the 21st century. Senckenberg-Buch, 74: 31–36.

MacLeod, N., Diver, P., Guralnick, R., Lazarus, D. & Malmgren, B., 2000: Computers, quantification, and databases. In: Lane, R.H., Steininger, F., Kaesler, R.L., Ziegler, W., Lipps, J.H. (Eds.): Fossils and the future; paleontology in the 21st century. Senckenberg-Buch, 74: 191–201.

Manuscript received April 2004
Revision accepted February 2005

0012-9402/06/01S067-12
DOI 10.1007/s00015-006-0608-z
Birkhäuser Verlag, Basel, 2006

Eclogae geol. Helv. 99 (2006) Supplement 1, S67–S78

A new genus of Entactiniidae (Radiolaria) from the Upper Permian of South China

QINGLAU FENG[1], YOUYAN MENG, WEIHONG HE & SONGZHU GU

Key words: Taxonomy, Radiolarian, Late Permian, South China

ABSTRACT

Four new species and one indeterminate species of radiolarians from the Changxiangian (Upper Permian) of southern Guangxi, South China are introduced. They all belong to one new genus: *Megaporus* n. gen. This new genus is characterized by two concentric spherical shells with pentagonal and hexagonal pore frames composed of thick but narrow bars, which is very similar in morphology to some genera within the Mesozoic subfamily Pantanelliinae PESSAGNO (1977), but differs from the latter in possessing initial spicules within medullary shell.

Introduction

The studies on the mass extinction of Permian radiolarians indicated that the Changxingian was a main extinction interval for the Permian radiolarian fauna and the main extinction event was a gradual process (Kakuwa 1996; Vishnevskaya 1997; Kuwahara & Yao 1998). Our investigations in South China, however, prove that some taxa, such as Entactinaria (Kozur & Mostler 1982), had high diversity during the late Changxingian, and then experienced rapid extinction at the end of the Changxingian. In the high diversity stage, some new taxonomic units developed. One of these units, described herein as *Megaporus* n. gen., is characterized by having two concentric spherical shells with large pores, which is very similar to some genera of the Mesozoic subfamily Pantanelliinae PESSAGNO (1977) in morphology, but differs from the subfamily in possessing initial spicules within the medullary shell.

Stratigraphy

The fauna described herein was collected from an area located between Dongmen and Liuqiao Towns, Fusui County, Guangxi Zhuang Autonomous Region (Fig. 1) and the strata in this region mainly consist of Upper Permian and Lower Triassic successions (BGMRGZAR 2001). The Upper Permian is divided in ascending order into three formations: the Heshan, Changxing and Dalong formations (Fig. 1). All specimens described in the present paper were collected from the Dongpan Section near Dongpan Village (22°16.196'N, 107°41.505'E). In this section, the Dalong Formation overlies Changxingian limestone of the Changxing Formation and is overlain by mudstones of the Lower Triassic Luolou Formation containing *Ophiceras tingi* TIEN and *Claraia dieneri* NAKAZAWA. Therefore, the Dalong Formation is upper Changxingian. The lower part of the Dalong Formation is mainly composed of mudstones and siliceous rocks; the middle part, of bedded siliceous rocks with shales; and the upper part, of siliceous mudstones, bentonites and muddy siliceous rocks with Permian ammonites, bivalves and brachiopods, for example: *Huananoceras* cf. *perornatum* CHAO & LIANG, *Laibinoceras* cf. *compressum* YANG, *Qiangjiangoceras* sp., *Euchondria jiangxianensis* GU & LIU, *Euchondria dalongensis* YIN, *Leptochondria intermedia* YIN, *Paracrurithyris pigmae* (LIAO), *Martinia* sp., *Spinomarginifera* and so on (Fig. 2). Abundant well-preserved radiolarians were obtained from the bedded siliceous rocks, including *Neoalbaillella optima* ISHIGA, KITO & IMOTO, *Albaillella triangularis* ISHIGA, KITO & IMOTO, *Albaillella yaoi* KUWAHARA, and so on together with the radiolarians described in this paper, which belong to the upper Changxingian *Neoalbaillella optima* Zone (Kuwahara et al. 1998; Yao et al. 2001) (Fig. 2).

[1] Faculty of Earth Sciences, China University of Geosciences, Wuhan 430074, China. E-mail: fengqlye@public.wh.hb.cn

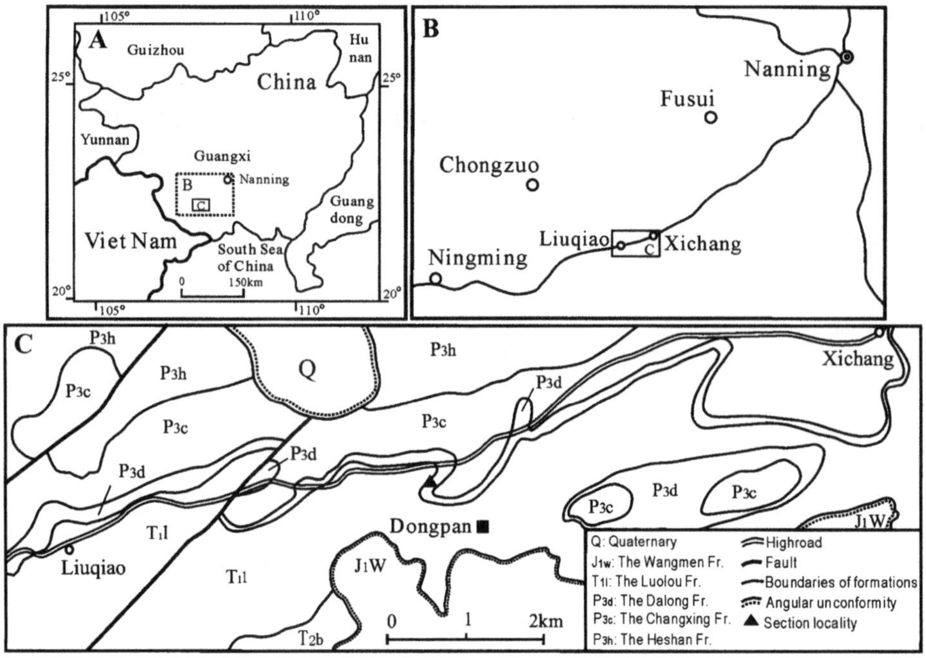

Fig. 1. A and B: Locality map of the studied area; C: Geologic map of the studied area and locality of the studied section.

Systematic Paleontology

All specimens described in this paper are deposited in Museum of China University of Geosciences, Wuhan, People's Republic China.

Class ACTINOPODA
Subclass RADIOLARIA
Superorder POLYCYSTIDA EHRENBERG 1838, *emend.* RIEDEL 1967
Order ENTACTINARIA KOZUR & MOSTLER 1982
Family Entactiniidae RIEDEL 1967

Type genus. – *Entactinia* RIEDEL 1967

Diagnosis. – Spherical to subspherical Entactinaria with commonly six to eght-spined initial spicule. Shell latticed, with one or more shells (De Wever et al. 2001).
Occurrence. – Ordovician to Lower Jurassic.

Genus ***Megaporus*** FENG new genus
Type species. – *Megaporus jini* FENG n. sp.

Description. – Test small, with two concentric shells: cortical and medullary shells. Cortical shell spherical, and composed of pentagonal and hexagonal pore frames with two to six three-bladed primary spines; pore frames thick and narrow; pores very large. Medullary shell small, consisting of pentagonal, hexagonal and irregular pore frames, and connected with cortical shell by some three-bladed and cylindrical beams. There is an initial spicule inside the medullary shell, and the spicule is relatively strong compared to the small medullary shell.

Discussion. – The structure of this new genus resembles that of *Entactinosphaera* FOREMAN (1963) from which it differs by the presence of very large pores on the cortical shell, and by irregular arrangement of the primary pines on the cortical shell. It is also similar to some genera of Pantanelliinae PESSAGNO (1977), such as *Cana* MEKIK (2000), *Gorgansium* PESSAGNO & BLOME (1980), and *Pantanellium* PESSAGNO (1977), but differs from the latter by having initial spicule in the medullary shell. The new genus is assigned to the Entactiniidae RIEDEL (1967) because it has two latticed shells and an initial spicule.

Etymology. – mega (Latin, adj.) = large; porus (Latin, adj.) = pore.

Occurrence. – Upper Permian, Dalong Formation in southern Guangxi, China.

Megaporus jini FENG new species
Plate 1, Figures 1–19; Plate 2, Figures 1–3, 8, 9, 12
?*Cryptostephanidium* sp., SHANG, CARIDROIT & WANG 2001, pl. 4, fig. 20.

Diagnosis. – A species of *Megaporus* with five to six three-bladed primary spines irregularly arranged on the cortical shell.

Description. – Test small, spherical, and with five to six approximately equal three-bladed primary spines. Spines gradually decrease in width distally. They are irregularly distributed on the cortical shell, the angle between two close spines being variable and not always at right-angle. Cortical shell is composed of pentagonal and hexagonal pore frames with massive but short secondary spines at the pore frame vertices besides the primary spines; pore frame thick and narrow; pores very large, about seven to ten pores present on outer surface of hemisphere. Medullary shell small, consisting of pentagonal, hexagonal and

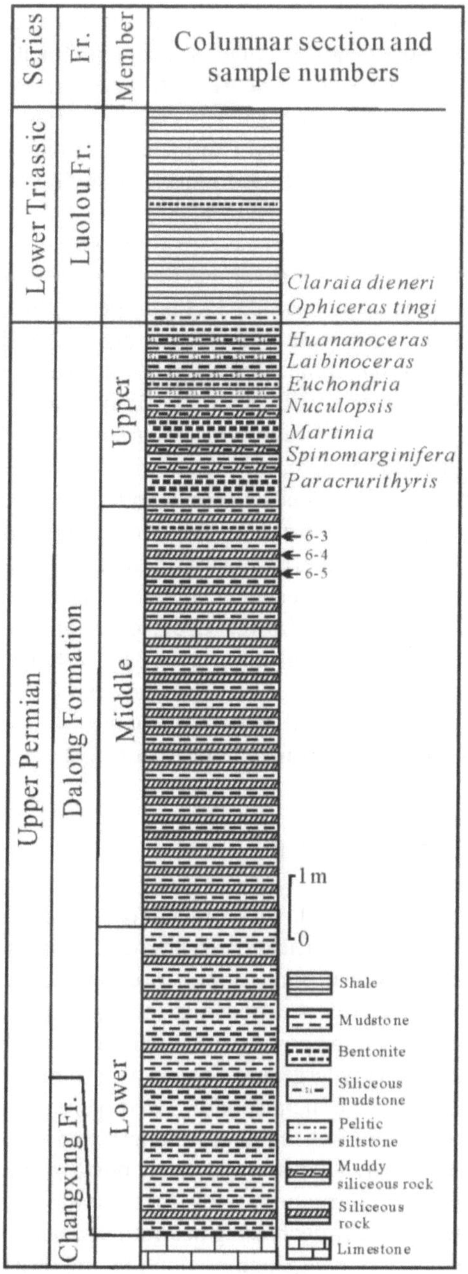

Fig. 2. Columnar section showing the stratigraphic position of the radiolarian fauna described in this paper.

irregular pore frames, and connected with cortical shell by some three-bladed and cylindrical beams. There are initial spicules inside the medullary shell, but the number of spicules is unclear.

Etymology. – This species is named for Prof. Yugan Jin of the Nanjing Institute of Geology and Palaeontology in honor of his many contributions to the study of the Permian stratigraphy.

Type. – Holotype, Plate 1 Figure 14, sample C6-3, catalog number X0301-14.

Measurements (μm). – Based on 24 specimens, of which only a few possess a medullary shell. Diameter of cortical shell 48-87 (average 72, holotype 70), diameter of pore of cortical shell 19-32 (average 26, holotype 29); diameter of medullary shell 25-27 (average 26, holotype 28), width of pores on medullary shell 5-9; width of primary spines near base 12-25 (average 20, holotype 21); length of primary spines 48-106 (average 80, holotype 56).

Occurrence. – This species is only known from the middle part of the Dalong Formation (upper Changxingian) in the Dongpan Section, southern Guangxi, China.

Discussion. – This species is similar to *Cana elegans* MEKIK (2000) from the Lower Cretaceous of Northwest Turkey in the characteristics of the cortical shell, but differs from the latter in having approximately equal primary spines irregularly distributed on the cortical shell and an initial spicule.

Megaporus yini FENG new species

Plate 2, Figures 4–7, 10, 11; Plate 3, Figures 16–20; Plate 4, Figures 1–9

Diagnosis. – A species of *Megaporus* with three unequal, coplanar primary spines and well-developed nodes.

Description. – Test is small, with two spherical concentric shells: cortical and medullary shells. Cortical shell composed of pentagonal and hexagonal pore frames with well-developed nodes at the pore frame vertices. Thickness of bars on the pore frames about 3 times their width. Pore very large, about six to nine pores present on out surface of hemisphere. Medullary shell with polygonal pore frames, and connected with the nodes on the cortical shell by several beams. Three coplanar primary spines are similar in configuration: three-bladed and gradually decreasing in width toward distal end, but unequal in length and asymmetrical in arrangement: two spines (basal spines) nearly equal in length and closer together, often shorter than third spine (polar spine). There are initial spicules inside the medullary shell.

Etymology. – This species is named for Prof. Hongfu Yin of China University of Geosciences (Wuhan) in honor of his many contributions to the study of the Permian-Triassic boundary.

Type. – Holotype, Plate 4, Figure 1, sample C6-5, catalog number X0301-15.

Measurements(μm). – Based on 16 specimens. Length of polar spine 37-63 (average 52, holotype 38), width of polar spine near base 11-24 (average 17, holotype 11); length of basal spine 35-53 (average 42, holotype 35), width of basal spine near base 11-23 (average 16, holotype 11); diameter of cortical shell 57-88 (average 71, holotype 57); diameter of pore of cortical shell 19-35 (average 29, holotype 19).

Occurrence. – This species is only collected from the middle part of the Dalong Formation (upper Changxingian) in the Dongpan Section, southern Guangxi, China.

Discussion. – This species resembles some species of the Mesozoic *Gorgansium* PESSAGNO & BLOME (1980) in outline, but differs from the latter in having a smaller test, larger pores on cortical shell and an initial spicule.

***Megaporus yangi* Feng new species**
Plate 3, Figures 1–6, 12–14

Diagnosis. – *Megaporus* with two unequal polar spines and well-developed nodes on cortical shell.

Description. – Test small, with two concentric shells: cortical and medullary shells. Cortical shell spherical, composed of pentagonal and hexagonal pore frames (predominantly hexagonal) with well-developed nodes at the pore frame vertices; nodes long, massive, and triradiate. Thickness of bars on the pore frames approximately four times their width. Pore very large, about seven to ten pores present on outer surface of hemisphere. Two polar spines are similar in configuration, but unequal in length and obliquely opposed to each other. Both spines are made up of three narrow, rounded longitudinal ridges alternating with three wide longitudinal grooves, and gradually decrease in width at distal direction. Medullary shell with polygonal pore frames, and connected with the nodes on the cortical shell by several triradiate and cylindrical beams. There is an initial spicule inside medullary shell.

Etymology. – This species is named for Prof. Qun Yang of Nanjing Institute of Geology and Palaeontology in honor of his many contributions to the study of the Jurassic Radiolaria.

Type. – Holotype, Plate 3, Figure 1, sample C6-3, catalog number X0301-16.

Measurements (μm). – Based on 9 specimens. Length of long polar spine 53-83 (average 75, holotype 77), width of long polar spine near base 19-37 (average 28, holotype 30); length of short polar spine 46-57 (average 51, holotype 50), width of short polar spine near base 28-40 (average 32, holotype 30); diameter of cortical shell 64-79 (average 71, holotype 71).

Occurrence. – This species is only collected from the middle part of the Dalong Formation (upper Changxingian) in the Dongpan Section, southern Guangxi, China.

Discussion. – The morphology of this species is similar to some species of the genus *Pantanellium* Pessagno (1977) from Upper Triassic-Lower Cretaceous in morphology, but differs from the latter in having an initial spicule inside medullary shell and longer nodes on the cortical shell.

***Megaporus unicum* Feng new species**
Plate 3, Figures 7–9

Diagnosis. – *Megaporus* with two unequal polar spines between which the minimum angle is 103°–140 °.

Description. – Test small with two concentric shells: cortical and medullary shells. Cortical shell spherical, composed of pentagonal and hexagonal pore frames (predominantly pentagonal) with well-developed nodes at the pore frame vertices; nodes long, massive, and triradiate. Thickness of bars on the pore frames about three times their width. Pore very large, about eight pores present on outer surface of hemisphere. Medullary shell small, with polygonal pore frames, and connected with the nodes on the cortical shell by several triradiate and cylindrical beams. Two polar spines are similar in configuration: consisting of three rounded longitudinal ridges alternating with three wide longitudinal grooves, and gradually decreasing in width at distal direction, but they are unequal in length and asymmetrical in arrangement: long primary spine is nearly two times the length of the short spine, and minimum angle between them about 125°.

Etymology. – *Unicus-a-um* (Latin, adj.) = unique.

Type. – Holotype, Plate 3, Figure 7, sample C6-5, catalog number X0301-17.

Measurements (μm). – Based on three specimens. Length of long polar spine 57-83 (average 70, holotype 83), width of long polar spine near base 23-30 (average 28, holotype 30); length of short polar spine 43-52 (average 48, holotype 43), width of short polar spine near base 18-27 (average 22, holotype 20); diameter of cortical shell 80-90 (average 85, holotype 83); minimum angle between both polar spines 105°-140° (average 125°, holotype 103°).

Occurrence. – This species is only collected from the middle part of the Dalong Formation (upper Changxingian) in the Dongpan Section, southern Guangxi, China.

Discussion. – This species is similar to *Denize magnifica* Mekik (2000) from the Lower Cretaceous of northwestern Turkey in morphology, but different from the latter in having an obtuse angle between both polar spines and well-developed nodes. It is distinguished with *Megaporus yangi* n. sp. by strongly asymmetrical arrangement of both polar spines.

***Megaporus* sp.**
Plate 2, Figure 13; Plate 3, Figures 10, 11

Description. – Test small, with two concentric shells: cortical and medullary shells. Cortical shell spherical, composed of pentagonal and hexagonal pore frames (predominantly pentagonal) with poorly developed nodes at the pore frame vertices. Thickness of bars on the pore frames approximately three times their width. Pore very large and about seven to nine pores present on outer surface of hemisphere. Two polar spines are similar in configuration, but unequal in length and obliquely opposed to each other. They are with three narrow, rounded longitudinal ridges alternating with three wide longitudinal grooves, and gradually decreasing in width at distal direction.

Occurrence. – Changxingian, the Dalong Formation in southern Guangxi, China.

Discussion. – It differs from *Megaporus yangi* n. sp. and *Megaporus unicum* n. sp. in having poorly developed nodes at the pore frame vertices.

Acknowledgments

This study has been supported by NSFC (Project No. 49972002, 40232025). We wish to express our sincere thanks to Prof. Qun Yang and Dr. Hui Luo of Nanjing Institute of Geology and Palaeontology, Chinese Academy of Sciences, Prof. Haoruo Wu of Institute of Geology and Geophysics, Chinese Academy of Sciences, and Prof. Jonathan Aichison of the University of Hong Kong for their valuable suggestions. We are grateful to Prof. Dr. Helmcke and Dr. R. Ingavat-Helmcke for their help during the course of the study.

REFERENCES

BUREAU OF GEOLOGY AND MINERAL RESOURCES OF GUANGXI ZHUANG AUTONOMOUS REGION, 2001: 1:50,000 Liuqiao geologic map and explanatory notes (in Chinese).

DE WEVER£ P., DUMITRIC, P., CAULET, J.P., NIGRINI, C. & CARIDROIT, M. 2001: Radiolarians in the sedimentary record. 178–180 p., Gordon and Breach Science Publishers.

EHRENBERG, C.G. 1838: Über die Bildung der Kreidefelsen und des Kreidemergels durch unsichtbare Organismen. Abhandlungen der Koeniglischen Akademie der Wissenschaften, zu Berlin, Jahrgang 1838, 59–147.

FOREMAN, H.P. 1963: Upper Devonian Radiolaria from the Huron member of the Ohio shale. Micropaleontology 9, 267–304.

KAKUWA, K. 1996: Correlation between the bedded chert sequence of southwest Japan and $\delta f^{13}C$ excursion of carbonte sequence, and its significance to the Permian-Triassic mass extinction. The Island Arc 5, 194–202.

KOZUR, H. & MOSTLER, H. 1982: Entactinian suborder, a new radiolarian suborder. Geologische-Palaeontologische Mitteilungen, Innsbruck 11, 399–414.

KUWAHARA, K. & YAO, A. 1998: Diversity of Late Permian radiolarian assemblages. News of Osaka Micropaleontologists, Special Volume 11, 33–46. (in Japanese with English abstract)

KUWAHARA, K., YAO, A. & YAMAKITA, S. 1998: Reexamination of Upper Permian radiolarian biostratigraphy. Earth Science 52(5), 391–404.

Mekik, F.A. 2000: Early Cretaceous Pantanelliidae (Radiolaria) from Northwest Turkey. Micropaleontology, 46 (1), 1–30.

Pessagno, E.A. Jr. 1977: Lower Cretaceous radiolarian biostratigraphy of the Great Valley Sequence and Franciscan Complex, California Coast Ranges. Contributions to the Cushman Foundation of Foraminiferal Research, Special Publication 15, 1–87.

PESSAGNO, E.A. JR., & BLOME, C.D. 1980: Upper Triassic and Jurassic Pantanellidae from California, Oregon and British Columbia. Micropaleontology 26 (3), 225–273.

RIEDEL, W.R. 1967: Subclass Actinopoda. In: HARLAND, W.B. et al. (Eds.): Protozoa, the Fossil Record, 291–298. Geological Society of London Publication.

SHANG, Q.H., CARIDROIT, M. & WANG Y.J. 2001: Radiolarians from the Uppermost Permian Changxingian of southern Guangxi. Acta Micropalaeontologica Sinica 18(3), 229–240.

VISHNEVSKAYA, V. 1997: Development of Paleozoic-Mesozoic Radiolaria in the Northwestern Pacific Rim. Marine Micropaleontology 30, 79–95.

YAO, J., YAO, A. & KUWAHARA, K. 2001: Upper Permian biostratigrapgic correlation between conodont and radiolarian zones in the Tamba-Mino Terrane, Southwest Japan. Journal of Geosciences, Osaka City University 44(5), 97–119.

Manuscript received January 2004
Revision accepted March 2005

Plate 1

Scanning electron photomicrographs of the Changxingian radiolarians from the Dalong Formation in southern Guangxi, Southw China (1–14, 16, 17: bar=50μm; 15, 18: bar=20μm; 19: bar=10μm). 1–19 *Megaporus jini* n. sp.; 1, C6-4/1698; 2, C6-4/1693; 3, C6-5/1723,; 4, C6-4/1696; 5, C6-4/1699; 6, C6-5/1722; 7, C6-5/1718; 8, C6-4/1710; 9, C6-4/1712; 10, C6-4/1700; 11, C6-5/1725; 12, C6-5/2055; 13, C6-5/1727; 14, C6-3/1140, holotype; 15, C6-3/1255, magnification of holotype, showing medullary shell; 16, C6-4/1716; 17, C6-4/1695; 18, C6-5/1717, magnification of figure 16, showing medullary shell and initial spicules in the medullary shell; 19, C6-5/1739, magnification of figure 16, showing initial spicules in the medullary shell.

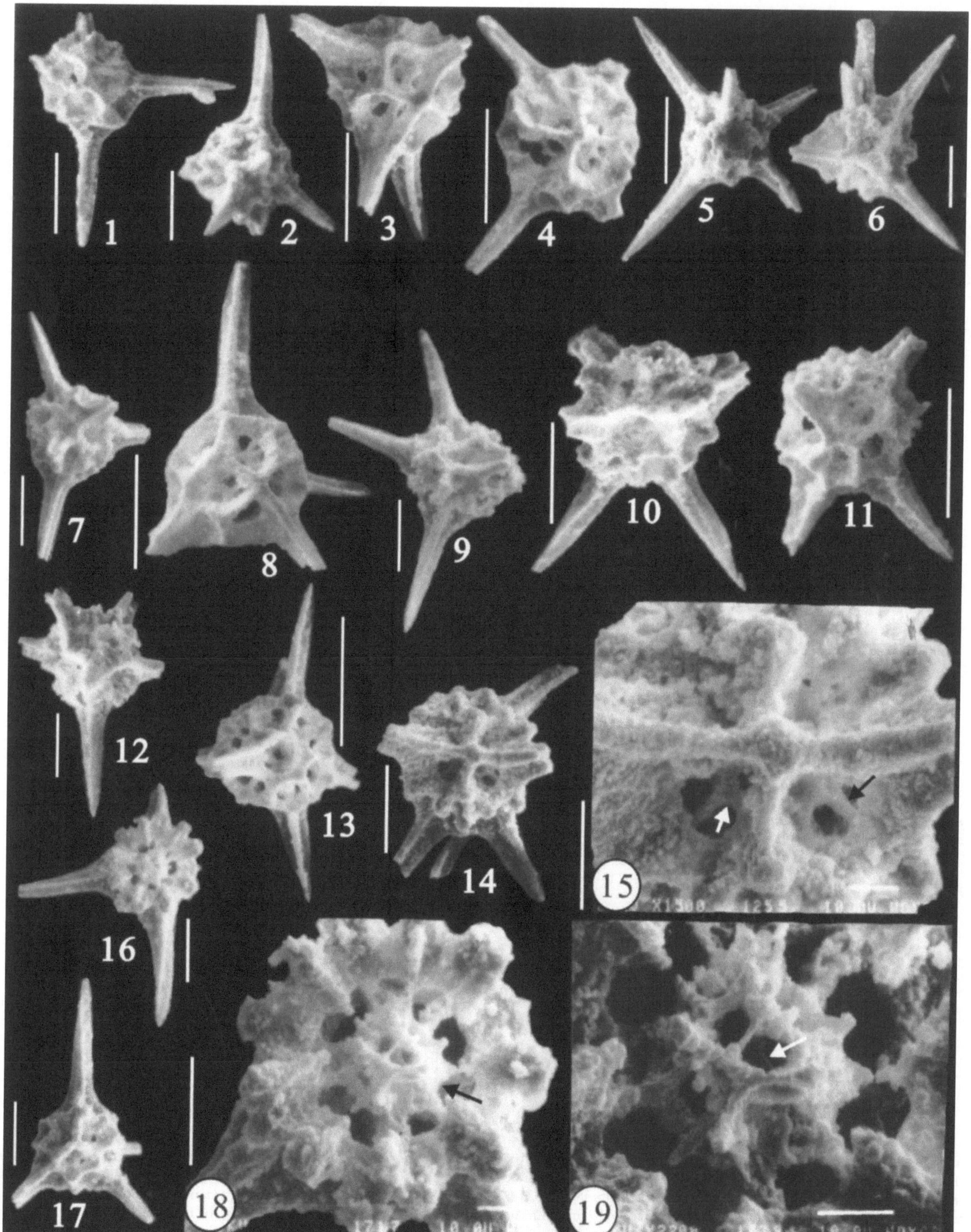
1
2
3
4
5
6
7
8
9
10
11
12
13
14
15
16
17
18
19

Plate 2

Scanning electron photomicrographs of the Changxingian radiolarians from the Dalong Formation in southern Guangxi, South China (1–3: bar=1μm; 4–8: bar=10μm; 9–13: bar=50μm). 1–3, 8, 9, 12 *Megaporus jini* n. sp.; 1–3, magnification of plate 1 figure 16, showing initial spicules in the medullary shell; 8, magnification of the figure 12, showing initial spicules in the medullary shell; 9, C6-5/3116; 12, C6-5/2500. 4–7, 10, 11, *Megaporus yini* n. sp.; 4 and 5, magnification of the figure 10, showing initial spicules in the medullary shell; 6 and 7, magnification of the figure 11, showing initial spicules in the medullary shell; 10, C6-5/3112; 11, C6-5/2498. 13, *Megaporus* sp., C6-5/3105.

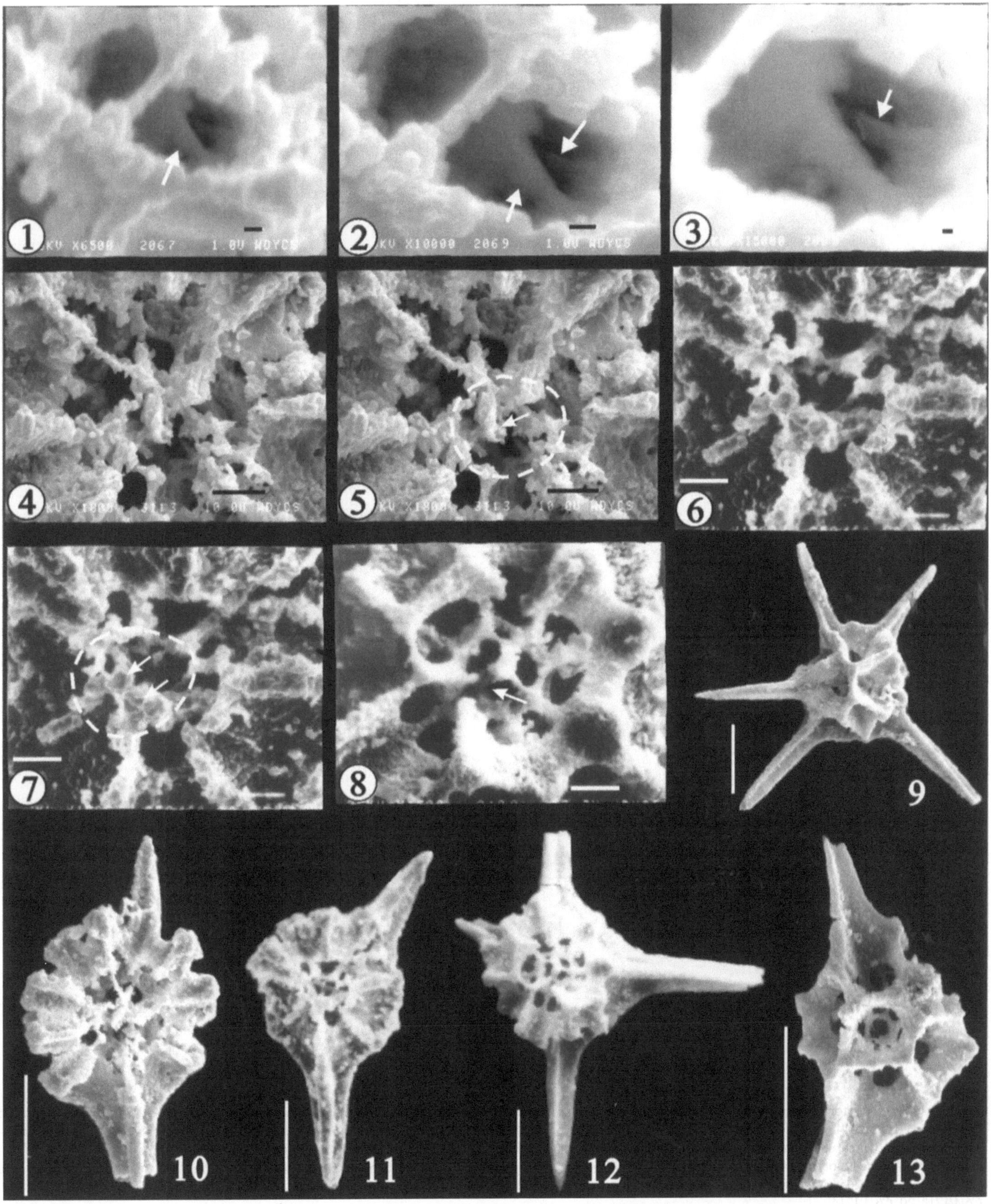
1
2
3
4
5
6
7
8
9
10
11
12
13

Plate 3

Scanning electron photomicrographs of the Changxingian radiolarians from the Dalong Formation in southern Guangxi, South China (1–14, 16–20: bar=50μm; 15: bar=100μm). 1–6, 12–14 *Megaporus yangi* n. sp.; 1, C6-3/1142, holotype; 2, C6-3/1145; 3, C6-5/1730; 4, C6-5/2057; 5, C6-5/1726; 6, C6-5/1737; 12, C6-3/1059, showing thickness of cortical shell, and beams between cortical and medullary shells; 13, C6-3/1252, magnification of figure 2, showing medullary shell; 14, C6-3/1254, magnification of the holotype, showing medullary shell. 7–9 *Megaporus unicum* n. sp.; 7, C6-5/2060, holotype; 8, C6-4/1686; 9, C6-3/1238. 10, 11 *Megaporus* sp.; 10, C6-3/1127; 11, C6-5/2058. 15, *Neoalbaillella optima* Ishiga, Kito and Imoto, C6-3/1093. 16–20 *Megaporus yini* n. sp.; 16, C6-5/1728; 17, C6-4/1706; 18, C6-5/1720; 19, C6-4/1713; 20, C6-4/1714.

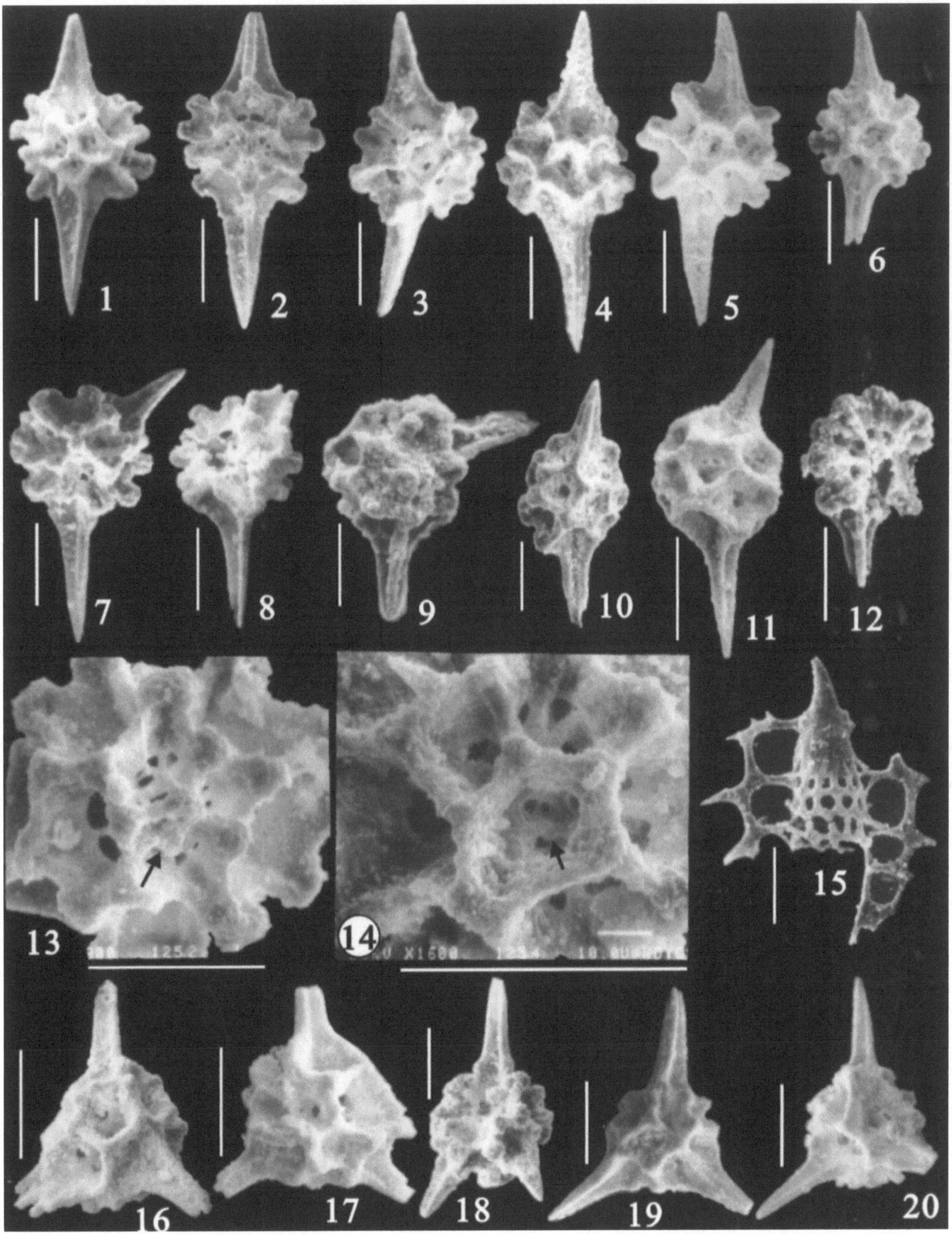
1
2
3
4
5
6
7
8
9
10
11
12
13
14
15
16
17
18
19
20

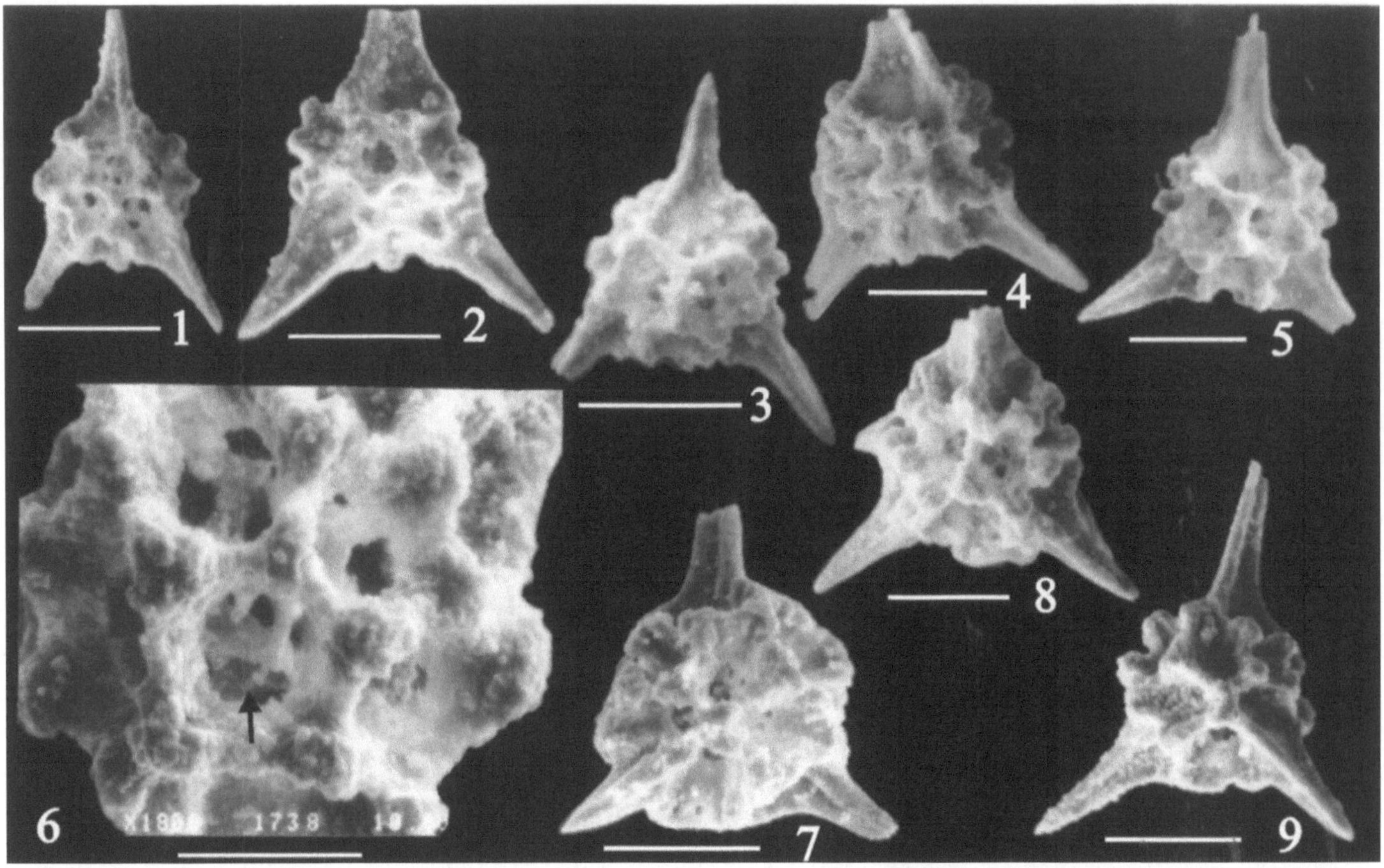

Plate 4

Scanning electron photomicrographs of the Changxingian radiolarians from the Dalong Formation in southern Guangxi, South China (1–5, 7–9: bar=50μm; 6: bar=20μm). 1–9 *Megaporus yini* n. sp.; 1, C6-5/1732, holotype; 2, C6-5/1735; 3, C6-4/1711; 4, C6-5/1731; 5, C6-5/1729; 6, C6-5/1738, magnification of the holotype, showing medullary shell; 7, C6-3/1143; 8, C6-5/1736; 9, C6-5/1734.

0012-9402/06/01S079-21
DOI 10.1007/s00015-006-0604-3
Birkhäuser Verlag, Basel, 2006

Eclogae geol. Helv. 99 (2006) Supplement 1, S79–S99

Radiolarian faunal turnover through the Paleocene-Eocene transition, Mead Stream, New Zealand

CHRISTOPHER J. HOLLIS[1]

Key words: South Pacific, Paleogene, Radiolaria, diatoms, paleoecology, global change

ABSTRACT

The Mead Stream section, northern Clarence Valley, is the most complete Paleocene-early Eocene record of pelagic sedimentation in the mid-latitude (~55° S paleolatitude) Pacific Ocean. Integrated studies of sediments, siliceous and calcareous microfossils and carbon isotopes have shown that major global climate events are recorded by distinct changes in lithofacies and biofacies. The consistent and often abundant occurrence of siliceous microfossils in the section provides a rare opportunity to undertake quantitative analysis of high-latitude radiolarian population changes through the late Paleocene and early Eocene. Late Paleocene assemblages are dominated by spumellarians, although the nassellarian species *Buryella tetradica* is the most abundant species. The Paleocene-Eocene boundary (= base of Paleocene-Eocene thermal maximum) in the Mead Stream section is marked by major faunal turnover, including an abrupt decrease in *B. tetradica*, first occurrences of several low-latitude species (e.g. *Amphicraspedum prolixum* s.s., *Lychnocanium auxilla*, *Podocyrtis papalis*, *Phormocyrtis turgida*, *Theocorys*? *phyzella*) and increased abundance of large, robust spumellarians relative to small actinommids. Above an 18-m thick, lowermost Eocene interval in which radiolarians are abundant to common, radiolarian abundance declines progressively, falling to <10 individuals per gram in the marl-dominated unit that is correlated with the early Eocene climatic optimum. These trends in siliceous microfossil populations signal major changes in watermass characteristics along the northeastern New Zealand margin in the earliest Eocene. Assemblages typical of cool, eutrophic, watermasses that dominated the Marlborough Paleocene were replaced in the early Eocene by assemblages more characteristic of oligotrophic, stratified, subtropical-tropical watermasses.

ZUSAMMENFASSUNG

Das Mead Stream Profil, nördliches Clarence Tal, ist die vollständigste Paläozän-Früh Eozän Aufzeichnung pelagischer Sedimentation in mittleren Breiten (~55° S Paläobreite) des Pazifischen Ozeans. Integrierte Studien von Sedimenten, kieseligen und kalkigen Mikrofossilien und Kohlenstoffisotopen haben gezeigt, dass bedeutende globale Klimaereignisse durch deutliche Veränderungen in Litho- und Biofazies gekennzeichnet sind. Das beständige und oft sehr häufige Vorkommen kieseliger Mikrofossilien im Profil bietet eine seltene Gelegenheit, quantitative Studien von Veränderungen innerhalb von Radiolarienpopulationen hoher Breiten während des späten Paläozäns und frühen Eozäns durchzuführen. Spätpaläozäne Vergesellschaftungen werden von Spumellarien dominiert, obwohl die Nassellarienart *Buryella tetradica* die häufigste Art ist. Die Paläozän-Eozän Grenze (=Basis des Paläozän-Eozän Wärmemaximums) im Mead Stream Profil ist durch bedeutende faunistische Erneuerungen gekennzeichnet, einschließlich der abrupten Abnahme von *B. tetradica*, dem ersten Auftreten zahlreicher Arten niederer Breiten (z. B. *Amphicraspedum prolixum* s.s., *Lychnocanium auxilla*, *Podocyrtis papalis*, *Phormocyrtis turgida, Theocorys*? *phyzella*) und der zunehmenden Häufigkeit von großen, robusten Spumellarien im Verhältnis zu kleinen Actinommida. Oberhalb eines 18 Meter mächtigen Intervals des untersten Eozäns, in dem Radiolarien sehr zahlreich bis häufig sind, nimmt die Radiolarienhäufigkeit progressive ab, bis auf eine Anzahl von weniger als zehn Individuen pro Gramm innerhalb der Mergel-dominierten Einheit, die dem früheozänen Klimaoptimum entspricht. Diese Trends innerhalb von Populationen kieseliger Mikrofossilien signalisieren bedeutende Veränderungen in Wassermassencharakteristika entlang des nordöstlichen neuseeländischen Kontinentalrandes im frühesten Eozän. Vergesellschaftungen typisch für kühle, eutrophische Wassermassen, die das Marlborough Paläozän dominierten, wurden im frühen Eozän durch Vergesellschaftungen ersetzt, die eher für oligotrophische, stratifizierte, subtropische-tropische Wassermassen charakteristisch sind.

1. Introduction

Radiolaria are the most diverse and widely distributed group of marine plankton preserved in the fossil record (Lipps 1993; De Wever et al. 2001). High evolutionary turnover and considerable geographic variation in populations make radiolarian fossils invaluable research tools in biostratigraphy and paleoceanography. Consequently, detailed census studies through sedimentary successions have the potential to reveal much about biosphere-geosphere interactions in the marine realm during critical episodes in Earth history. Whereas quantitative studies are common for late Cenozoic radiolarians (e.g. Boltovskoy 1987; Caulet et al. 1992), very few such studies have been made of early Cenozoic or Mesozoic assemblages.

[1] Institute of Geological and Nuclear Sciences, PO Box 30-368, Lower Hutt, New Zealand. Email: c.hollis@gns.cri.nz

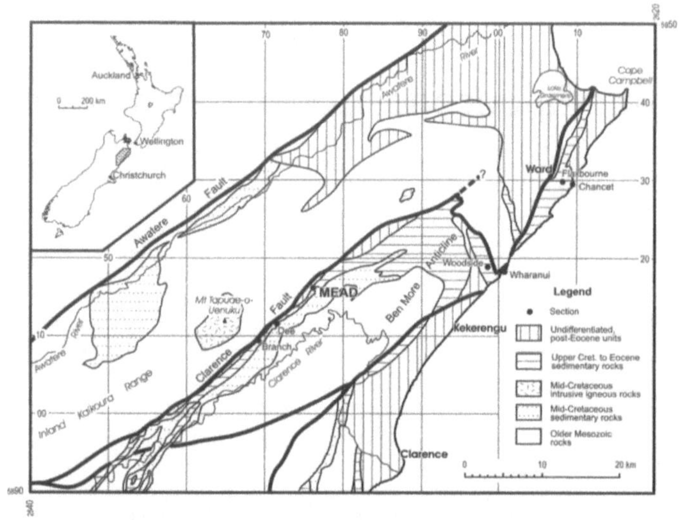

Fig. 1. Simplified geological map of eastern Marlborough (after Crampton et al. 2003) showing location of Mead Stream and other stratigraphic sections mentioned in the text.

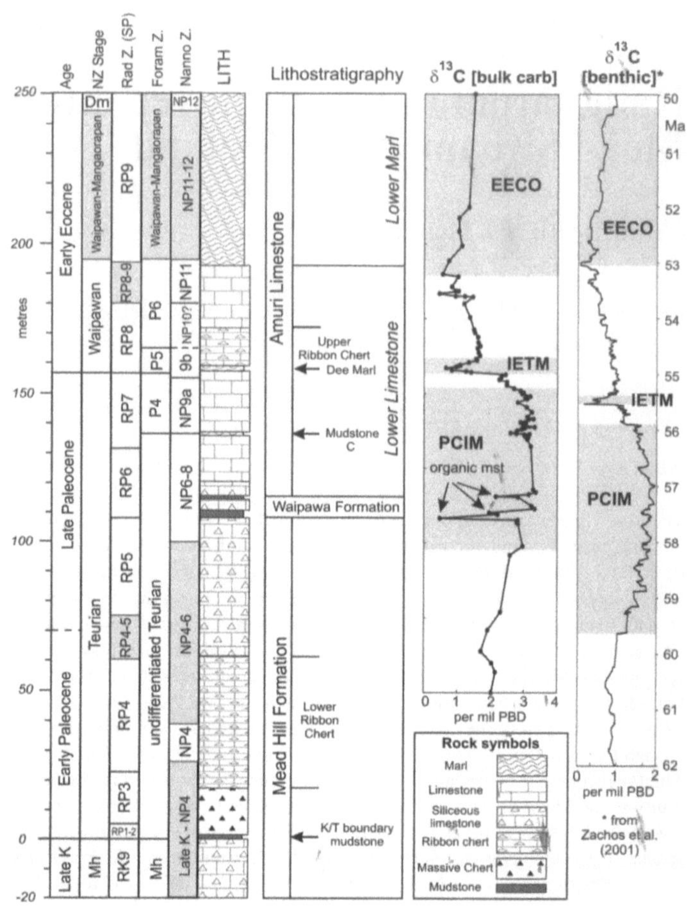

Fig. 2. Lithostratigraphy, biostratigraphy and carbon isotope stratigraphy of Upper Cretaceous-lower Eocene strata at Mead Stream (after Hollis et al. 2005). For age and biostratigraphy columns, intervals of uncertain age or biozone are shaded. Three episodes of major global change are identified from bulk carbonate $\delta^{13}C$ stratigraphy: late Paleocene carbon isotope maximum (PCIM), Paleocene-Eocene thermal maximum (PETM) and early Eocene climatic optimum (EECO). For comparison, the global compilation of benthic foraminiferal $\delta^{13}C$ data of Zachos et al. (2001) is also shown; the plotted values are running five-point averages calibrated to the geochronometric time scale of Berggren et al. (1995) following Hollis et al. (2005).

Notable exceptions are studies of Cretaceous-Tertiary boundary radiolarians in New Zealand and Ecuador (Hollis 1996; Keller et al. 1997; Hollis et al. 2003a, b). In this article we report the results of the first quantitative study of radiolarian assemblage variation through the Paleocene-Eocene (P-E) transition (~60-50 Ma).

In addition, this is the first study of radiolarian faunal change through the P-E transition in the mid-latitude South Pacific; eastern Marlborough being situated at ~55° S in the latest Paleocene (King et al. 1999; Sutherland et al. 2001). Sanfilippo & Nigrini (1998a) reviewed DSDP and ODP records of late Paleocene-early Eocene radiolarians and established a composite biostratigraphy from well-preserved, low-latitude assemblages in 12 sites between 40°N and 30°S. Most of these sites are in the western central and North Atlantic, with two in the South Atlantic, one in the western central Indian Ocean and one in the North Pacific. Subsequently, Nigrini & Sanfilippo (2000) and Sanfilippo & Blome (2001) reported on radiolarian assemblages in well-constrained P-E boundary records in Caribbean and western North Atlantic ODP holes, respectively.

The section exposed in the southern branch of Mead Stream, Clarence River valley (Fig. 1), is the most complete known record of South Pacific pelagic-hemipelagic sedimentation from Late Cretaceous to middle Eocene (Strong et al. 1995; Hollis et al. 2005). Rich radiolarian assemblages make it a key reference section for early Paleogene radiolarian biostratigraphy and it is the type section for eight zones (RP6-13) in the South Pacific radiolarian zonation (Strong et al. 1995; Hollis 1997, 2002; Hollis et al. 2005). Detailed stratigraphic and paleoenvironmental studies have been completed on the Cretaceous/Tertiary (K/T) boundary (Hollis et al. 2003a) and the Paleocene-Eocene (P-E) transition (Hollis et al. 2005). In the latter study, carbon isotope stratigraphy was integrated with radiolarian, foraminiferal and calcareous nannofossil biostratigraphy in order to relate lithofacies changes to major perturbations in the early Paleogene global carbon cycle and climate system. Three of these episodes of perturbations have been identified in upper Paleocene and lower Eocene strata at Mead Stream (Fig. 2): the late Paleocene carbon isotope maximum (PCIM – Corfield & Cartlidge 1992, Thompson & Schmitz 1997; Kurtz et al. 2003), the Paleocene-Eocene thermal maximum (PETM – Zachos et al. 2001, 2003), and the early Eocene climatic optimum (EECO – Zachos et al. 2001). The base of the PETM corresponds with the Paleocene-Eocene boundary (Zachos et al. 2003).The aim of the current study is to use changes in radiolarian assemblages through the P-E transition at Mead Stream to further elucidate the influence of these global events on oceanic conditions in the Southwest Pacific.

2. Material and Methods

This study is based on 42 radiolarian assemblages recovered from rock samples between 100 and 250 m above the well-constrained K/T boundary (Hollis et al. 2003a) in the southern and main branch of Mead Stream. Sampling methods are described by Hollis et al. (2005). For radiolarian microfossil extraction, samples were crushed into 5–10 mm chips and leached in 15% hydrochloric acid (HCl) until reaction ceased. Siliceous rocks with weak reaction to HCl digestion were then rinsed and leached in 5% hydrofluoric acid for 2–4 hours. Samples were then washed through a 63 µm screen and the residues cleaned by gentle heating in a 1:1 solution of 10% hydrogen peroxide and calgon, $(NaPO_3)_6$.

Taxonomic nomenclature is summarised in Table 1. Selected species are illustrated in Plates 1–3. Census data (Hollis et al. 2005, supplementary data) have been derived from vertical traverses of strewn slides under transmitted light. For sparse samples (<300 radiolarians per slide), all radiolarians on one, two or more slides were counted. For richer samples, the following double count method was utilised. First, all specimens were counted until a total count of ~300 was achieved. The proportion of the slide examined to this point was determined and the abundance of common taxa (>15 radiolarians or >5% of the total fauna) was estimated for the remainder of the slide. The remainder of the slide was then examined and occurrences of rare taxa (<5% in initial count) recorded. This method has the advantage of establishing reliable abundance estimates for rare taxa without expending great amounts of time counting many hundreds of specimens. One disadvantage is that the total count is highly variable, ranging in this dataset from <60 to >7000 individuals. Abundance of larger (>63 µm) diatoms was also recorded during the initial count (Hollis et al. 2005).

In order to obtain an accurate estimate of relative abundance, all specimens have been assigned a taxonomic category. Although these counting groups range from the level of subspecies to family, they are given equal weighting in diversity analysis.

3. Stratigraphy

The lithostratigraphy and foraminiferal, radiolarian and dinoflagellate cysts biostratigraphy of Upper Cretaceous to middle Eocene strata at Mead Stream was described by Strong et al. (1995). Subsequently, the radiolarian biostratigraphy was reviewed and revised by Hollis (1997). Hollis et al. (2003a) described the stratigraphy of the K/T boundary transition in detail, as part of a paleoenvironmental study of this interval at Mead Stream and nearby Branch Stream (Fig. 1). The K/T boundary is defined as zero datum in the current study. As a forerunner to the present study, Hollis et al. (2005) completed a detailed stratigraphic study of upper Paleocene-lower Eocene strata at Mead Stream, incorporating lithostratigraphy, foraminiferal, calcareous nannofossil and radiolarian biostratigraphy and carbon isotope stratigraphy (Fig. 2).

Paleocene-Eocene strata exposed along Mead Stream consist of well-bedded units of chert, siliceous limestone with chert nodules, siliceous mudstone, limestone and marl, which form three formations within the Muzzle Group (Reay 1993). The Mead Hill Formation consists of dm-bedded chert and siliceous limestone. It extends from a faulted base 170 m below the K/T boundary to a sharp contact with overlying the Waipawa Formation at 108 m above the K/T boundary. The Waipawa Formation is a 7.5 m thick interval (108–115.5 m) that consists of lower and upper siliceous mudstone units with an intervening unit of siliceous limestone. The two mudstone units carry the same distinctive geochemical signature of the Waipawa Formation as in its type area in Hawkes Bay, North Island (Killops et al. 2000).

The overlying formation, Amuri Limestone, has been subdivided into several lithotypes and members (Reay 1993; Hollis et al. 2005). The basal lithotype is Lower Limestone, which includes two members, Dee Marl and Upper Ribbon Chert. Lower Limestone (116–192 m) consists of dm-bedded siliceous limestone with thin marl interbeds and rare chert nodules. One unusually thick marl bed provides a useful reference point (136.5–136.6 m) within Lower Limestone and is referred to as Mudstone C (following Hollis et al. 2005). Dee Marl is a 2.4 m thick unit (157–159.4 m) of dm-bedded alternating marl and marly limestone. This unit marks the base of the Eocene at Mead Stream and at the type section, Dee Stream (Hancock et al. 2003). Upper Ribbon Chert, the highest occurrence of chert in the section, is a 12.6 m thick interval (159.4–172 m) of dm-bedded siliceous limestone with narrow chert stringers in the centre of the beds. The abundance of chert decreases gradually up-section, whereas the abundance and thickness of marl interbeds increases. The top of Lower Limestone is placed at the base of the first of a series of thick marl beds intercalated with limestone beds. The overlying lithotype, Lower Marl, consists of 110 m of alternating marl and marly limestone beds, similar to the lithology of Dee Marl. The stratigraphy of the upper part of Amuri Limestone, which is not considered here, is described by Strong et al. (1995).

4. Siliceous microfossil occurrence

Of 134 samples processed for radiolarians in upper Paleocene and lower Eocene strata at Mead Stream (~75–300 m), 42 samples contained radiolarian assemblages of adequate preservation and abundance for species level identification and quanti-

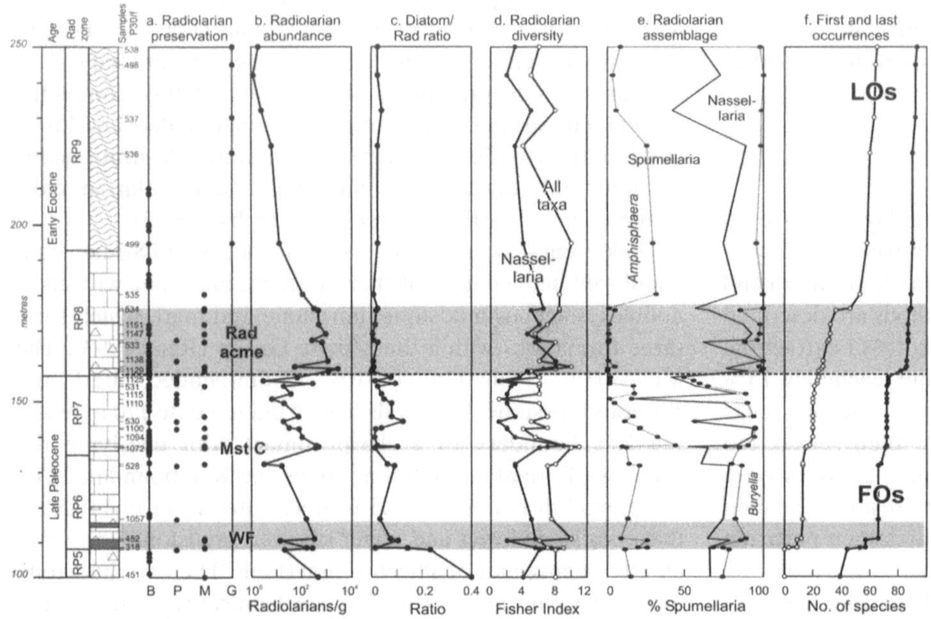

Fig. 3. Variation (a) radiolarian preservation, (b) radiolarian abundance, (c) diatom/radiolarian ratio, (d) radiolarian diversity (Fisher α Index), (e) relative abundance of major radiolarian groups, and (f) radiolarian first and last occurrences in upper Paleocene-lower Eocene siliceous microfossil assemblages at Mead Stream. Radiolarian preservation is recorded as **B**arren, **P**oor (internal and surface features obscured by recrystallisation), **M**oderate (internal features obscured by infilling or recrystallisation, surface features preserved), **G**ood (test morphology not affected by recrystallisation, no infilling, breakage minor to moderate). Shaded intervals represent two distinctive units (WF = Waipawa Formation, Mst C = Mudstone C) and a 21 m-thick lowermost Eocene interval characterised by peak radiolarian abundances (rad acme).

tative analysis (Fig. 3). Radiolarians are generally common but poorly preserved in the upper Paleocene where strong recrystallisation, including siliceous infilling and overgrowth, is evident in siliceous limestone samples (Fig. 3a–b). In the first 21 m of Eocene strata, radiolarians are generally abundant and moderately preserved. Recrystallisation is less pronounced in this interval. This probably reflects reduced silica content in Dee Marl. In the overlying Upper Ribbon Chert, chertification is restricted to a central zone in each bed and samples taken from the outer parts of beds contained relatively well-preserved siliceous microfossils. In the overlying marl-rich Eocene interval, radiolarians are increasingly rare but preservation is moderate to good.

Diatoms are common in radiolarian residues in the Paleocene but are rare to very rare in the Eocene (Fig. 3c). The original change in diatom abundance across the P/E boundary is likely to have been even more pronounced than is described here. Owing to their smaller average size and more delicate tests, diatoms are more prone to diagenetic destruction than radiolarians. Therefore, diatoms are likely to be under-represented in Paleocene siliceous limestone samples relative to the better preserved Eocene material.

Radiolarian diversity is highly variable. Most assemblages contain 20–50 taxa whereas three assemblages have <20 taxa and seven assemblages have >50 taxa. The Fisher α Index is used to assess diversity trends in census data (Fig. 3d). The diversity curve for all taxa is considered to be less reliable than the curve for nassellarians. Most nassellarians have been identified to species level, whereas many spumellarians remain undifferentiated below family level. This results in an underestimate of overall diversity in spumellarian-dominated samples, notably in the earliest Eocene. Nassellarian α values indicate relatively low diversity in the Paleocene, apart from a short-lived increase in the marl bed (Mudstone C) near the base of Zone RP7. A rapid increase in diversity in the earliest Eocene is followed by a gradual decline with relatively low α values through the Lower Marl.

Paleocene radiolarian assemblages are dominated by spumellarians (Fig. 3e) and, in order of abundance, the most common species and species groups are: *Buryella tetradica*, *Haliomma* spp. gr. b, *Amphisphaera coronata* gr, *A. goruna*, *Lithelius minor* gr. and *Buryella granulata*. In the lowermost 21 m of Eocene strata, spumellarian dominance is greater than in the Paleocene and the most common species are: *Haliomma* gr. b., *Amphisphaera coronata* gr. *Lithelius minor* gr. and *Theocorys*? cf. *phyzella*. In overlying Eocene strata, spumellarians dominate to a similar extent as in the Paleocene and the most common species are: *Haliomma* gr. b, *Amphisphaera coronata* gr. *Lithelius minor* gr., *Calocycloma ampulla*, *Phormocyrtis striata striata*, *Podocyrtis papalis*, *Sethocytris babylonis*, *Axoprunum pierinae* and *Periphaena heliasteriscus*.

5. Radiolarian faunal turnover through the Paleocene-Eocene transition

In contrast to low-latitude regions where there is little change in radiolarian assemblages across the Paleocene-Eocene boundary (Sanfilippo & Nigrini 1998a; Nigrini & Sanfilippo 2000; Sanfilippo & Blome 2001), four episodes of significant faunal turnover are observed in the upper Paleocene-lower Eocene succession at Mead Stream (Fig. 3f, Table 2).

The first episode occurs in the lower Waipawa Formation and is associated with the RP5/RP6 zone boundary. The first occurrences (FOs) of 22 species and the last occurrences (LOs) of 12 species are recorded within 3 m of strata (~108–111 m). Some of the LOs are isolated records of Cretaceous-early Pa-

Table 1. Annotated list of radiolarian species or species groups encountered in this study. Species ranges are based on the South Pacific zonation for Mead Stream and other South Pacific ("Sth Pac") sections (Hollis, 1997, 2002; Hollis et al. 1997), and on the tropical zonation (Sanfilippo and Nigrini, 1998a) for low latitude ("Low Lat") records; LK = Late Cretaceous, UZP = unzoned Paleocene, EOC = Eocene, MIO = Miocene. Reference to taxon concept: CC42 = Clark and Campbell, 1942; F73 = Foreman, 1973; H02 = Hollis, 2002; H97a = Hollis, 1997; H97b = Hollis et al. 1997; K99 = Kozlova, 1999; N92 = Nishimura, 1992; S85 = Sanfilippo et al. 1985; SR92 = Sanfilippo and Riedel, 1992; S95 = Strong et al. 1995; SR73 = Sanfilippo & Riedel, 1973.

Taxon name	**Zonal range**			**Taxon concept**	
	Mead	**Sth Pac**	**Low Lat.**	**Ref.**	**Strong et al. 1995:**
Amphicraspedum murrayanum HAECKEL	RP8	RP8	RP6-8	SR73	
Amphicraspedum prolixum form. gracilis (LIPMAN)	RP8-9	RP8-9	?	K99	
Amphicraspedum prolixum SANFILIPPO & RIEDEL gr.	RP7-10	RP7-10	RP7-13	SR73	
Amphicraspedum prolixum SANFILIPPO & RIEDEL *s.s.*	RP8-10	RP8-10	RP7-9	RP7-9	208
Amphipternis alamedaensis (CAMPB. & CL.)	RP3-9	RK9-RP9	LK-RP7	H97a	
Amphipyndax stocki (CAMPB. & CLARK)	RK9-RP7	RK9-RP7	LK-UZP	H97a	
Amphisphaera aff. magnaporulosa CL. & CAMPB.	RP5-9	RP4-6	?	H02	
Amphisphaera coronata s.l. (EHRENBERG)	RP4-9	RP4-6	UZP-EOC	H97a	208 (*A. radiosa*)
Amphisphaera goruna (SANFILIPPO & RIEDEL)	RP3-8	RP2-6	UZP-RP7	H97a	208, fig. 8G, 9A
Amphisphaera kina HOLLIS	RP2-7	RP2-6	?	H97a	208, fig. 8F
Amphisphaera macrosphaera (NISHIMURA)	RP3-9	RP3-14	UZP-RP6	H97a	208 (*A. radiosa*)
Amphymenium splendiarmatum CLARK & CAMPB.	RP6-8	RP2-6	RP8-20	H97a	
Artostrobus pusillus (EHRENBERG)	RK9-RP8	RK9-RP8	?	H97a	
Aspis murus NISHIMURA	RP6	RP5-6	RP6	N92	
Axoprunum pierinae (CLARK & CAMPB.)	RP5-9	RP6-14	UZP-EOC	SR73	208, fig. 10C
Axoprunum ? aff. *bispiculum* (POPOFSKY)	RP6-9	RP4-6	?	H02	
Bathropyramis magnifica (CLARK & CAMPB.)	RP5-13	RP4-13	?	H02	
Bathropyramis sanjoaquinesis (CAMPB. & CLARK)	RK9-RP8	RK9-RP8	LK-UZP	H97a	
Bekoma bidartensis RIEDEL & SANFILIPPO	RP8	RP8	RP7-8	S95	208, fig. 9G
Bekoma campechensis FOREMAN	RP6	RP6	RP6-7	S95	208, fig. 9F
Bekoma divaricata FOREMAN	RP6-8	RP6-8	RP7	S95	208, fig. 9H
Buryella dumitricai PETRUSHEVSKAYA	RP4-7	RP4-7	?	H97a	208, fig. 8L-M
Buryella foremanae PETRUSHEVSKAYA	RP4-7	RP4-7	?	H97a	208, fig. 8K
Buryella granulata (PETRUSHEVSKAYA)	RP3-8	RP4-8	?	H97a	208, fig. 8J
Buryella helenae O'CONNOR	RP6	RP4-6	?	H02	
Buryella kaikoura HOLLIS	RP5-6	RP5-6	?	H97a	
Buryella pentadica FOREMAN	RP6	RP6	UZP-RP7	N92	
Buryella petrushevskayae O'CONNOR	RP6	RP5-6	?	H02	
Buryella tetradica tetradica FOREMAN	RP5-9	RP5-8	UZP-RP9	H02	208, fig. 8N, 9Q-R
Buryella tetradica tridica O'CONNOR	RP6	RP5-6	?	H02	
Calocycloma ampulla (EHRENBERG)	RP5-10	RP5-10	RP6-14	S95	208. fig. 9V-X
Cassideus aff. mariae NISHIMURA	RP5-9	RP5-9	?	H02	209, fig. 9C (*Microsciadiocapsa* ? sp.)
Clathrocyclas australis HOLLIS	RP3-6	RP2-6	?	H97a	P. 208, 8P (*Clathrocycloma* sp. A)
Clathrocyclas universa gr.	RP5-13	RP5-13	?	H97b	
Conoactinomma stilloformis (LIPMAN)	RP9	RP9	?	K99	206 (*Conocaryomma* sp.)
Cornutella californica CAMPB. & CLARK	RK9-RP13	RK9-RP13	LK-UZP	H97a	
Corythomelissa adunca (SANFILIPPO & RIEDEL)	RP5-8	RP4-8	RP6-8	H02	208, fig. 9B
Corythomelissa spp.	RP6-8	RP6-8	?	H02	
Cryptocarpium ? cf. *ornatum* (EHRENBERG)	RP6-13	RK9-RP13	?	H97a	
Cycladophora aff. *cosma* LOMBARI AND & LAZARUS	RP6-8	RP4-8	?	H02	
Cycladophora cf. *bicornis* (POPOFSKY)	RP6-8	RP6-8	?	H02	208, fig. 9D (*Clathrocycl. humerus*)
Dictyomitra andersoni (CAMPB. & CLARK)	RK9-RP6	RK9-RP6	LK-UZP	H97a	
Haliomma gr. B	RP4-9		H97a		
Lamptonium fabaeforme fabaeforme (KRASHEN.)	RP8-9	RP8-9	RP7-9	S85	
Lamptonium pennatum FOREMAN	RP7-9	RP7-9	UZP-RP8	S95	208, fig. 9I
Lamptonium ? aff. *colymbus* FOREMAN	RP9		F73		
Lithelius ? *minor* gr. JÖRGENSEN	RK9-RP9	RK9-RP14	?	H97a	
Lithocampe wharanui HOLLIS	RP3-6	RK9-RP6	?	H97a	
Lithomespilus coronatus SQUINABOL	RK9RP7	RK9-RP8	UZP-RP8	H97a	
Lithostrobus longus GRIGORJEVA	RP1-7	RP1-6	?	H02	209, fig. 8H-I (*Stichomitra wero*)
Lophophaena mugaica (GRIGORJEVA)	RP5-8	RP2-8	?	H02	

Table 1. (cont.)

Taxon name	**Zonal range**			**Taxon concept**	
	Mead	**Sth Pac**	**Low Lat.**	**Ref.**	**Strong et al. 1995:**
Lychnocanium aff. *carinatum* Ehrenberg	RP8	RP8	RP6-7	N92	209, fig. 9M (*Rhopalocanium*? sp.)
Lychnocanium aff. *tripodium* Ehrenberg	RP7-8	RP7-8	?	K99	
Lychnocanium amphitrite (Foreman)	RP9-13	RP9-14	RP8-20	F73	
Lychnocanium auxilla (Foreman)	RP8	RP8	RP6-7	F73	
Lychnocanium bellum Clark & Campb.	RP9-13	RP9-14	RP8-14	F73	209, fig. 11I-J (as *Lychnocanoma*)
Lychnocanium satelles (Kozlova)	RP5-9	RP5-9	?	H02	208, fig. 9L (L. *auxilla*)
Lychnocanium sp. B	RP7-8		?	S95	209, fig. 9J
Lychnocanium zhamoidai Kozlova	RP7-9		?	K99	209, fig. 9K (*Lychnocanoma* sp. A)
Middourium regulare	RP6-8	RP6-8	RP6-7	K99	
Mita regina (Campb. & Clark)	RK9-RP6	RK9-RP6	LK-UZP	H97a	
Mita sp. A (? = *Siphocampe?* "eilizabethae")	RP9-10	RP9-RP15	?	S95	209, fig. 10X
Monobrachium irregulare	RP7-8	RP7-8	RP6-7	K99	
Myllocercion acineton Foreman	RK9-RP6	RK9-RP4	LK-UZP	H97a	
Orbiculiforma renillaeformis gr. (Campb. & Cl.)	RK9-RP6	RK9-RP6	LK-UZP	H02	
Patagospyris confluens (Ehrenberg)	RP4-RP8	RP4-8	UZP-RP10	K99	
Periphaena alveolata (Lipman)	RP5-9	RP5-9	?	K99	
Periphaena heliasteriscus (Clark & Campb.)	RP8-9	RP6-9	UZP-EOC	SR73	
Peritiviator? dumitricai Nishimura	RP5-7	RP5-6	UZP-RP6	N92	
Petalospyris fovealata Ehrenberg	RP5-9	RP5-9	?	K99	209, fig. 8Q-R
Petalospyris senta (Kozlova)	RP6-9	RP6-9	?	K99	209, fig. 8S
Phacostaurus? quadratus Nishimura	RP6-7	RP6	RP6-7	N92	
Phormocyrtis cubensis (Riedel & Sanfilippo)	RP8	RP8	RP6-8	F73	
Phormocyrtis striata exquisita (Kozlova)	RP7-8	RP7-8	RP6-9	S95	209, fig. 9N
Phormocyrtis striata striata (Brandt)	RP7-10	RP7-10	RP7-14	S95	209, fig. 9O-P
Phormocyrtus turgida (Krasheninnikov)	RP8	RP8	RP6-8	F73	
Plectodiscus circularis (Clark & Campb.)	RP6-8	RP6-8	RP6-EOC	PK72	
Podocyrtis acalles Sanfilippo & Riedel	RP8-9	RP8-9	RP8-9	S92	209, fig. 9T-U (P. *aphorma*)
Podocyrtis papalis Ehrenberg	RP8-10	RP8-14	RP7-18	S92	209, fig. 9S
Pterocodon poculum Nishimura	RP6-7	RP6-7	RP6-7	H02	
Saturnalis kennetti Dumitrica	RP3-8	RP1-8	?	H97a	
Sethochytris babylonis (Clark & Campb.) gr.	RP6-13	RP6-15	RP6-20	H02	208, fig. 11A-B (as *Lychnocanoma*)
Siphocampe nodosaria (Haeckel)	RP6-13	RP4-15	RP6-14	H02	209 (S. *arachnaea*)
Siphocampe quadrata (Petrushev. & Kozlova)	RP8-13	RP5-15	RP6-20	H02	209, fig. 10R
Spongurus bilobatus gr. Clark & Campb.	RP5-9	RP4-9	?	H02	
Stichomitra carnegiense Campb. & Clark	RK9-RP6	RK9-RP6	LK-UZP	H97a	
Stylodictya targaeformis Clark & Campb.	RP5-9	RP5-9	?	CC42	
Stylosphaera minor Clark & Campb.	RP3-9	RP3-6	UZP-MIO	H97a	
Theocampe urceolus (Campb. & Clark)	RP9-13	RP9-15	RP8-20	S95	209, fig. 10Q
Theocampe vanderhoofi (Campb. & Clark)	RK9-RP6	RK9-RP6	LK-UZP	H97a	
Theocorys? aff. *phyzella* Foreman	RP6-8	RP6-8	?	H02	208, fig. 9E (*Calocyclas asperum*)
Theocorys? cf. *phyzella* Foreman	RP7-8	?	RP7	SB01	(as *Theocorys* aff. *phyzella*)
Theocorys? *phyzella* Foreman	RP8		RP7-8	F73	
Tholodiscus densus (Kozlova)	RP6-8	RK9-RP8	?	H97a	

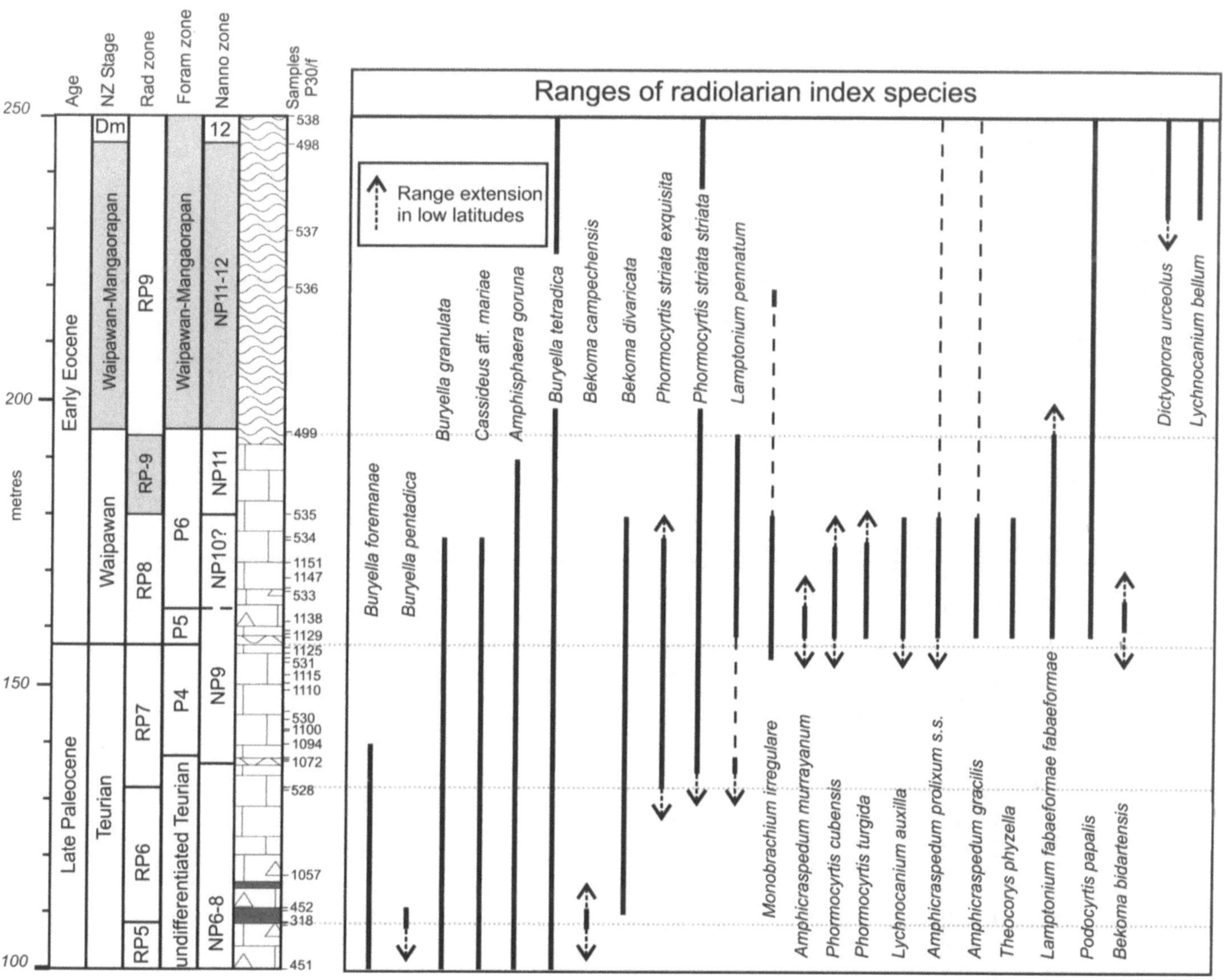

Fig. 4. Ranges of radiolarian index species in upper Paleocene-lower Eocene strata at Mead Stream. For age and biostratigraphy columns, intervals of uncertain age or biozone are shaded. Low-latitude ranges are based on (Sanfilippo & Nigrini 1998a, b) and Sanfilippo & Blome (2001).

leocene species (e.g. *Lithocampe wharanui*, *Mita regina*, *Myllocercion acineton*) that may be reworked into the siliciclastic Waipawa Formation. Other events are isolated records of taxa that are largely restricted to two particularly rich and diverse assemblages (P30/f318, f452), e.g. *Buryella helenae*, *B. petrushevskayae*, *Cryptocarpium* cf. *ornatum*. As most of the FOs recorded in this interval relate to species that are known to have earlier local FOs (Hollis 1997, 2002), this episode is not thought to reflect a time of high evolutionary turnover. Instead, the large number of events at the base of the Waipawa Formation suggests that either the unit is highly condensed or that there is a basal unconformity.

The second episode occurs within or directly below the 10 cm thick marl bed (Mudstone C). The FOs of 6 species and the LOs of 5 species occur within 5 m of strata. Most of the LOs are the last record of Cretaceous-Paleocene species that are generally very rare in the upper Paleocene but are present in one or more of three relatively rich samples in this interval (P20/f1072, f1076, f1081); two of these species may be reworked into Mudstone C (*Amphipyndax stocki*, *Theocampe vanderhoofi*). Three species with FOs in this interval are key index species that have much earlier FOs in low latitudes (*Lamptonium pennatum*, *Phormocyrtis striata exquisita*, *P. s. striata* – Fig. 4). All are very rare in overlying Paleocene strata but occur consistently in the lowermost Eocene.

The third faunal turnover episode occurs at the base of the Dee Marl (Fig. 3f, Table 2). The FOs of 13 species are recorded within the first 3 m of Eocene strata at Mead Stream, with 9 of these species in the basal Eocene sample. Four of these species appear to be global markers for the P/E boundary

Table 2. Radiolarian species in order of (A) first and (B) last occurrence at Mead Stream. Species with first occurrences lower in the section (Hollis et al. 2003a) or last occurrences higher in the section (Strong et al. 1995) are not shown. (MHF = Mead Hill Formation, WF = Waipawa Formation, DM = Dee Marl, URC = Upper Ribbon Chert)

A. IN ORDER OF FIRST OCCURRENCE

International Age	Late Paleocene																								Early Eocene																	
South Pacific Radiolarian Zone	5		RP6						RP7																RP8											RP9						
Lithologic Unit	MHF			WF			Lower Limestone																		DM		URC				Lower Lst					Lower Marl						
NZ Fossil Record No. P30/f	451	526	527	318	503	452	1057	528	529	1072	1076	1081	1094	1099	1100	530	1101	1110	1105	1115	531	1122	1123	1125	1129	1132	532	1138	533	1146	1147	1151	1223	534	535	499	536	537	498	538	539	497
DEPTH (metres above KTB)	99.0	107.7	108.2	108.3	110.3	110.8	116.4	131.5	132.0	136.6	136.9	137.2	139.6	141.9	142.2	144.0	145.4	149.1	150.3	151.8	154.0	154.8	155.6	156.6	158.3	159.0	159.6	161.2	166.5	167.1	168.8	171.5	175.5	176.0	180.0	194.5	220.0	230.0	245.0	250.0	280.0	292.5
Axoprunum pierinae	*	*	*			*	*	*	*	*	*	*	*		*	*	*	*		*	*	*				*	*	*	*	*	*	*	*	*	*	*	*	*	*	*	*	*
Axoprunum ? aff. *bispiculum*	*		*	*	*	*	*	*	*	*	*	*	*	*	*	*	*	*		*	*	*			*	*	*	*	*	*	*	*	*	*		*						
Bathropyramis magnifica	*	*	*	*	*	*	*	*		*	*	*	*		*	*	*	*	*	*	*	*		*		*	*	*	*	*	*	*		*	*	*			*		*	*
Buryella tetradica	*	*	*	*	*	*	*	*	*	*	*	*	*	*	*	*	*	*	*	*	*	*	*	*	*	*	*	*	*	*	*	*	*	*	*	*	*	*		*		
Cassideus aff. mariae	*	*	*		*	*	*	*			*	*					*	*			*			*	*	*	*	*	*	*	*	*	*									
Conocaryomma stilloformis	*																																									*
Corythomelissa adunca	*	*	*			*					*	*					*				*					*	*		*				*	*	*							
Heliostylus spp.	*					*												*																*								*
Peritiviator ? *dumitricai*	*					*						*																														
Periphaena alveolata	*	*	*	*		*	*		*	*	*	*	*	*	*	*	*			*	*	*			*	*	*	*	*	*	*	*	*	*	*		*				*	
Stylodictya targaeformis	*	*				*																																*	*			
Calocycloma ampulla		*	*	*		*	*	*		*	*	*													*	*		*	*		*		*		*	*		*	*	*	*	*
Cycladophora aff. *cosma*		*				*					*															*																
Lophophaena mugaica		*	*			*					*																			*												
Patagospyris confluens		*	*	*						*	*		*		*										*	*		*		*	*											
Petalospyris fovealata		*	*	*		*					*	*	*				*	*							*	*	*		*			*		*		*						
Amphisph. aff. *magnaporulosa*			*	*	*	*	*	*		*	*	*	*	*	*	*	*	*		*	*	*	*	*		*	*	*	*	*	*		*		*		*					
Amphymenium splendiarmatum			*	*		*				*	*													*				*			*	*		*	*							
Aspis murus			*																																							
Bekoma campechensis			*			*																																				
Burella pentadica			*	*																																						
Buryella tridica			*	*																																						
Cycladophora cf. *bicornis*			*							*	*						*												*				*	*								
Petalospyris senta			*			*					*															*	*	*	*			*	*	*	*	*						
Siphocampe nodosaria			*																							*		*			*		*	*								
Amphicraspedum prolixum gr.				*																				*	*	*		*	*		*	*	*	*	*	*		*	*			
Buryella helenae				*																																						
Buryella petrushevskayae				*																																						
Cryptocarpium cf. *ornatum*				*		*																																*	*		*	

Table 2 (cont.)

International Age	Late Paleocene																								Early Eocene																	
South Pacific Radiolarian Zone	5		RP6						RP7																RP8											RP9						
Lithologic Unit	MHF			WF			Lower Limestone																		DM		URC				Lower Lst					Lower Marl						
NZ Fossil Record No. P30/f	451	526	527	318	503	452	1057	528	529	1072	1076	1081	1094	1099	1100	530	1101	1110	1105	1115	531	1122	1123	1125	1129	1132	532	1138	533	1146	1147	1151	1223	534	535	499	536	537	498	538	539	497
DEPTH (metres above KTB)	99.0	107.7	108.2	108.3	110.3	110.8	116.4	131.5	132.0	136.6	136.9	137.2	139.6	141.9	142.2	144.0	145.4	149.1	150.3	151.8	154.0	154.8	155.6	156.6	158.3	159.0	159.6	161.2	166.5	167.1	168.8	171.5	175.5	176.0	180.0	194.5	220.0	230.0	245.0	250.0	280.0	292.5
Bekoma divaricata						*	*																		*	*			*					*	*							
Lychnocanium zhamoidai						*	*	*		*	*	*																	*				*	*	*	*				*	*	
Middourium regulare						*		*		*	*	*	*	*	*	*	*			*	*				*	*	*	*	*	*	*	*		*								
Phacostaurus ? quadratus						*				*	*																															
Plectodiscus circularis						*	*			*	*	*	*																*													
Sethochytris babylonis gr.						*	*		*	*	*	*		*	*						*	*		*	*	*	*	*	*		*	*	*	*	*	*	*	*	*	*	*	*
Theocorys ? aff. *phyzella*						*	*	*	*	*	*											*																				
Theocorys ? cf. *phyzella*						*	*		*	*	*	*	*									*			*	*	*	*	*	*	*	*	*	*	*							
Tholodiscus densus						*					*															*		*						*								
Lychnocanium sp. B									*	*	*							*		*	*	*							*													
Phormocyrtis striata exquisita									*	*	*	*	*				*			*		*			*	*	*	*	*	*	*	*	*	*								
Lychnocanium aff. *tripodium*										*	*	*													*	*	*															
Phormocyrtis striata striata										*	*	*										*		*	*	*	*		*	*	*	*	*	*	*	*		*		*	*	*
Pterocodon poculum										*	*	*																														
Lamptonium pennatum											*	*														*	*	*	*	*		*		*	*	*						
Monobrachium irregulare																						*	*		*	*	*	*	*	*	*	*	*	*	*		*					
Amphicraspedum murrayanum																									*	*																
Amphicraspedum prolixum s.s.																									*	*	*	*	*		*		*	*	*							*
A. prolixum f. *gracilis*																									*	*			*					*	*	*						*
Lamptonium f. fabaeformae																									*	*			*			*		*	*	*					*	
Lychnocanium auxilla																									*	*	*	*	*	*		*	*	*	*							
Phormocyrtis cubensis																									*	*		*		*			*									
Phormocyrtis turgida																									*	*					*			*								
Theocorys ? phyzella																									*	*	*	*	*	*	*	*	*	*	*							
Podocyrtis papilis																									*	*	*	*	*	*		*	*	*	*	*			*	*	*	*
Clathrocyclas universa gr.																										*	*	*	*	*	*	*		*	*			*				
Lychnocanium aff. *carinatum*																										*			*					*								
Podocyrtis acelles																										*	*		*									*		*		*
Bekoma bidartensis																											*		*													
Siphocampe quadrata																													*			*	*									
Periphaena heliasteriscus																																	*		*	*	*		*	*	*	*
Mita sp. A																																		*	*							*
Lamptonium ? aff. *columbus*																																				*	*	*				
Theocampe urceolus																																						*		*		*
Lychnocanium bellum																																						*				*
Lychnocanium amphitrite																																								*	*	
First Occurrences	10	5	9	4	0	9	0	0	2	3	1	0	0	0	0	0	0	0	0	0	0	1	0	0	9	3	1	0	1	0	0	0	1	1	0	1	0	2	0	1	0	0

Table 2 (cont.)
B. IN ORDER OF LAST OCCURRENCE

International Age	Late Paleocene																								Early Eocene																	
South Pacific Radiolarian Zone	5		RP6						RP7																RP8											RP9						
Lithologic Unit	MHF			WF			Lower Limestone																		DM		URC				Lower Lst					Lower Marl						
NZ Fossil Record No. P30/f	451	526	527	318	503	452	1057	528	529	1072	1076	1081	1094	1099	1100	530	1101	1110	1105	1115	531	1122	1123	1125	1129	1132	532	1138	533	1146	1147	1151	1223	534	535	499	536	537	498	538	539	497
DEPTH (metres above KTB)	99.0	107.7	108.2	108.3	110.3	110.8	116.4	131.5	132.0	136.6	136.9	137.2	139.6	141.9	142.2	144.0	145.4	149.1	150.3	151.8	154.0	154.8	155.6	156.6	158.3	159.0	159.6	161.2	166.5	167.1	168.8	171.5	175.5	176.0	180.0	194.5	220.0	230.0	245.0	250.0	280.0	292.5
Aspis murus			*																																							
Lithocampe wharanui	*		*																																							
Mita regina			*																																							
Buryella helenae				*																																						
Buryella kaikoura	*		*	*																																						
Burella pentadica			*	*																																						
Buryella petrushevskayae				*																																						
Buryella tetradica tridica			*	*																																						
Myllocercion acineton		*		*																																						
Bekoma campechensis			*			*																																				
Clathrocyclas australis	*	*	*			*																																				
Dictyomitra andersoni			*	*	*	*																																				
Stichomitra carnegiense			*	*		*	*																																			
Amphipyndax stocki	*		*	*	*	*				*											?																					
Theocampe vanderhoofi			*	*						*																																
Phacostaurus ? quadratus						*				*	*																															
Peritiviator ? dumitricai	*					*						*																														
Pterocodon poculum										*	*	*																														
Buryella dumitricai	*	*	*	*	*	*	*			*		*	*																													
Buryella foremanae	*	*	*	*	*	*	*	*			*	*	*																													
Amphisphaera kina	*	*	*	*	*	*	*	*	*	*	*	*	*	*	*		*			*																						
Lithomespilus coronatus			*	*		*		*	*	*	*	*	*	*	*	*	*	*		*	*	*																				
Theocorys ? aff. *phyzella*						*	*	*	*	*	*											*																				
Lithostrobus longus	*	*	*	*		*	*			*		*				*		*				*		*																		
Amphicraspedum murrayanum																									*	*																
Cycladophora aff. *cosma*		*				*					*															*																
Lychnocanium aff. *tripodium*										*	*	*													*	*	*															
Bekoma bidartensis																											*		*													
Lychnocanium sp. B									*	*	*							*		*	*	*							*													
Plectodiscus circularis						*	*			*	*	*	*																*													
Lophophaena mugaica		*	*			*					*																			*												
Patagospyris confluens		*	*	*						*	*		*		*										*	*		*		*	*											
Cassideus aff. mariae	*	*	*		*	*	*	*			*	*					*	*			*			*	*	*	*	*	*	*	*	*	*									
Orbiculiforma renillaeformis gr.	*					*																								*			*									
Phormocyrtis cubensis																									*	*		*		*			*									
Siphocampe quadrata																													*			*	*									

Table 2 (cont.)

International Age	Late Paleocene																								Early Eocene																	
South Pacific Radiolarian Zone	5		RP6						RP7																RP8											RP9						
Lithologic Unit	MHF			WF			Lower Limestone																		DM		URC				Lower Lst					Lower Marl						
NZ Fossil Record No. P30/f	451	526	527	318	503	452	1057	528	529	1072	1076	1081	1094	1099	1100	530	1101	1110	1105	1115	531	1122	1123	1125	1129	1132	532	1138	533	1146	1147	1151	1223	534	535	499	536	537	498	538	539	497
DEPTH (metres above KTB)	99.0	107.7	108.2	108.3	110.3	110.8	116.4	131.5	132.0	136.6	136.9	137.2	139.6	141.9	142.2	144.0	145.4	149.1	150.3	151.8	154.0	154.8	155.6	156.6	158.3	159.0	159.6	161.2	166.5	167.1	168.8	171.5	175.5	176.0	180.0	194.5	220.0	230.0	245.0	250.0	280.0	292.5
Artostrobus pusillus	*		*			*		*		*	*	*	*		*		*	*		*		*		*								*		*								
Buryella granulata	*	*	*	*	*	*	*		*	*	*	*	*		*			*		*	*	*		*	*	*	*	*	*	*	*	*	*	*								
Cycladophora cf. *bicornis*			*							*	*						*												*				*	*								
Lychnocanium aff. *carinatum*																										*			*					*								
Middourium regulare						*		*		*	*	*	*	*	*	*	*			*	*				*	*	*	*	*	*	*	*		*								
Phormocyrtis striata exquisita									*	*	*	*	*				*			*		*			*	*	*	*	*	*	*	*	*	*								
Phormocyrtis turgida																									*	*					*			*								
Saturnalis kennetti	*				*						*	*			*												*							*								
Siphocampe nodosaria			*																							*		*			*		*	*								
Tholodiscus densus						*					*															*		*						*								
Amphisphaera goruna	*	*	*	*	*	*	*	*	*	*	*	*	*	*	*	*	*				*	*		*		*	*	*	*	*	*	*	*	*	*							
Amphymenium splendiarmatum			*	*		*				*	*													*				*			*	*		*	*							
Bekoma divaricata						*	*																		*	*			*					*	*							
Corythomelissa adunca	*	*	*			*					*	*					*				*					*	*		*				*	*	*							
Lychnocanium auxilla																									*	*	*	*	*	*		*	*	*	*							
Theocorys ? cf. *phyzella*						*	*		*	*	*	*	*									*			*	*	*	*	*	*	*	*	*	*	*							
Theocorys ? *phyzella*																									*	*	*	*	*	*	*	*	*	*	*							
Amphipternis alamadaensis	*	*				*																												*	*	*						
Axoprunum ? aff. *bispiculum*	*		*	*	*	*	*	*	*	*	*	*	*	*	*	*	*	*		*	*	*			*	*	*	*	*	*	*	*	*	*		*						
Lamptonium pennatum											*	*														*	*	*	*	*		*		*	*	*						
Petalospyris fovealata		*	*	*		*					*	*	*				*	*							*	*	*		*			*		*		*						
Petalospyris senta			*			*					*															*	*	*	*			*	*	*	*	*						
Amphisph. aff. *magnaporulosa*			*	*	*	*	*	*		*	*	*	*	*	*	*	*	*		*	*	*	*	*		*	*	*	*	*	*		*		*		*					
Monobrachium irregulare																						*	*		*	*	*	*	*	*	*	*	*	*	*		*					
Clathrocyclas universa gr.																										*	*	*	*	*	*	*		*	*			*				
Lamptonium ? aff. *columbus*																																				*	*	*				
Lychnocanium satelles	*	*	*	*		*	*	*	*	*	*	*	*	*	*	*					*					*		*	*				*	*	*	*		*				
Stylodictya targaeformis	*	*				*																																*	*			
Buryella tetradica	*	*	*	*	*	*	*	*	*	*	*	*	*	*	*	*	*	*	*	*	*	*	*	*	*	*	*	*	*	*	*	*	*	*	*	*	*	*		*		
Cryptocarpium cf. *ornatum*				*		*																																*	*		*	
Lamptonium f. fabaeformae																									*	*			*			*		*	*	*					*	
Lychnocanium amphitrite																																								*	*	
Lychnocanium zhamoidai						*	*	*		*	*	*																	*				*	*	*	*				*	*	
Periphaena alveolata	*	*	*	*		*	*		*	*	*	*	*	*	*	*	*			*	*	*			*	*	*	*	*	*	*	*	*	*	*		*				*	
Podocyrtis acelles																										*	*		*									*		*		*
Last Occurrences	10	0	3	6	0	3	1	0	0	2	1	2	0	0	0	0	0	0	0	1	0	2	0	1	0	2	1	0	3	1	1	0	4	10	7	5	2	3	1	1	5	1

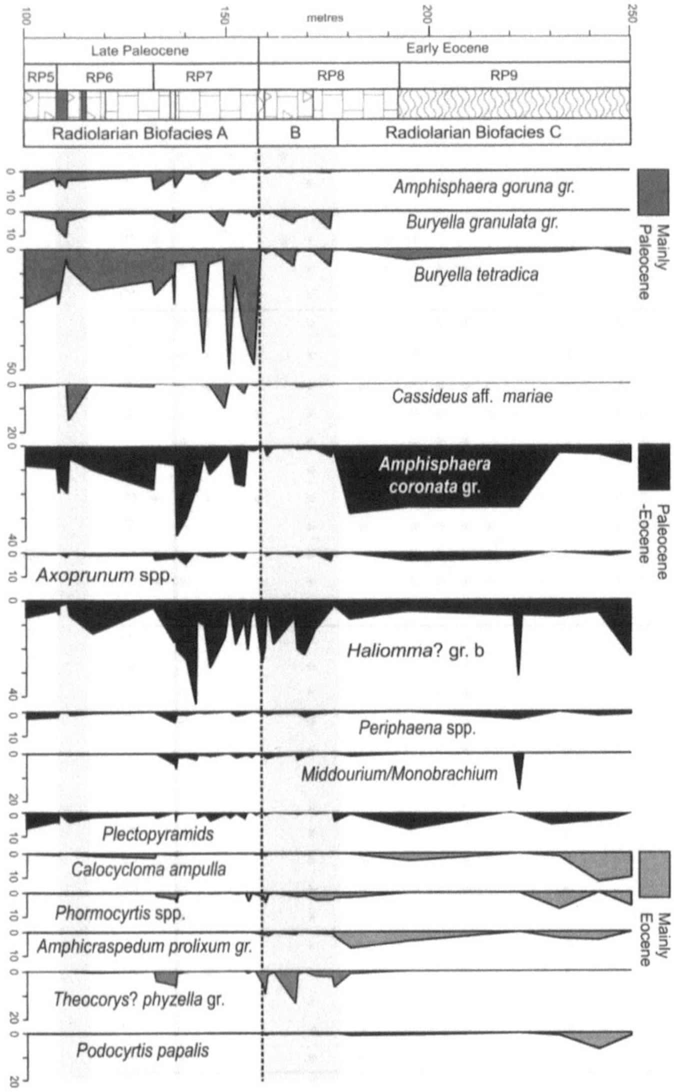

Fig. 5. Relative abundance of common radiolarian species and species groups in upper Paleocene-lower Eocene strata at Mead Stream.

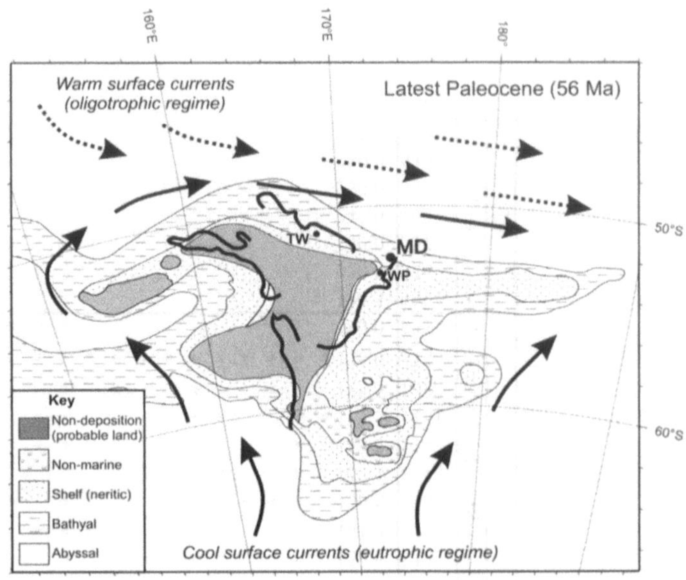

Fig. 6. Paleogeographic reconstruction of New Zealand region during the latest Paleocene (adapted from King et al. 1999) showing major current systems based on coupled climate model simulations for the Eocene (Huber 2002)

(Sanfilippo & Nigrini 1998a; Sanfilippo & Blome 2001): *Lamptonium fabaeforme fabaeforme, Podocyrtis papalis, Phormocyrtis turgida* and *Theocorys*? *phyzella* (Fig. 4). Five of the species are index species with late Paleocene FOs in low latitudes (Sanfilippo & Nigrini 1998a, b): *Amphicraspedum murrayanum, A prolixum* s.s., *Bekoma bidartensis, Lychnocanium auxilla, Phormocyrtis cubensis* (Fig. 4).

There are relatively few significant LOs associated with the P/E boundary (Fig. 3f). The LOs of *Lithomespilus coronatus, Lithostrobus longus* and *Theocorys*? aff. *phyzella* occur within the uppermost 3 m of Paleocene strata. Of the three LOs that occur within the basal Eocene Dee Marl, two are species that are restricted to the unit (*Amphicraspedum murrayanum, Lychnocanium* aff. *tripodium*) while one is the LO of a sporadically occurring late Paleocene species (*Cycladophora* aff. *cosma*). In contrast to the faunal changes observed at the base of the Waipawa Formation, the low number of LOs relative to FOs in the P/E boundary interval indicates that the faunal turnover is not due to a stratigraphic discontinuity. The abrupt appearance of many low-latitude species indicates that warming at the PETM promoted pole-ward migration of warm-water radiolarians. Similar, but less extreme, short-lived warming may explain the FOs of several low-latitude species around the time of deposition of Mudstone C (Fig. 3f, Table 2).

The final episode of marked faunal change in the studied section occurs ~20 m above the P/E boundary (Fig. 3f, Table 2). Within the same 5 m interval in which radiolarian abundance begins to decrease (175–180 m), the LOs of 21 species are recorded. These species include many that thrived through much of the late Paleocene as well as many that appeared in the earliest Eocene (Fig. 4). Only two FOs are recorded in this interval, one of these being the FO of the common Eocene phacodiscid *Periphaena heliasteriscus*. The high number of LOs, coupled with the onset of long term decline in radiolarian abundance and diversity, indicates that this episode of faunal change reflects a oceanographic changes unfavourable to all radiolarians.

6. Radiolarian biofacies

To simplify the discussion and presentation of census data (Fig. 5) in this section, some species are combined into larger groups that reflect a close phylogenetic relationship or morphological similarity:

1. *Amphisphaera coronata* gr. (*A. macrosphaera, A. coronata* s.l, *A.* aff. *magnaporulosa*).

2. *Amphisphaera goruna* gr. (*A. kina*, *A. goruna*)
3. *Axoprunum* spp. (*Axoprunum pierinae*, *A. aff. bispiculum*)
4. *Buryella granulata* gr. (*B. granulata*, *B. foremanae*, *B. dumitricai*)
5. *Middourium/Monobrachium* (*Middourium regulare*, *Monobrachium irregulare*)
6. *Periphaena* spp. (*Periphaena alveolata*, *P. heliasteriscus*)
7. *Phormocyrtis* spp. (*Phormocyrtis cubensis*, *P. striata exquisita*, *P. striata striata*, *P. turgida*)
8. Plectopyramidins (*Bathropyramis* and *Cornutella* spp.)
9. *Theocorys*? *phyzella* gr. (*Theocorys*? cf. *phyzella*, *Theocorys*? *phyzella*)

Three radiolarian biofacies are identified in the Paleocene-Eocene section at Mead Stream: (A) upper Paleocene (100–157 m), (B) lowermost Eocene (157–178 m) and (C) lower Eocene (178–250 m).

Biofacies A is characterised by abundant *Buryella tetradica*, common to abundant *Amphisphaera coronata* gr. and *Haliomma* gr. b, and few to common *Amphisphaera goruna* gr., *Buryella granulata* gr. and *Cassideus* aff. *mariae* (Fig. 5). Diatoms are common in this biofacies (Fig. 3c) and sediments are moderately siliceous, with a highly siliceous facies of the organic-rich Waipawa Formation (Killops et al. 2000) occurring in the lower part of the interval. Faunal similarities with diatom-rich, biosiliceous Paleocene lithofacies lower in the Marlborough sequence (Hollis et al. 1995, 2003a, b) and on the eastern margin of Campbell Plateau (Hollis 2002) suggest moderately eutrophic conditions, consistent with weak to moderate upwelling in outer shelf to upper slope depths (Fig. 6).

This biofacies spans an interval correlated with the PCIM, which is interpreted as a 3 m.y. episode of enhanced global carbon burial, either through increased marine or terrestrial productivity (Corfield & Cartlidge 1992; Thompson & Schmitz 1997; Kurtz et al. 2003). The regional occurrence of organic-rich mudstone in the lower part of the interval (Killops et al. 2000; Hollis et al. 2005) and abundant siliceous microfossils (Hollis 2002) indicates that this event is associated with enhanced marine productivity in the New Zealand region.

Biofacies B occurs in the lowermost 21 m of Eocene strata (157–178 m). It is characterised by abundant *Haliomma* gr. b and few to common *Phormocyrtis* spp. and *Theocorys*? *phyzella* gr. (Fig. 5). The base of the biofacies coincides with the P/E boundary and the onset of the PETM. A marked increase in radiolarian abundance and diversity is associated with a sharp decline in diatom abundance and in the abundance of *Buryella tetradica* – the dominant radiolarian species in the upper Paleocene (Fig. 3b–d, f). This species and *B. granulata* s.s. are rare in this interval, apart from two samples in which both species are common (P30/f533, f1223). The high abundance of large, robust spumellarians, namely *Haliomma* gr. b, *Middourium/Monobrachium* gr. and *Periphaena alveolata*, is a distinctive feature of Biofacies B. These taxa tend to be most abundant in marl samples of Dee Marl and are also common to abundant in underlying Paleocene marly layers, notably at ~137 and ~142 m, and overlying Eocene marl layers (Fig. 5). Despite the significant lithologic change from Dee Marl to Upper Ribbon Chert, there is little change in siliceous microfossil assemblages. A slight increase in biosiliceous productivity in Upper Ribbon Chert may be inferred from the increase in silica content and a small increase in diatom abundance, although overall radiolarian abundance and diversity declines (Fig. 3b–d).

Biofacies B has unusual characteristics that make interpretation difficult. The presence of many low-latitude species indicates relatively warm climatic conditions. The low abundance of diatoms and of radiolarian taxa characteristic of the eutrophic Paleocene Biofacies A, suggests conditions of relatively low productivity. However, radiolarian abundance and diversity maxima are recorded in Dee Marl and in underlying Mudstone C, which also contains an assemblage similar to Biofacies B. A very similar assemblage is present in lowermost Eocene carbonate ooze at high-latitude DSDP Site 277, southwest Campbell Plateau (Hollis et al. 1997). The three features of high total abundance, high diversity and high numbers of warm-water species are consistent with late Cenozoic tropical radiolarian assemblages (Casey 1993). This suggests that the PETM may have promoted the expansion of weakly eutrophic subtropical water into the Southwest Pacific (Fig. 6). Further biogeographic study of the distinctive spumellarians that dominate this biofacies is required to confirm this interpretation.

Biofacies C occurs in the upper part of the lower Eocene sequence (178 to 250 m). It is characterised by abundant *Amphisphaera coronata* gr., *Haliomma* gr. b, common *Calocycloma ampulla*, few to common *Phormocyrtis striata striata* and few to rare *Amphicraspedum prolixum* gr., *Axoprunum* spp., *Buryella tetradica*, *Periphaena heliasteriscus*, plectopyramidins, *Podocyrtis papalis* and *Sethochytris babylonis* gr. An upwards decrease in radiolarian abundance and diversity, coupled with low diatom abundance, indicates that this biofacies represents relatively oligotrophic conditions. Similar assemblages are found in lower Eocene carbonate ooze at DSDP site 277 (Hollis et al. 1997). As this biofacies is correlated with the EECO, it seems likely that global warming associated with this event promoted the expansion of oligotrophic subtropical watermasses into the Southwest Pacific.

7. Conclusions

The first quantitative study of radiolarian assemblages from an upper Paleocene-lower Eocene bathyal succession has identified three distinct biofacies that can be used to determine the influence of major changes in the global carbon cycle and climate system on Southwest Pacific oceanic conditions.

Biofacies A is associated with the PCIM, a 3 m.y. episode of maximum $\delta^{13}C$ values that is inferred to represent high rates of carbon burial due to enhanced marine or terrestrial productivity (Corfield & Cartlidge 1992; Thompson & Schmitz 1997; Kurtz et al. 2003). Hollis et al. (2005) argued that Mead Stream lithofacies, including organic mudstone and siliceous limestone, are indicative of enhanced marine productivity in the Marlborough basin during the PCIM. Siliceous microfossil assemblages also indicate enhanced marine productivity. Biofacies A is associated with common diatoms and is dominated by the same or similar species that are found in biogenic silica-rich lithologies in the lower Paleocene in Marlborough (Hollis et al. 2003a, b) and in the upper Paleocene of the Campbell Plateau margin (Hollis 2002). Deposition of radiolarian and diatom-rich sediments in the Marlborough basin, as well as on the northwest and southeast continental margin of New Zealand, in the late Paleocene between ~60–56 Ma indicates that the PCIM was associated with enhanced marine productivity, at least in the southern Pacific Ocean. This is consistent with ocean circulation models for the South Pacific (Huber 2002).

Biofacies B is associated with the PETM, a 100–200,000 yr episode of extreme global warmth recorded as pronounced negative $\delta^{13}C$ and $\delta^{18}O$ excursions (Zachos et al. 2001), which at Mead Stream is associated with a lithofacies change from siliceous limestone to marly limestone and marl (Dee Marl). A marl bed (Mudstone C) ~20 m below the PETM may represent a precursor event. Not only is the bed also associated with a negative $\delta^{13}C$ excursion, but radiolarian assemblages are very similar to those of Biofacies B. Biofacies B is characterised by high radiolarian abundance and diversity, but very low diatom abundance. It is dominated by large, robust smooth-walled spumellarians of uncertain biogeographic affinity, and contains the only local record of several low-latitude species. The biofacies records an abrupt warm-water incursion in the East Coast Basin. The combination of an incursion of subtropical water and waning cool-water upwelling may have promoted an unusual bloom of warm water species. The occurrence of similar earliest Eocene radiolarian assemblages at New Zealand's southern continental margin (Hollis et al. 1997) indicates that warm subtropical waters may have extended as far south as 60°.

Biofacies C is associated with the onset of the EECO, a 3 m.y. episode of extreme global warming (Zachos et al. 2001), which at Mead Stream is associated with the transition from limestone- to marl-dominated sedimentation. Biofacies C is characterised by declining radiolarian abundance and diversity and low diatom abundance, implying a marked decrease in biosiliceous productivity. Significantly, many low-latitude as well as high-latitude radiolarian species disappear at the onset of the EECO. With similar early Eocene radiolarian assemblages recorded from the southern Campbell Plateau (Hollis et al. 1997), progressive warming is inferred to have promoted increasingly oligotrophic oceanic conditions around New Zealand.

Acknowledgements

Sample collection and stratigraphic integration was carried out with the assistance of Jerry Dickens, Brad Field, Craig Jones and Percy Strong. Samples were processed and slides were prepared by Kellee Anderson, Neville Orr and John Simes. The manuscript benefited for careful reviews by Annika Sanfilippo, Akiko Nishimura, James Crampton, Erica Crouch and Percy Strong. This research was supported by the Foundation for Research Science and Technology (GNS Global Change Through Time Programme, Contract CO5X0202).

REFERENCES

BERGGREN W., KENT D., SWISHER C. & AUBRY M.-P. 1995: A revised Cenozoic geochronology and chronostratigraphy. In: BERGGREN, W. et al. (Eds): Geochronology, time scales and global stratigraphic correlation, 129–212, SEPM Special Publication 54.

BOLTOVSKY D. 1987: Sedimentary record of radiolarian biogeography in the equatorial to Antarctic western Pacific Ocean. Micropaleontology 33, 267–281.

CASEY R. 1993: Radiolaria. In: LIPPS, J.H. (Ed.): Fossil Prokaryotes and Protists, 249–284. Blackwell Scientific Publications, Oxford/London, UK.

CAULET J., VENEC-PEYRE M.-T., VERGNAUD-GRAZZINI C. & NIGRINI C. 1992: Variation of South Somalian upwelling during the last 160 ka: radiolarian and foraminifera records in core MD 85674. In: SUMMERHAYES, C. et al. (Eds): Upwelling Systems: Evolution since the Early Miocene, Geological Society of London, Special Publication 64, 379–389.

CORFIELD R. M. & CARTLIDGE J. E. 1992: Oceanographic and climatic implications of the Palaeocene carbon isotope maximum. Terra Nova 4, 443–455.

CRAMPTON J., LAIRD M., NICOL A., TOWNSEND D. & VAN DISSEN R. 2003: Palinspastic reconstructions of southeastern Marlborough, New Zealand, for Late Cretaceous to Eocene times. New Zealand Journal of Geology and Geophysics 46, 153–175.

DE WEVER P., DUMITRICA P., CAULET J.-P., NIGRINI C. & CARIDROIT M. 2001: Radiolarians in the sedimentary record. 533 p., Gordon and Breach Science Publishers, The Netherlands.

HANCOCK H., DICKENS G., STRONG C., HOLLIS C. & FIELD B. 2003: Foraminiferal and carbon isotope stratigraphy through the Paleocene-Eocene transition at Dee Stream, Marlborough, New Zealand. New Zealand Journal of Geology and Geophysics 46, 1–19.

HOLLIS C. 1996: Radiolarian faunal change through the Cretaceous-Tertiary transition of eastern Marlborough, New Zealand. In: MACLEOD, N. & KELLER, G. (Eds.): Cretaceous-Tertiary Mass Extinctions: Biotic and Environmental Changes, 173–204, Norton Press, New York.

– 1997: Cretaceous-Paleocene Radiolaria from eastern Marlborough, New Zealand. Institute of Geological and Nuclear-Sciences Monograph 17, 152 p.

– 2002: Biostratigraphy and paleoceanographic significance of Paleocene radiolarians from offshore eastern New Zealand. Marine Micropaleontology 46, 265–316.

HOLLIS C., DICKENS G., FIELD B., JONES C. & STRONG C. 2005: The Paleocene-Eocene transition at Mead Stream, New Zealand: a southern Pacific record of early Cenozoic global change. Palaeogeography, Palaeoclimatology, Paleoecology 215, 313–343.

HOLLIS C., RODGERS K. & PARKER R. 1995: Siliceous plankton bloom in the earliest Tertiary of Marlborough, New Zealand. Geology (Boulder) 23, 835–838.

HOLLIS C., WAGHORN D., STRONG C. & CROUCH E. 1997: Integrated Paleogene biostratigraphy of DSDP site 277 (Leg 29): foraminifera, calcareous nannofossils, Radiolaria, and palynomorphs. Institute of Geological & Nuclear Sciences Science Report 97/7, 1–73.

HOLLIS C., RODGERS K., STRONG C., FIELD B. & ROGERS K. 2003a: Paleoenvironmental changes across the Cretaceous/Tertiary boundary in the northern Clarence Valley, southeastern Marlborough, New Zealand. New Zealand Journal of Geology and Geophysics 46, 209–234.

HOLLIS C., STRONG C., RODGERS K. & ROGERS K. 2003b: Paleoenvironmental changes across the Cretaceous/Tertiary boundary at Flaxbourne River and Woodside Creek, eastern Marlborough, New Zealand. New Zealand Journal of Geology and Geophysics 46, 177–197.

HOLLIS C., DICKENS G., FIELD B., JONES C. & STRONG C. 2005: The Paleocene-Eocene transition at Mead Stream, New Zealand: a southern Pacific record of early Cenozoic global change. Palaeogeography, Palaeoclimatology, Paleoecology 215, 313–343.

HUBER M. 2002: Straw man 1; preliminary view of the tropical Pacific from a global coupled climate model simulation of the early Paleogene. Proceedings of the Ocean Drilling Program, initial reports, Paleogene equatorial transect, Leg 199. Proceedings of the Ocean Drilling Program, Part A: Initial Reports 199, 30.

KELLER G., ADATTE T., HOLLIS C., ORDONEZ M., ZAMBRANO I., JIMENEZ N., STINNESBECK E., ALEMAN A. & HALE-ERLICH W. 1997: The Cretaceous/Tertiary boundary event in Ecuador: reduced biotic effects due to eastern boundary current setting. Marine Micropaleontology 31, 97–133.

KILLOPS S., HOLLIS C., MORGANS H., SUTHERLAND R., FIELD B. & LECKIE D. 2000: Paleoceanographic significance of Late Paleocene dysaerobia at the shelf/slope break around New Zealand. Palaeogeography, Palaeoclimatology, Palaeoecology 156, 51–70.

KING P., NAISH T., BROWNE G., FIELD B. & EDBROOKE S. 1999: Cretaceous to Recent sedimentary patterns in New Zealand. Institute of Geological and Nuclear Sciences Folio Series 1. 35 p.

KURTZ A., KUMP L., ARTHUR M., ZACHOS J. & PAYTAN A. 2003: Early Cenozoic decoupling of the global carbon and sulfur cycles. Paleoceanography 18(4), 1090, doi:10.1029/2003PA000908

LIPPS J. H. 1993: Fossil prokaryotes and protists. 342 p., Blackwell Scientific, Boston.

NIGRINI C. & SANFILIPPO A. 2000: Paleogene radiolarians from Sites 998, 999, and 1001 in the Caribbean. Proceedings of the Ocean Drilling Program, Scientific Results 165, 57–81.

REAY M. 1993: Geology of the middle part of the Clarence Valley. Institute of Geological and Nuclear Sciences Geological Map 10. 144 p. + 1 map.

SANFILIPPO A. & BLOME C. 2001: Biostratigraphic implications of mid-latitude Palaeocene-Eocene radiolarian faunas from Hole 1051A, ODP Leg 171B, Blake Nose, western North Atlantic. In: KROON, D. et al. (Eds.): Western North Atlantic Palaeogene and Cretaceous palaeoceanography. Geological Society Special Publication 183, 185–224.

SANFILIPPO A. & NIGRINI C. 1998a: Upper Paleocene-Lower Eocene deep-sea radiolarian stratigraphy and the Paleocene/Eocene Series boundary. In: AUBRY, M.-P., et al. (Eds.): Late Paleocene-Early Eocene climatic and biotic events in the marine and terrestrial records, 244–276, Columbia University Press, New York.

– 1998b: Code numbers for Cenozoic low latitude radiolarian biostratigraphic zones and GPTS conversion tables. Marine Micropaleontology 33, 109–156.

STRONG C., HOLLIS C. & WILSON G. 1995: Foraminiferal, radiolarian, and dinoflagellate biostratigraphy of Late Cretaceous to middle Eocene pelagic sediments (Muzzle Group), Mead Stream, Marlborough, New Zealand. New Zealand Journal of Geology and Geophysics 38, 171–209.

SUTHERLAND R., KING P. & WOOD R. 2001: Tectonic evolution of Cretaceous rift basins in south-eastern Australia and New Zealand: implications for exploration risk assessment. In: HILL K. & BERNECKER T. (Eds.): Eastern Australasian basins symposium: a refocused energy perspective for the future. Petroleum Exploration Society of Australia Special Publication 1, 3–13.

THOMPSON E. & SCHMITZ B. 1997: Barium and the late Paleocene $\delta^{13}C$ maximum: evidence of increased marine surface productivity. Paleoceanography 12, 239–254.

ZACHOS J., PAGANI M., SLOAN L., THOMAS E. & BILLUPS K. 2001: Trends, rhythms, and aberrations in global climate 65 Ma to present. Science 292, 686–693.

ZACHOS J., WARA M., BOHATY S., DELANEY M., PETRIZZO M., BRILLA., BRALOWER T. & PREMOLI-SILVA I. 2003: A transient rise in tropical sea surface temperature during the Paleocene-Eocene thermal maximum. Science 302, 1551–1554.

Manuscript received January 2004
Revision accepted February 2005

Plate 1

Scanning electron micrographs of spumellarian radiolarians from lower Eocene strata, Mead Stream. Scale bars = 100 μm.

1. *Amphisphaera coronata* EHRENBERG gr. CH12076, P30/f534, 176 m, RP8.
2. *Axoprunum* aff. *bispiculum* (POPOFSKY). CH12086, P30/f534, 176 m, RP8.
3. *Axoprunum pierinae* (CLARK AND CAMPBELL). CH12079, P30/f534, 176 m, RP8.
4. *Stylosphaera minor* CLARK AND CAMPBELL. CH12078, P30/f534, 176 m, RP8.
5. *Haliomma* gr. b. (cf. *Theocosphaerella rotunda* (BORISSENKO)). CH12005, P30/f1129, 158.3 m, RP8.
6. *Haliomma* gr. b. CH12001, P30/f1129, 158.3 m, RP8.
7. *Periphaena*? *duplus* (KOZLOVA). CH12020, P30/f1129, 158.3 m, RP8.
8. *Periphaena*? *duplus* (KOZLOVA). CH12019, P30/f1129, 158.3 m, RP8.
9. *Periphaena*? *duplus* (KOZLOVA). CH12016, P30/f1129, 158.3 m, RP8.
10. *Periphaena*? *duplus* (KOZLOVA). CH12018, P30/f1129, 158.3 m, RP8.
11. *Middourium regulare* (BORISSENKO). CH12301, P30/f1129, 158.3 m, RP8.
12. *Monobrachium irregulare* (NISHIMURA). CH12012, P30/f1129, 158.3 m, RP8.
13. *Monobrachium irregulare* (NISHIMURA). CH12011, P30/f1129, 158.3 m, RP8.
14. *Amphiscraspedum prolixum* SANFILIPPO AND RIEDEL CH12094, P30/f534, 176 m, RP8.
15. *Amphicraspedum gracilis* (LIPMAN). CH12093, P30/f534, 176 m, RP8.
16. *Amphicraspedum gracilis* (LIPMAN). CH12092, P30/f534, 176 m, RP8.
17. *Amphicraspedum gracilis* (LIPMAN). CH12091, P30/f534, 176 m, RP8.
18. *Amphicraspedum murrayanum* HAECKEL. CH12030, P30/f1129, 158.3 m, RP8.
19. *Amphicraspedum murrayanum* HAECKEL. CH12064, P30/f1129, 158.3 m, RP8.
20. *Amphiscraspedum prolixum* SANFILIPPO AND RIEDEL CH12059, P30/f1129, 158.3 m, RP8.
21. *Amphiscraspedum prolixum* SANFILIPPO AND RIEDEL CH12089, P30/f534, 176 m, RP8.
22. Spumellarian gen. et sp. indet. CH12085, P30/f534, 176 m, RP8.

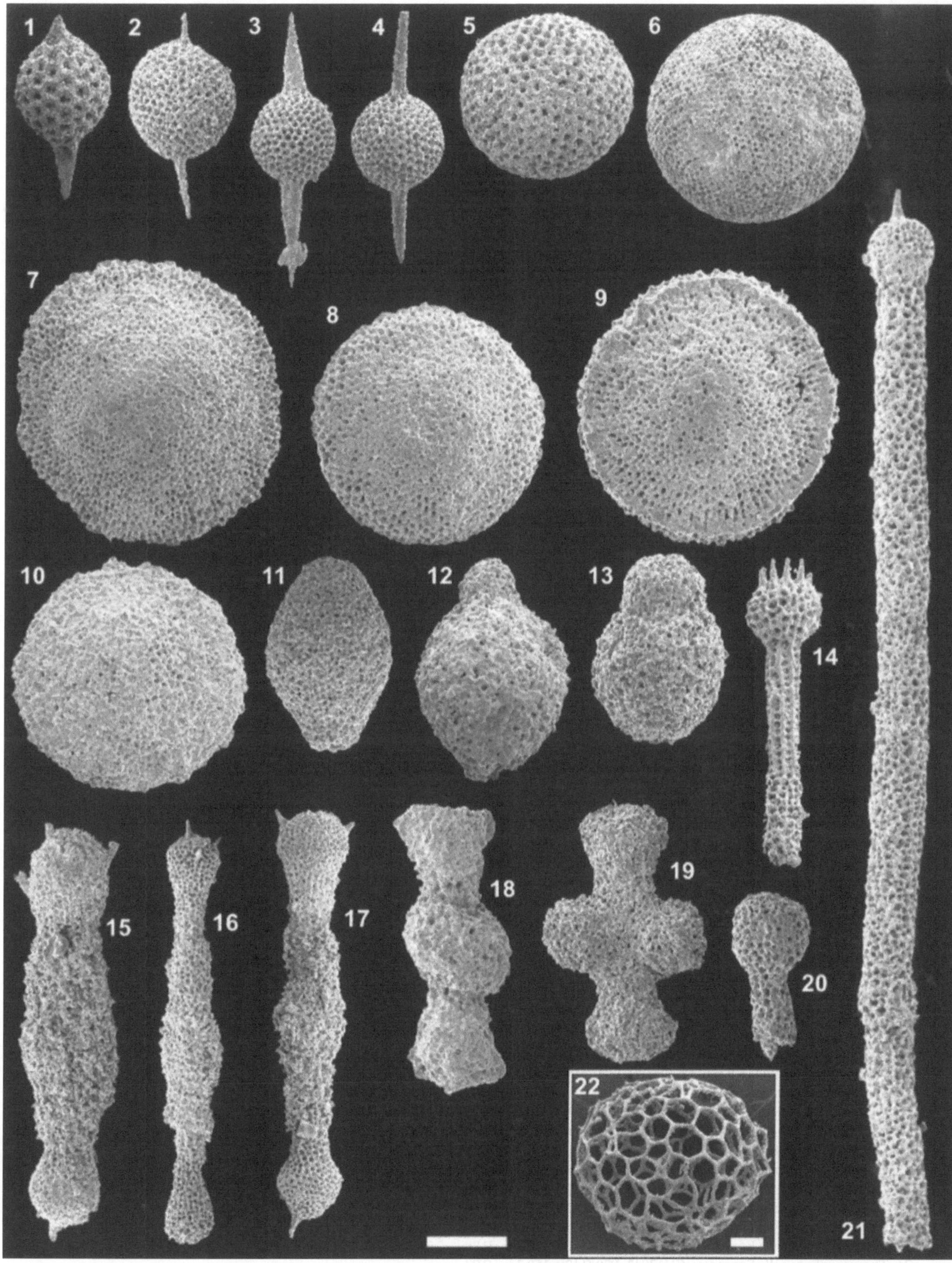
1
2
3
4
5
6
7
8
9
10
11
12
13
14
15
16
17
18
19
20
21
22

Plate 2

Scanning electron micrographs of nassellarian radiolarians from upper Paleocene (Fig. 4) and lower Eocene strata, Mead Stream. Scale bar = 100 µm.

1. *Patagopsyris confluens* (Ehrenberg). CH12074, P30/f1129, 158.3 m, RP8.
2. *Petalospyris foveolata* (Ehrenberg). CH12075, P30/f1129, 158.3 m, RP8.
3. *Clathrocyclas universa* Clark and Campbell. CH12103, P30/f534, 176 m, RP8.
4. *Buryella tetradica* Foreman. CH01028, P30/f318, 108.3, RP6.
5. *Buryella tetradica* Foreman. CH12040, P30/f1129, 158.3 m, RP8.
6. *Buryella tetradica* Foreman. CH12065, P30/f1129, 158.3 m, RP8.
7. *Bathropyramis magnifica* (Clark and Campbell). CH12101, P30/f534, 176 m, RP8.
8. *Bathropyramis magnifica* (Clark and Campbell). CH12100, P30/f534, 176 m, RP8.
9. *Lychnocanium ponderosa* (Kozlova). CH12067, P30/f1129, 158.3 m, RP8.
10. *Lychnocanium satelles* (Kozlova). CH12098, P30/f534, 176 m, RP8.
11. *Phormocyrtis striata exquisita* (Kozlova). CH12070, P30/f1129, 158.3 m, RP8.
12. *Phormocyrtis striata striata* Brandt. CH12038, P30/f1129, 158.3 m, RP8.
13. *Phormocyrtis turgida* (Krasheninnikov). CH12023, P30/f1129, 158.3 m, RP8.
14. *Phormocyrtis turgida* (Krasheninnikov). CH12056, P30/f1129, 158.3 m, RP8.
15. *Phormocyrtis cubensis* (Riedel and Sanfilippo). CH12057, P30/f1129, 158.3 m, RP8.
16. *Phormocyrtis cubensis* (Riedel and Sanfilippo). CH12029, P30/f1129, 158.3 m, RP8.
17. *Phormocyrtis cubensis* (Riedel and Sanfilippo). CH12028, P30/f1129, 158.3 m, RP8.
18. *Lamptonium fabaeforme fabaeforme* (Krasheninnikov). CH12055, P30/f1129, 158.3 m, RP8.
19. *Lamptonium fabaeforme fabaeforme* (Krasheninnikov). CH12048, P30/f1129, 158.3 m, RP8.
20. *Theocorys*? *phyzella* Foreman. CH12073, P30/f1129, 158.3 m, RP8.
21. *Theocorys*? *phyzella* Foreman. CH12042, P30/f1129, 158.3 m, RP8.
22. *Theocorys*? *phyzella* Foreman. CH12058, P30/f1129, 158.3 m, RP8.
23. *Theocorys*? *phyzella* Foreman. CH12045, P30/f1129, 158.3 m, RP8.
24. *Theocorys*? *phyzella* Foreman. CH12043, P30/f1129, 158.3 m, RP8.

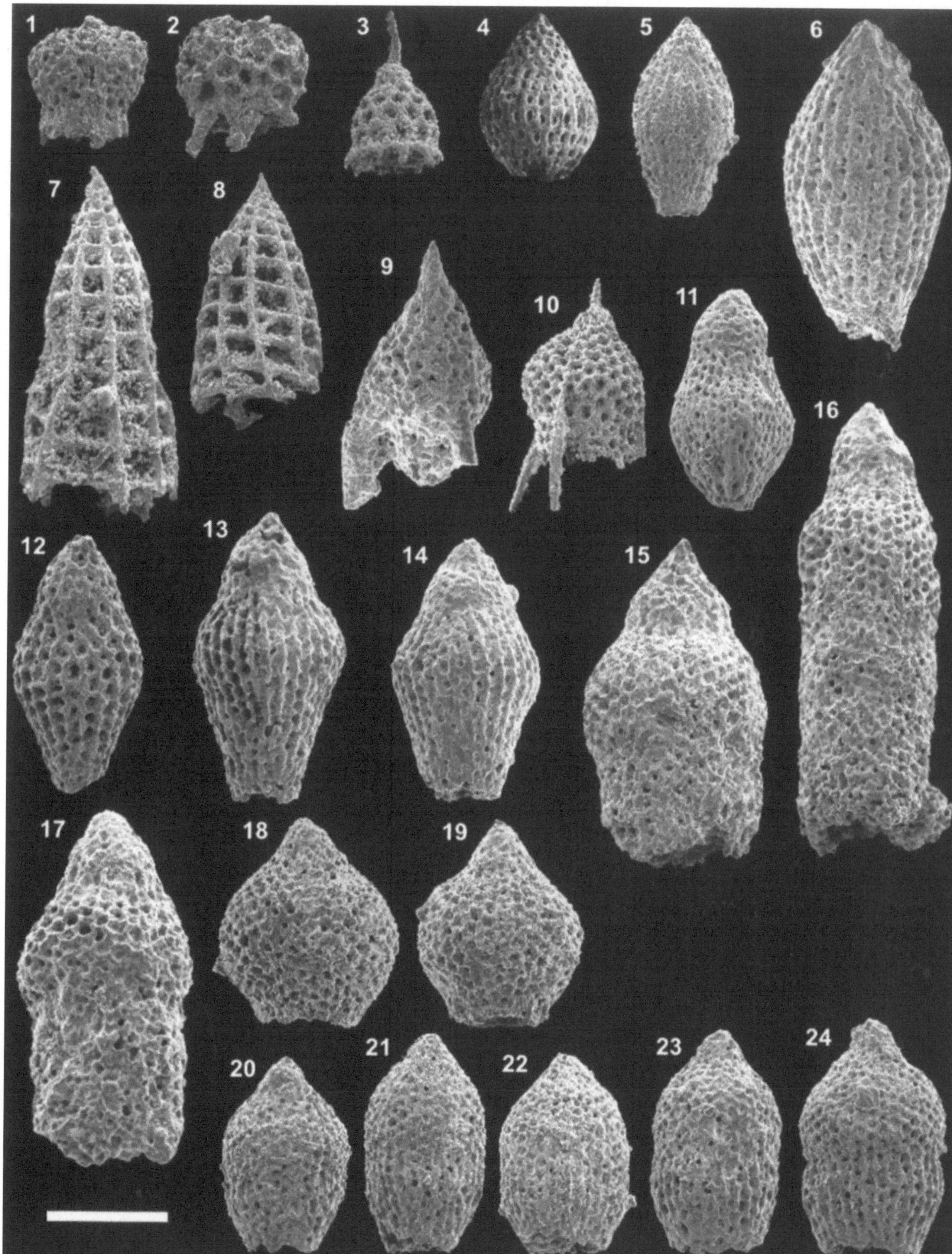
1
2
3
4
5
6
7
8
9
10
11
12
13
14
15
16
17
18
19
20
21
22
23
24

Plate 3

Transmitted light micrographs of radiolarians from upper Paleocene and lower Eocene strata, Mead Stream. Scale bars = 100 μm.

1. *Haliomma* gr. A-K47/4, P30/f451, 99 m, RP5.
2. *Haliomma* gr. A-L36/3, P30/f533, 166.5 m, RP8.
3. *Haliomma* gr. A-S28/0, P30/f533, 166.5 m, RP8.
4. *Haliomma* gr. A-D29/3, P30/f533, 166.5 m, RP8.
5. *Haliomma* gr. A-R46/1, P30/f533, 166.5 m, RP8.
6. *Periphaena*? *duplus* (KOZLOVA). A-032/2, P30/f451, 99 m, RP5.
7. *Periphaena*? *duplus* (KOZLOVA). A-S31/3, P30/f451, 99 m, RP5.
8. *Periphaena*? *duplus* (KOZLOVA). A-G29/1, P30/f451, 99 m, RP5.
9. *Middourium regulare* (BORISSENKO). A-W53/0, P30/f1076, 136.9 m, RP7.
10. *Dorcadospyris foveolata* (EHRENBERG). A-L49/3, P30/f527, 108.2 m, RP6.
11. *Dorcadospyris foveolata* (EHRENBERG). A-P42/0, P30/f527, 108.2 m, RP6.
12. *Dorcadospyris confluens* (EHRENBERG). A-W50/1, P30/f527, 108.2 m, RP6.
13. *Lophophaena mugaica* (GRIGORJEVA). A-B42/1, P30/f527, 108.2 m, RP6.
14. *Corythomelissa adunca* (SANFILIPPO AND RIEDEL). **A-?**, P30/f451, 99 m, RP5.
15. *Corythomelissa* sp. A. B-F36/0, P30/f527, 108.2 m, RP6.
16. *Buryella granulata* (PETRUSHEVSKAYA). D-P34/0, P30/f451, 99 m, RP5.
17. *Buryella foremanae* (PETRUSHEVSKAYA). B-U42/0, P30/f527, 108.2 m, RP6.
18. *Buryella tetradica* FOREMAN. A-F48/0, P30/f1076, 136.9 m, RP7.
19. *Buryella tetradica* FOREMAN. D-P32/0, P30/f451, 99 m, RP5.
20. *Buryella pentadica* FOREMAN. C-034/2, P30/f527, 108.2 m, RP6.
21. *Pterocodon poculum* (NISHIMURA). A-O47/2, P30/f1072, 136.6 m, RP7.
22. *Pterocodon poculum* (NISHIMURA). A-L37/3, P30/f1076, 136.9 m, RP7.
23. *Theocorys*? aff. *phyzella* FOREMAN. C-G39/0, P30/f452, 110.8 m, RP6.
24. *Theocorys*? aff. *phyzella* FOREMAN. C-E41/3, P30/f452, 110.8 m, RP6.
25. *Theocorys*? cf. *phyzella* FOREMAN. D-U29/3, P30/f452, 110.8 m, RP6.
26. *Theocorys*? *phyzella* FOREMAN. B-V33/1, P30/f1072, 136.6 m, RP7.
27. *Theocorys*? *phyzella* FOREMAN. A-H34/4, P30/f1072, 136.6 m, RP7.

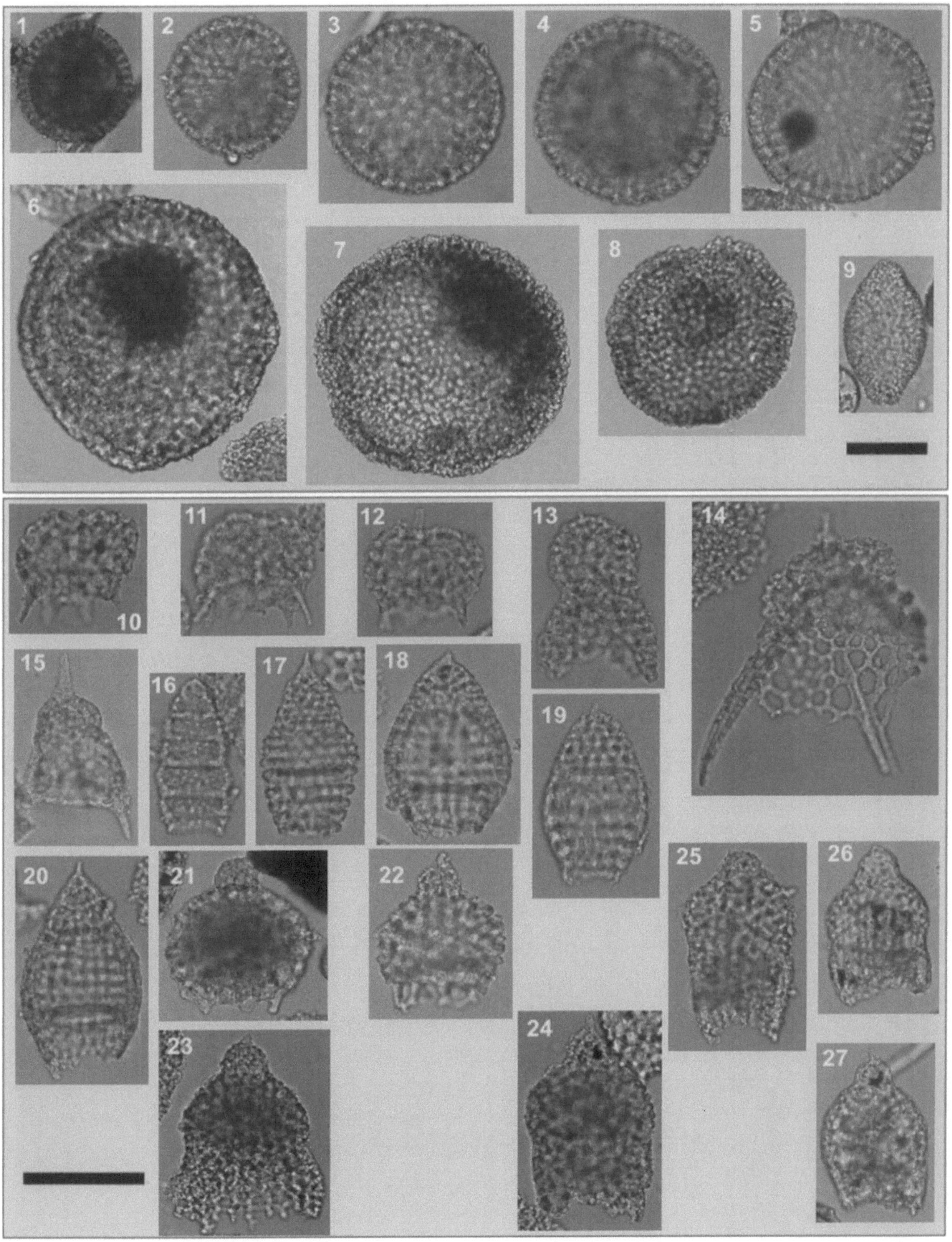

1
2
3
4
5
6
7
8
9
10
11
12
13
14
15
16
17
18
19
20
21
22
23
24
25
26
27

0012-9402/06/01S101-25
DOI 10.1007/s00015-006-0606-1
Birkhäuser Verlag, Basel, 2006

Eclogae geol. Helv. 99 (2006) Supplement 1, S101–S125

Upper Permian to Middle Jurassic radiolarian assemblages of Busuanga and surrounding islands, Palawan, Philippines

EDANJARLO J. MARQUEZ[1,2], JONATHAN C. AITCHISON[2] & LAWRENCE R. ZAMORAS[3]

Key words: North Palawan Block, radiolarians, biostratigraphy, Triassic, Permian, Jurassic, East Asian accretion

ABSTRACT

The North Palawan Block is regarded as the southernmost continuation of a Late Mesozoic accretionary complex, which developed along the length of the East Asian margin. It records a long (up to ~ 100 My) period of pelagic deposition on an oceanic plate from Late Permian to Late Jurassic when subduction resulting in the disappearance of the plate by the Early Cretaceous began. Subduction-accretion resulted in the development of three lithotectonic belts. From thirteen localities within these belts, radiolarian investigations yielded 173 species belonging to 92 genera and 45 families. Most of the samples contain Middle to Upper Triassic faunas. Several sections containing upper Permian, Lower Jurassic, and Middle Jurassic assemblages were also found. The presence of benthonic and planktonic foraminifers at some localities indicates that portions of the Liminangcong Formation were deposited in environments above the carbonate compensation depth (CCD). Manganese deposits found in some areas (e.g. Busuanga and Dabatonay) suggest that parts of the depositional area experienced a very low average sedimentation rate. Examination of the North Palawan accretionary complex reveals the ghosted history of the Izanagi Plate and constrains the timing of subduction beneath the eastern margin of Asia.

Introduction

Palawan is a NE-SW trending ridge located between the South China Sea and the Sulu Sea. The northern part of Palawan, together with the Calamian Island Group, Cuyo Island Group, Reed Bank, Spratly Islands, Dangerous Grounds, western Mindoro, Tablas and northwest Panay, is interpreted as a microcontinent, called the North Palawan Block (NPB) (Figure 1). The block is inferred to have rifted from the eastern margin of Eurasia as a result of the opening of the South China Sea at around 32 Ma (Hamilton 1979; Taylor & Hayes 1980; Holloway 1982; Briais et al. 1993; Almasco et al. 2000) and collided with the Philippine mobile belt during the middle Miocene (Hall 2002). Several investigations (e.g., Isozaki et al. 1988; Kojima 1989; Faure & Ishida 1990) considered the NPB as the southernmost continuation of the accretionary complex along the Eurasian margin that extends southwards from the Russian Far East, through NE China, NE and SW Japan and the Ryukyu islands.

In the last 30 years, radiolarian research has contributed tremendously in the deciphering and understanding of the pre-Cenozoic geology of the North Palawan Block. The Calamian Island Group is an integral part of this block and has been the subject of such investigations (Isozaki et al. 1988; Cheng 1989, 1992; Tumanda et al. 1990; Tumanda 1991a, 1991b, 1994; Yeh 1992; Yeh and Cheng 1996a, 1996b, 1998; Tumanda-Mateer et al. 1996; Zamoras & Matsuoka 2000, 2001 and Zamoras 2001). These studies produced radiolarian biostratigraphic zonations of the Middle Permian to Jurassic chert sequences. Zamoras (2001) and Zamoras & Matsuoka (2001) established that the chert-clastic succession youngs from north to south, and discriminated three different belts – the Northern Busuanga (NBB), Middle Busuanga (MBB) and Southern Busuanga (SBB) belts. They reported that the NBB accreted to the Asian margin in the late Middle Jurassic, whereas, the MBB and SBB accreted in Late Jurassic and Early Cretaceous, respectively.

The present research was undertaken in order to attempt to locate the Permo-Triassic boundary in the region and to expand the knowledge of the Triassic radiolarians in Southeast Asia, in particular those from the Calamian Island Group, in-

[1] Department of Physical Sciences and Mathematics, College of Arts and Sciences, University of the Philippines-Manila, Padre Faura St., Ermita, Manila, Philippines. Email: ejmarquez@mail.upm.edu.ph; ejmarque@graduate.hku.hk
[2] Department of Earth Sciences, The University of Hong Kong, Pokfulam Road, Hong Kong SAR, China. Email: jona@hku.hk
[3] Graduate School, University of the East, Manila, 1008 Philippines

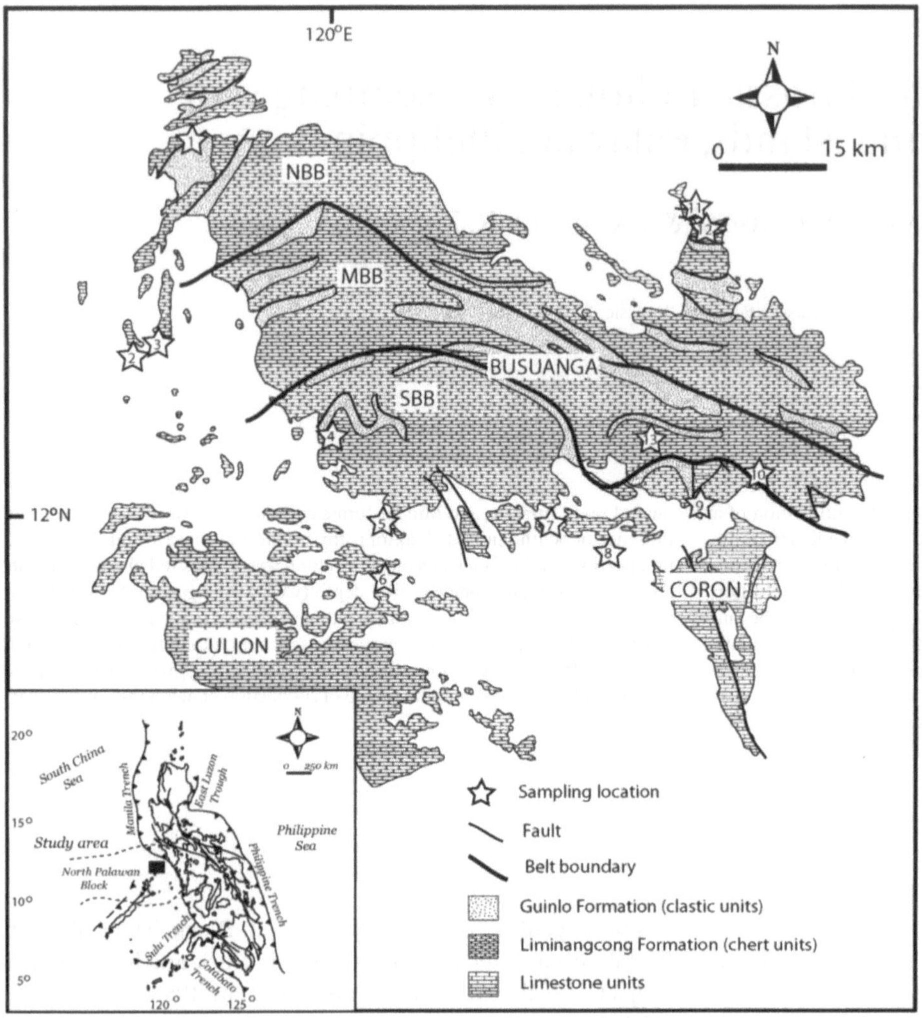

Fig.1. Geologic map of Busuanga and surrounding islands modified after Zamoras (2001). SBB = Southern Busuanga Belt, MBB = Middle Busuanga Belt, NBB = Northern Busuanga Belt. Locations of the sections where the cherts belonging to the Liminangcong Formation were collected are indicated. Numbers refer to the following areas: 1. Calauit-Illultuk Bay, 2. Dabatonay, 3. Mapadolo, 4. Dipuyay-Decalangwang River, 5. Lusong, 6. Marily, 7. Mayanpayan, 8. Dimanglet, 9. Road cut, township of Coron, 10. Calindo, 11. Sitio Tingil, 12. Mabudyen, 13. Mabintangin River.

cluding Busuanga and surrounding islands. Results presented herein provide information from many new localities, which complement the previous works in the description of Upper Permian and Middle Jurassic radiolarians. The data presented herein help to constrain the age of the oceanic materials accreted along the eastern margin of Eurasia during the Late Mesozoic.

Geologic Background

Busuanga is the largest island in the Calamian Island Group (Figure 1). As with the rest of the group, the island is part of a Late Mesozoic subduction complex that extended from the Russian Far East southwards at least as far as the north Palawan mainland. The subduction complex mainly consists of accreted limestones and cherts intercalated with trench-fill clastic sediments. The location of the sampling sites and a geologic map of the study area are given in Figure 1.

Various authors have proposed different stratigraphic subdivisions. Isozaki et al. (1988) interpreted the rock suite comprising the islands together with the north Palawan mainland to be a subduction-related complex and named this suite the North Palawan Complex. Faure & Ishida (1990) considered these rocks to be olistostromes. Zamoras (2001) and Zamoras & Matsuoka (2001) referred to the accretionary complex on Busuanga as the Malampaya Sound Group.

Several investigators considered the chert as part of the Liminancong Formation (Hashimoto & Sato 1973; Bureau of Mines and Geosciences 1984; Wolfart et al. 1986; Zamoras 2001; Zamoras & Matsuoka 2001) and assigned various ages: Middle Triassic (Hashimoto & Sato 1973; Fontaine 1979; Wolfart 1984); Lower to Middle Jurassic (Bureau of Mines and Geosciences 1984) and Middle-Upper Permian to Upper Jurassic (Wolfart et al. 1986; Isozaki et al. 1988; Tumanda 1991a; 1991b; Zamoras 2001; Zamoras and Matsuoka 2001). Japan International Cooperation Agency-Metal Mining Agency of Japan (1990) subdivided the chert into Liminangcong Chert (Triassic) and Busuanga Chert (Jurassic). Wolfart et al. (1986) adopted the name Bacuit Formation for the Permian sequence of chert, clastic rocks and limestone.

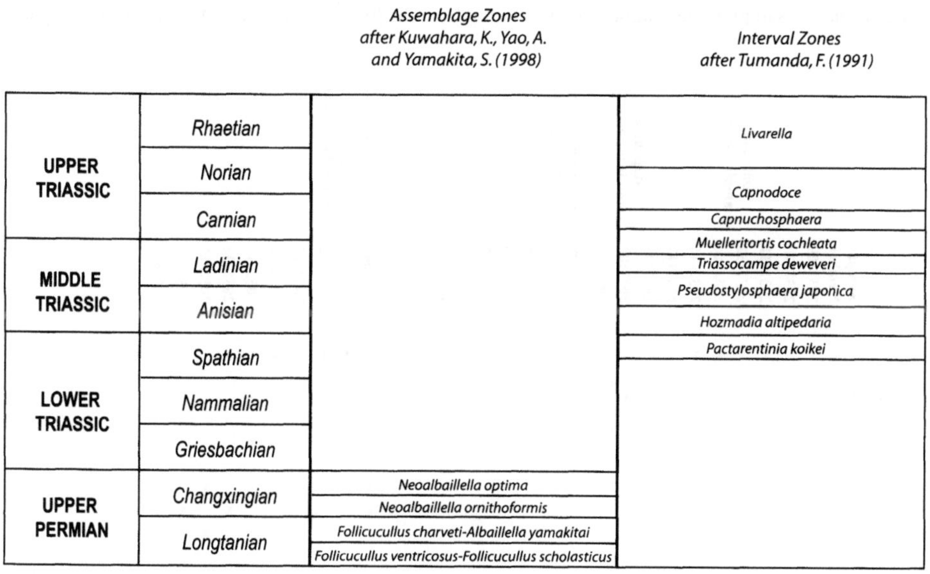

Fig. 2. Permian and Triassic radiolarian zonation schemes used in this study.

The limestone is distributed sporadically. Based on foraminifers, algae, conodonts, Porifera and Cnidaria (Fontaine 1979; Hashimoto et al. 1980; Fontaine et al. 1983, 1986; Wolfart et al. 1986), they range from Middle Permian to Upper Jurassic. The Bureau of Mines and Geosciences (1984) discriminated two limestone units: the Malajon Limestone (Upper Triassic) and the Coron Formation (Upper Triassic to Lower Jurassic). Kiessling & Flügel (2000) described the limestone outcrops on and around Busuanga, assigning them to the Coron Formation. They interpreted the carbonates as reefs (Malajon) and carbonate platforms (Kalampisanan, Busuanga and Coron).

On geologic maps prepared by the Bureau of Mines and Geosciences, clastic rocks are separated into the King Ranch and Liminangcong formations. Several workers used the term Guinlo Formation for the clastic succession (Hashimoto & Sato 1973; Zamoras 2001; Zamoras & Matsuoka 2001). Wolfart et al. (1986) subdivided the clastic rocks into the Coron Formation (Triassic) and the Guinlo Formation (Jurassic). Stratigraphic assignments vary from Middle Jurassic (Tumanda 1991a, 1991b), Middle Jurassic to Upper Jurassic (Tumanda-Mateer et al. 1996) and Middle Jurassic to Lower Cretaceous (Zamoras 2001; Zamoras & Matsuoka 2001). The reported ages of the clastic rocks were established based on radiolarians.

Zamoras (2001) and Zamoras &Matsuoka (2001) introduced the name Bicatan mélange for the minor mélange bodies units they mapped at Buyod Point, Bicatan Peninsula. They recognized two types; limestone-basalt-chert mélange and sandstone-chert-mudstone mélange.

Radiolarian biostratigraphy

Ninety-four samples yielding identifiable radiolarians comprise part of a suite of 261 samples collected from the Calamian Group of Islands, Palawan. Moderately to well-preserved faunas were recovered from the chert on Busuanga and surrounding islands (Calauit, Dabatonay, Mapadolo, Lusong, Marily, Mayanpayan and Dimanglet). Foraminifers were also noted in samples from Mabudyen and on the islands of Marily, Dimanglet, Mapadolo and Calauit. The conodont *Epigondollela* sp. was found in one sample collected along the Concepcion-Dipuyay road on Busuanga.

In order to establish the age of the radiolarian assemblages, we used the zonation schemes proposed by Kuwahara et al. (1998) for the Upper Permian, and by Tumanda (1991a, 1991b) for the Triassic (Figure 2). The Jurassic biostratigraphic ranges of the radiolarian species discussed in this study have been determined from synthesis of occurrences recorded in the literature.

In the Mabintangin River section (sample location 13), three late Permian zones have been recognized. The presence of *Albaillella cavitata* Kuwahara was used to delineate the upper Longtanian interval in the Liminangcong Formation. *A. cavitata* Kuwahara is a common species occurring in the *Follicucullus charveti-Albaillella yamakitai* assemblage zone (Kuwahara et al. 1998). Most of the Mabintangin samples also contain radiolarian assemblages belonging to the succeeding *Neoalbaillella ornithoformis* assemblage zone. *Follicucullus* sp. cf. *F. charveti* Caridroit & De Wever, *Foremanhelena musashiensis* (Sashida & Tonishi), *Albaillella*? *protolevis* Kuwahara and *Nazarovella gracilis* De Wever & Caridroit are some of the species found in these samples.

F. musashiensis (Sashida & Tonishi) and *Albaillella triangularis* Ishiga, Kito & Imoto are found in one sample. *F. musashiensis* (Sashida & Tonishi) is a common species of the *Neoalbaillella ornithoformis* assemblage zone and the *Neoalbaillella optima* assemblage zone whereas *Albaillella triangularis* Ishiga, Kito & Imoto, considering its occurrence, belongs to the *Neoalbaillella optima* zone. Based on this, the sample is assigned to the *Neoalbaillella optima* zone.

Table 1. Radiolarian assemblages from the Liminangcong Formation samples. The samples were collected from the following sections: Dabatonay and Mapadolo, Calauit-Illultuk Bay and Dipuyay-Decalangwang River.

Sample Location	**Dabatonay (sample location 2)**							**Mapadolo (sample location 3)**			**Calauit-Illultuk Bay (sample location 1)**		**Dipuyay-Decalangwang River (sample location 4)**																	
Species / **Sample Number**	**2031109**	**2031110**	**2031111**	**2031112**	**2031114**	**2031116**	**2031117**	**2031101**	**2031102**	**2031103**	**2031904**	**2031905**	**2032008**	**2032009**	**2032010**	**2032011**	**2032013**	**2032040**	**2032057**	**2032049**	**2032050**	**2032014**	**2032016**	**2032018**	**2032019**	**2032020**	**2032021**	**2032022**	**2032023**	**2032024**
Acanthosphaera (?) *mocki* KOZUR & MOSTLER																							x							
Archaeocenosphaera spp.	x	x	x	x	x	x	x		x		x		x	x		x	x	x	x	x	x	x	x	x	x	x		x	x	x
Bulbocyrtium spp.																				x	x									
Capnuchosphaera deweveri KOZUR & MOSTLER, emend. BLOME									x																					
Capnuchosphaera missionensis CORDEY									x																					
Celluronta sp.																									x					
Crucella sp.								x																						
Cryptostephanidium cornigerum DUMITRICA	x																			x							x			
Eptingium manfredi DUMITRICA	x			x	x							x							x	x						x		x	x	x
Eptingium nakasekoi KOZUR ET AL.																											x			
Eptingium sp.																														x
Haeckelicyrtium? sp.										x																				
Haliomma sp.																				x										
Hindeosphaera spinulosa (NAKASEKO & NISHIMURA)																			x							x			x	
Hozmadia reticulata DUMITRICA, KOZUR & MOSTLER																														x
Hozmadia rotunda (NAKASEKO & NISHIMURA)	x																											x		
Hozmadia spinifera SUGIYAMA																			x											
Hozmadia spp.	x	x	x														x		x	x		x	x			x			x	
Ladinocampe japonica (NAKASEKO & NISHIMURA)																		x		x										
Ladinocampe sp.																				x										
Oertlispongus inaequispinosus DUMITRICA, KOZUR & MOSTLER		x																												
Pantanellium sp.																							x							
Pararuesticyrtium spp.	x																	x	x							x	x		x	
Parasepsagon spp.	x		x		x										x							x				x		x		
Paroertlispongus spp.												x		x	x		x													
Pentactinocapsa awaensis (NAKASEKO & NISHIMURA)																				x								x		x
Pentactinocarpus fusiformis DUMITRICA	x	x	x	x	x															x										
Pentactinocarpus tetracanthus DUMITRICA					x																									
Pentactinorbis kozuri DUMITRICA																			x											
Poulpus spp.		x			x																									
Praeconocaryomma sp.																								x						
Pseudostylosphaera compacta (NAKASEKO & NISHIMURA)	x																	x	x								x	x	x	x
Pseudostylosphaera japonica (NAKASEKO & NISHIMURA)			x	x	x															x								x	x	x
Pseudostylosphaera longispinosa KOZUR & MOSTLER	x	x			x	x														x						x			x	
Pseudostylosphaera magnispinosa YEH	x		x																x										x	
Pseudostylosphaera spp.	x				x		x												x	x	x					x	x		x	x
Pseudostylosphaera timorensis SASHIDA & KAMATA																			x							x	x			
Rikivatella spp.											x												x				x			x
Sarla sp.																													x	
Spinotriassocampe annulata (NAKASEKO & NISHIMURA)						x														x	x									
Spinotriassocampe hungarica KOZUR	x				x													x												
Spongosilicarmiger sp.																														x
Spongostephanidium japonicum (NAKASEKO & NISHIMURA)																			x							x		x	x	
Spongostephanidium longispinosum SASHIDA	x																													
Stauracontium spp.																x							x			x				x
Staurolonche fluegeli KOZUR & MOSTLER																										x			x	
Staurolonche spp.	x	x					x				x									x										x
Staurolonche trispinosa (KOZUR & MOSTLER)																											x		x	
Tiborella florida (NAKASEKO & NISHIMURA)																	x		x								x		x	
Tiborella magnidentata DUMITRICA, KOZUR & MOSTLER																			x											
Tiborella sp.	x																													
Triassocampe coronata BRAGIN	x	x	x	x	x			x					x						x	x	x						x	x	x	x
Triassocampe deweveri (NAKASEKO & NISHIMURA)	x	x	x	x	x	x								x			x	x		x								x	x	x
Triassocampe spp.	x	x			x					x							x	x	x	x								x	x	x
Triassocampe tulbuanensis TUMANDA												x																		
Trilonche sp. cf. *T. chiakensis* (SASHIDA & IGO)	x																													
Trilonche spp.															x		x										x			x
Range	**upmst. Ans-Lad**	**u. Ans.-Lad**	**u. Ans.-Lad**	**u. Ans.-Lad**	**u. Ans.-Lad**	**u. Ans.-Lad**	**u. Ans.-Lad?**	**u. Ans-Lad**	**Crn**	**u. Nor-Rht?**	**u. Ans-Lad**	**u. Ans-Lad**	**u. Ans-Lad?**	**u. Ans-Lad?**	**u. Ans-Lad?**	**u. Ans-Lad**	**u. Ans-l. Lad**	**u. Ans-l. Lad**	**u. Ans-l. Lad**	**u. Ans-Lad**	**u. Ans-Lad**	**u. Ans-Lad**	**u. Ans-Lad**	**u. Ans-Lad**	**u. Ans-Lad**	**u. Ans-Lad**	**u. Ans-Lad**	**Lad**	**u. Ans-Lad**	**u. Ans-Lad**

Table 2. Radiolarians occurring at Dipuyay-Decalangwang River.

Sample Location	Dipuyay-Decalangwang River (sample location 4)																													
Species / **Sample Number**	**2032025**	**2032026**	**2032027**	**2032028**	**2032029**	**2032030**	**2032031**	**2032032**	**2032033**	**2032034**	**2032035**	**2032036**	**2032037**	**2032038**	**2032039**	**2032041**	**2032042**	**2032043**	**2032044**	**2032045**	**2032046**	**2032047**	**2032048**	**2032051**	**2032052**	**2032054**	**2032055**	**2032059**	**2032058**	**2032060**
Acanthosphaera (?) *mocki* KOZUR & MOSTLER	x		x					x					x																	
Archaeocenosphaera spp.	x	x	x	x	x	x	x	x	x	x	x	x		x	x	x	x	x	x	x	x	x	x	x	x	x	x	x	x	x
Cryptamphorella sp.											x																			
Cryptostephanidium cornigerum DUMITRICA																							x							
Eptingium manfredi DUMITRICA	x	x	x	x	x		x	x		x	x	x	x	x	x	x	x	x	x	x		x				x	x	x	x	x
Eptingium ramovsi KOZUR ET AL.	x																													
Eptingium spp.	x		x	x			x													x							x			
Haliomma spp.			x									x				x														
Hexalonche sp.			x																											
Hindeosphaera spinulosa (NAKASEKO & NISHIMURA)				x			x																			x		x		x
Hozmadia reticulata DUMITRICA, KOZUR & MOSTLER			x					x	x	x						x							x							
Hozmadia rotunda (NAKASEKO & NISHIMURA)										x																x	x		x	x
Hozmadia sp. cf. *H. reticulata* DUMITRICA, KOZUR & MOSTLER												x																		
Hozmadia spinifera SUGIYAMA																														
Hozmadia spp.		x	x	x			x	x	x	x	x	x		x			x	x	x			x			x	x				
Katorella bifurcata KOZUR & MOSTLER																										x				
Ladinocampe japonica (NAKASEKO & NISHIMURA)																									x					
Ladinocampe sp.										x																				
Muelleritortis sp.	x																													
Napora spp.																x		x							x					
Nassellaria gen. et. sp. indet.												x																		
Pantanellium sp.			x																											
Pararuesticyrtium spp.	x	x	x	x	x		x		x	x			x		x	x	x	x			x					x	x		x	
Parasepsagon spp.		x											x	x							x					x	x	x	x	x
Parasepsagon variabilis (NAKASEKO & NISHIMURA)			x																											
Pentactinocapsa awaensis (NAKASEKO & NISHIMURA)	x	x	x	x	x	x	x		x	x	x	x		x		x			x			x				x	x	x		x
Pentactinocarpus fusiformis DUMITRICA												x	x	x	x			x	x	x	x	x	x							
Pentactinocarpus tetracanthus DUMITRICA									x		x																			
Pentaspongodiscus spp.			x	x								x								x										
Plafkerium abboti PESSAGNO			x																											
Poulpus spp.												x													x					
Pseudostylosphaera compacta (NAKASEKO & NISHIMURA)	x	x	x	x	x	x	x	x			x		x		x	x			x	x	x	x		x		x	x	x	x	x
Pseudostylosphaera goricanae KOZUR ET AL.			x																											
Pseudostylosphaera japonica (NAKASEKO & NISHIMURA)	x	x	x	x	x	x	x	x		x	x	x	x	x		x		x				x				x	x	x		x
Pseudostylosphaera longispinosa KOZUR & MOSTLER			x			x	x	x	x	x			x	x	x	x		x	x		x	x							x	
Pseudostylosphaera spp.	x	x			x	x	x	x	x	x	x	x	x					x			x	x	x		x	x			x	x
Pseudostylosphaera timorensis SASHIDA & KAMATA						x	x	x			x							x												
Rikivatella spp.																										x		x		
Sarla spp.					x		x		x			x							x											x
Sepsagon sp.																									x					
Spinotriassocampe annulata (NAKASEKO & NISHIMURA)											x		x	x	x			x												
Spongostephanidium japonicum (NAKASEKO & NISHIMURA)		x	x	x			x		x																	x		x		x
Stauracontium spp.												x										x								
Staurolonche spp.	x		x	x	x																x									
Staurolonche trispinosa (KOZUR & MOSTLER)			x	x	x																				x		x	x		x
Thaisphaera spp.	x																													
Tiborella florida (NAKASEKO & NISHIMURA)	x																	x												
Tiborella sp. cf. *T. anisica* KOZUR & MOSTLER				x																										
Tiborella spp.			x		x					x															x	x				x
Triassistephanidium laticorne DUMITRICA					x			x																						
Triassocampe campanilis (KOZUR & MOSTLER)																														
Triassocampe coronata BRAGIN	x		x		x	x	x	x	x	x	x	x	x		x	x	x	x	x			x	x	x	x	x		x		x
Triassocampe deweveri (NAKASEKO & NISHIMURA)	x			x	x	x	x					x		x	x				x	x			x	x						
Triassocampe scalaris DUMITRICA, KOZUR & MOSTLER													x																	
Triassocampe spp.	x	x	x	x	x			x	x	x	x	x		x		x		x				x		x	x	x	x	x		x
Trilonche spp.		x				x	x		x									x	x		x	x								
Range	u. Ans-Lad	u. Ans-Lad	u. Ans-Lad	u. Ans-Lad	u. Ans-Lad	u. Ans-Lad	u. Ans-Lad	u. Ans-Lad	u. Ans-l. Lad	u. Ans-Lad	u. Ans-l. Lad	u. Ans-Lad	u. Ans-Lad	u. Ans-Lad	u. Ans-Lad	u. Ans-Lad	u. Ans-Lad	u. Ans-l. Lad	u. Ans-l. Lad	u. Ans-Lad	u. Ans-Lad	u. Ans-Lad	u. Ans-Lad	u. Ans-Lad	u. Ans-Lad	u. Ans-Lad	u. Ans-Lad	u. Ans-Lad	u. Ans-Lad	u. Ans-Lad

Triassic radiolarians were observed from samples collected from several localities. A number of species is common in the *Pseudostylosphaera japonica* interval zone (Tumanda 1991a, 1991b). This zone, considered as late Anisian-early Ladinian in age, is defined by the successive lowest occurrences of *Hindeosphaera spinulosa* (NAKASEKO & NISHIMURA) and *Eptingium* group at the base of the zone and *Triassocampe deweveri* (NAKASEKO & NISHIMURA) at the top. Some of the faunas present in our assemblage, which are commonly found in this zone, include *Hindeosphaera spinulosa* (NAKASEKO & NISHIMURA), *Eptingium* spp., *Pseudostylosphaera magnispinosa* YEH, *Tiborella florida* (NAKASEKO & NISHIMURA), *Katorella bifurcata* KOZUR & MOSTLER, *Pentactinocarpus tetracanthus* DUMITRICA, *Pentactinorbis kozuri* DUMITRICA, *Stauroloncha trispinosa* (KOZUR & Mostler), *Parasepsagon variabilis* (NAKASEKO & NISHIMURA) and *Tiborella magnidentata* DUMITRICA, KOZUR & MOSTLER.

The *Triassocampe deweveri* interval zone (Tumanda 1991a, 1991b) is represented by samples from Dipuyay-Decalangwang River (sample location 4) and Calindo (sample location 10). The following species common to this zone were found in our samples: *Triassocampe deweveri* (NAKASEKO & NISHIMURA), *Spongostephanidium japonicum* (NAKASEKO & NISHIMURA), *Oertlispongus inaequispinosus* DUMITRICA, KOZUR & MOSTLER, *Stauroloncha trispinosa* (KOZUR & MOSTLER), *Hozmadia rotunda* (NAKASEKO & NISHIMURA), *Pentactinorbis kozuri* DUMITRICA, *Pseudostylosphaera japonica* (NAKASEKO & NISHIMURA), *Pseudostylosphaera compacta* (NAKASEKO & NISHIMURA), *Eptingium manfredi* DUMITRICA, *Hindeosphaera spinulosa* (NAKASEKO & NISHIMURA), *Cryptostephanidium cornigerum* DUMITRICA and *Parasepsagon variabilis* (NAKASEKO & NISHIMURA).

The Ladinian to probably lower Carnian interval (*Muelleritortis cochleata* interval zone) is represented by sample HKU-PLWN 02031402. This sample is characterized by the presence of the diagnostic species *Muelleritortis cochleata* (NAKASEKO & NISHIMURA). The Carnian and upper Carnian-middle Norian were established based on the presence of *Capnuchosphaera deweveri* KOZUR & MOSTLER, emend. BLOME and *Capnodoce anapetes* DE WEVER, respectively. Based on the presence of *Livarella* spp., the upper Norian-Rhaetian was identified.

Jurassic radiolarians were also observed. Diagnostic species are very rare. However, based on the reported ranges of some species, the samples are assigned Jurassic ages. These include *Katroma westermanni* WHALEN & CARTER (upper Sinemurian), *Plessus* sp. aff. *P. aptus* YEH (lower Sinemurian) and *Transhsuum medium* (Unitary Association Zones 1–7, early-middle Aalenian to late Bathonian-early Callovian).

Following is the description of the localities investigated:

Calauit-Illultuk Bay (sample location 1)

Samples HKU-PLWN02031901 to HKU-PLWN02031908 were collected from two coastal sections located at 12°16.078'N/119°52.080'E and 12°17.484'N/119°51.960'E. The lithology present is recrystallized thinly bedded greenish-gray chert. Only two of these samples (HKU-PLWN02031904 and HKU-PLWN02031905) yielded identifiable radiolarians (Table 1). The assemblage present is upper Anisian-Ladinian.

Dabotonay (sample location 2)

Samples HKU-PLWN02031109 to HKU-PLWN02031120 were collected from a coastal section (12°6.186'N/119°50.758'E) composed of steeply dipping laminated to thinly bedded red chert. The beds are complexly folded and faulted. Some parts of the section contain manganese nodules. Occurrences of radiolarians in the area are shown on Table 1. Samples HKU-PLWN 02031109 to HKU-PLWN02031112, HKU-PLWN02031114, HKU-PLWN02031116 and HKU-PLWN02031117 contain upper Anisian-Ladinian radiolarians.

Mapadolo (sample location 3)

Samples HKU-PLWN02031101 to HKU-PLWN02031108 were collected from a coastal section (12°06.849'N/119° 52.006'E) of complexly folded, steeply dipping laminated to thinly bedded chert. The chert is recrystallized. Three samples yielded rare radiolarians. Samples HKU-PLWN02031101 and HKU-PLWN02031103 contain upper Anisian-Ladinian and upper Norian-Rhaetian assemblages, respectively. Sample HKU-PLWN02031102 yielded Carnian radiolarians (Table 1).

Dipuyay-Decalangwang River (sample location 4)

Samples HKU-PLWN02032007 to HKU-PLWN02032060 were collected from a section on the Concepcion-Dipuyay road (12°01.848'N/119°59.792'E). The lithology is predominantly chert with minor siliceous mudstone. Beds strike ~030 and dip steeply to the NW. Three radiolarian assemblages are recognized. The assemblages are diverse and contain well-preserved radiolarians. Samples HKU-PLWN02032008 to HKU-PLWN02032010, HKU-PLWN02032016, HKU-PLWN02032018 to HKU-PLWN 02032021, HKU-PLWN02032023 to HKU-PLWN02032032, HKU-PLWN02032034, HKU-PLWN02032036 to HKU-PLWN 02032039, HKU-PLWN02032041, HKU-PLWN02032042, HKU-PLWN02032045 to HKU-PLWN02032052, HKU-PLWN 02032054, HKU-PLWN02032055 and HKU-PLWN02032058 to HKU-PLWN02032060 contain upper Anisian-Ladinian radiolarians. Samples HKU-PLWN02032033, HKU-PLWN02032035, HKU-PLWN02032040, HKU-PLWN02032043, HKU-PLWN 02032044 and HKU-PLWN02032057 yielded upper Anisian-lower Ladinian radiolarians. Sample HKU-PLWN02032022 contains Ladinian radiolarians. Tables 1 and 2 show radiolarian occurrences in the area.

Lusong (sample location 5)

Samples HKU-PLWN02031802 to HKU-PLWN02031805 were collected from the western coast of Lusong (11°58.899'N/

Table 3. Radiolarian assemblages within the Liminangcong Formation. The samples were collected from Lusong, Marily, Mayanpayan and Dimanglet.

Sample Location	Lusong (sample location 5)		Marily (sample location 6)			Mayanpayan (sample location 7)				Dimanglet (sample location 8)																		
Species / Sample Number	2031802	2031803	2031302	2031303	2031304	2031709	2031710	2031715	2031713	2031705	2031706	2031707	2031413	2031414	2031418	2031419	2031420	2031425	2031427	2031430	2031402	2031405	2031407	2031408	2031409	2031410	2031702	2031701
Acanthosphaera? sp.																	x											
Archaeocenosphaera spp.				x	x								x			x				x	x	x	x					x
Archaeodictyomitra spp.			x						x											x								
Archicapsa sp.				x																								
Betraccium deweveri PESSAGNO & BLOME														x	x													
Betraccium maclearni PESSAGNO & BLOME														x														
Betraccium sp.cf. *B. inornatum* BLOME								x																				
Busuanga(?) sp.																											x	
Canoptum spp.												x				x												
Capnodoce anapetes DE WEVER																										x		
Capnodoce extenta BLOME													x															
Capnuchosphaera lea DE WEVER																						x	x	x				
Complexapora sp.																				x								
Corum sp. cf. *C. delgado* SUGIYAMA																												x
Eptingium sp.																												x
Gorgansium sp.				x	x					x																		
Haeckelicyrtium? sp.																						x						
Hagiastrum sp.?			x																									
Haliomma sp.					x																							
Japonocampe sp.	x																							x				x
Katroma coliforme (HORI)									x																			
Katroma sp.																		x		x								
Katroma sp. cf. *K. irvingi* WHALEN & CARTER									x																			
Katroma westermanni WHALEN & CARTER																				x								
Livarella magna TEKIN															x	x	x											
Minocapsa sp.							x																					
Muelleritortis cochleata (NAKASEKO & NISHIMURA)																					x						x	
Napora sp.																												
Pachus multinodosus TEKIN																							x					
Pantanellium sp.						x		x				x																
Pantanellium ultrasincerum BLOME					x																							
Parahsuum simplum YAO			x						x																			
Parahsuum spp.				x	x							x								x								
Pararuesticyrtium sp.																					x							
Paroertlispongus sp.										x																x		
Paronaella sp.																							x					x
Parvicingula sp.												x																
Pentaspongodiscus sp.	x																											
Pleesus sp. cf. *P. aptus* YEH									x																			
Podobursa sp.																				x								
Praemesosaturnalis huxleyensis (CARTER)														x														
Praemesosaturnalis rugosus (YEH)																									x			
Praemesosaturnalis sandspitense (BLOME)														x														
Praemesosaturnalis sp. cf. *P. finchi* (PESSAGNO)										x	x																	
Pseudostylosphaera compacta (NAKASEKO & NISHIMURA)																											x	
Pseudostylosphaera japonica (NAKASEKO & NISHIMURA)																					x							
Pseudostylosphaera spp.																											x	
Sarla sp.								x																				
Sethocapsa spp.			x		x										x					x								
Silicarmiger sp.																					x							
Spongostephanidium japonicum (NAKASEKO & NISHIMURA)																											x	
Stauroloncha spp.																											x	x
Stichocapsa spp.					x		x		x									x	x	x								
Triassocampe deweveri (NAKASEKO & NISHIMURA)																					x							
Trilonche sp.					x						x			x				x										
Tripocyclia sp.					x																							
Vinassaspongus subsphaericus KOZUR & MOSTLER																											x	
Williriedellum sp.												x																
Xipotheca sp. cf. *X. rugosa* BRAGIN, emend. TEKIN		x																										
Range	u. Crn-l. Nor	u. Crn-l. Nor	Het-Sin?	Het-Sin?	Het-Sin?	Het-Sin?	Het-Sin?	u. Nor-Rht	Sin	u. Nor.-Rht	u. Nor.-Rht	Baj?	u. Crn.-Nor	u. Nor.-Rht	u. Nor.-Rht	u. Nor.-Rht	u. Nor.-Rht	u. Nor.-Rht	u. Nor.-Rht	Sin	Lad-l. Crn	u. Crn-l. Nor	u. Crn-l. Nor	u. Crn-l. Nor	u. Nor-Rht	u. Crn-Nor	Lad-l. Crn	u. Crn-Nor

Table 4. Radiolarians occurring at Sitio Tingil, Mabudyen, a road cut 35 km east of the township of Coron (11°59.625'N/120°13.637'E) and Calindo.

Sample Location	**Sitio Tingil (sample location 11)**			**Mabudyen (sample location 12)**		**Road cut, township of Coron (sample location 9)**		**Calindo (sample location 10)**							
Species / **Sample Number**	**2031210**	**2031217**	**2031220**	**2031201**	**2031202**	**2032118**	**2032119**	**2032102**	**2032103**	**2032106**	**2032109**	**2032110**	**2032112**	**2032117**	**2032101**
Archaeocenosphaera spp.	x	x						x	x	x	x		x	x	
Astrocentrus sp. cf. *A. pulcher* KOZUR & MOSTLER	x														
Bulbocyrtium sp.												x			
Canoptum sp. aff. *C. unicum* PESSAGNO & WHALEN						x									
Canoptum spp.						x	x								
Capnodoce sp. cf. *C. anapetes* DE WEVER			x												
Celluronta sp.												x			
Deflandrecyrtium carterae YEH & CHENG						x									
Eptingium manfredi DUMITRICA	x							x	x	x	x	x		x	
Eptingium sp.									x						
Hexalonche sp.											x				
Hindeosphaera spinulosa (NAKASEKO & NISHIMURA)								x			x				
Hozmadia reticulata DUMITRICA, KOZUR & MOSTLER								x							
Hozmadia rotunda (NAKASEKO AND NISHIMURA)								x	x		x	x	x		
Hozmadia spp.								x			x			x	
Ladinocampe sp.														x	
Linaresia sp.															x
Livarella magna TEKIN						x	x								
Oertlispongus diacanthus Sugiyama	x														
Paleosaturnalis sp. cf. *P. dotti* (BLOME)					x										
Pararuesticyrtium sp.								x	x		x		x		
Parasepsagon spp.									x					x	
Parasepsagon variabilis (NAKASEKO & NISHIMURA)								x							
Paronaella sp.						x									
Pentactinocapsa awaensis NAKASEKO & NISHIMURA												x		x	
Pentaspongodiscus spp.										x			x	x	
Praemesosaturnalis gracilis (KOZUR & MOSTLER)					x										
Praemesosaturnalis huxleyensis (CARTER)					x										
Praemesosaturnalis rugosus YEH					x										
Praemesosaturnalis spp.		x		x	x										
Pseudostylosphaera compacta (NAKASEKO & NISHIMURA)								x		x	x	x		x	
Pseudostylosphaera japonica (NAKASEKO & NISHIMURA)									x			x	x		
Pseudostylosphaera longispinosa KOZUR & MOSTLER								x				x		x	
Pseudostylosphaera magnispinosa YEH														x	
Pseudostylosphaera spp.	x								x	x	x	x	x	x	
Pseudostylosphaera timorensis SASHIDA & KAMATA								x						x	
Sarla sp.		x													
Spinotriassocampe annulata (NAKASEKO & NISHIMURA)														x	
Spongosilicarmiger sp.											x				
Spongostephanidium japonicum (NAKASEKO & NISHIMURA)								x			x	x		x	
Spongostylus tricostatus KOZUR, KRAINER & MOSTLER								x							
Spumellaria gen. et. sp. indet.								x							
Stauracontium sp.										x					
Staurolonche spp.								x		x	x			x	x
Staurolonche trispinosa (KOZUR & MOSTLER)								x				x		x	
Stichocapsa sp.															x
Thaisphaera sp.										x					
Tiborella florida (NAKASEKO & NISHIMURA)	x											x		x	
Tiborella sp.												x			
Transhsuum medium TAKEMURA															x
Triassocampe coronata BRAGIN									x		x			x	
Triassocampe deweveri (NAKASEKO & NISHIMURA)												x		x	
Triassocampe spp.	x								x	x	x		x	x	
Trilonche spp.	x			x	x			x				x			
Range	**u. Ans-l. Lad**	**u. Nor-Rht**	**u. Crn-u. Nor**	**u. Nor-Rht**	**u. Nor-Rht**	**u. Nor-Rht**	**u. Nor-Rht**	**u. Ans-Lad**	**u. Ans.-Lad**	**Lad**	**u. Ans-Lad**	**u. Ans-Lad**	**Lad**	**u. Ans-Lad**	**u. Baj-u. m. Bth?**

120°00.509'E). The chert present at this locality is dull red and thinly bedded. Bedding is oriented 030/50NW. Two samples, HKU-PLWN02031802 and HKU-PLWN02031803, yielded rare upper Carnian-lower Norian radiolarians (Table 3).

Marily (sample location 6)

Samples HKU-PLWN02031302 to HKU-PLWN02031305 were collected from a coastal section consisting of reddish brown chert beds. The section is situated at the SE side of the island, 11°56.848'N/120°00.942'E. Bedding strikes 165 and dips steeply to the east. Three samples (HKU-PLWN02031302 to HKU-PLWN02031304) contain Lower Jurassic? (Hettangian-Sinemurian or younger) radiolarians. The occurrences of radiolarians are shown in Table 3.

Mayanpayan (sample location 7)

The coastal outcrop where samples HKU-PLWN02031709 to HKU-PLWN02031715 were collected is located on the SE side of the island, 11°58.816'N/120°07.382'E. It consists of folded thinly to medium bedded reddish brown chert. Samples HKU-PLWN02031709 and HKU-PLWN02031710 contain Lower Jurassic? (Sinemurian or younger) assemblages. Sample HKU-PLWN02031713 contains Sinemurian radiolarians whereas sample HKU-PLWN02031715 yielded upper Norian-Rhaetian radiolarians (Table 3).

Dimanglet (sample location 8)

Samples HKU-PLWN02031401 to HKU-PLWN02031435 and HKU-PLWN02031701 to HKU-PLWN02031708 were collected from a continuous chert section exposed along the SE coast of the island (12°01.848'N/119°59.792). The reddish brown chert is thinly to medium bedded. It is highly folded and faulted. Some horizons are massively recrystallized and appear to be thickly bedded. Four radiolarian assemblages are recognized. Samples HKU-PLWN02031410, HKU-PLWN02031413 and HKU-PLWN02031701 yielded upper Carnian-Norian radiolarians. Ladinian – lower Carnian faunas occur in samples HKU-PLWN02031402 and HKU-PLWN02031702 whereas upper Carnian – lower Norian radiolarians are found in samples HKU-PLWN02031405, HKU-PLWN02031407 and HKU-PLWN02031408. Samples HKU-PLWN02031409, HKU-PLWN 02031414, HKU-PLWN20301418 to HKU-PLWN02031420, HKU-PLWN02031425, HKU-PLWN02031427, HKU-PLWN 02031705 and HKU-PLWN02031706 contain upper Norian-Rhaetian faunas. Lower to Middle Jurassic radiolarians occur in samples HKU-PLWN02031430 (Sinemurian or younger) and HKU-PLWN02031707 (Bajocian or younger). Table 3 shows the radiolarians occurring in this area.

Road cut, 3.5 km east of the township of Coron (sample location 9)

Two samples, HKU-PLWN02032118 and HKU-PLWN 02032119, were collected from this roadside section (11°59.625'N/120°13.637'E). The chert isweathered and thinly bedded. Bedding strikes 160 and dips gently to the SW. The two samples contain upper Norian-Rhaetian radiolarians (Table 4).

Calindo (sample location 10)

Samples HKU-PLWN02032102 to HKU-PLWN02032117 were collected from a roadside section located at 12°00.939'N/ 120°15.681'E. The outcrop is composed of uniformly bedded reddish brown chert. Bedding strikes 030 and dips gently to the NW. Upper Anisian-Ladinian radiolarians occur in samples HKU-PLWN02032102, HKU-PLWN02032103, HKU-PLWN02032109, HKU-PLWN02032110 and HKU-PLWN 02032117 whereas Ladinian radiolarians are present in samples HKU-PLWN02032106 and HKU-PLWN02032112 (Table 4).

A spot sample (HKU-PLWN02032101) was collected from the top of a small hill (100 m high) located at 12°01.477'N/120°16.232'E. The sample, a chert, is upper Bajocian to uppermost middle Bathonian? (Table 4).

Sitio Tingil, 2.5 km SE of Coconogon Point (sample location 11)

Samples HKU-PLWN02031209 to HKU-PLWN02031220 and HKU-PLWN02031618 to HKU-PLWN02031621 were collected from a coastal section (12°12.570'N/120°13.282'E) composed of brown, dark gray to black, thinly bedded to laminated chert. The beds are folded and locally highly sheared. Sample HKU-PLWN02031210 yielded upper Anisian-lower Ladinian radiolarians whereas samples HKU-PLWN02031217 and HKU-PLWN02031220 contain upper Norian-Rhaetian and upper Carnian – upper Norian radiolarians, respectively (Table 4).

Mabudyen, 2 km west of Malawig (sample location 12)

The coastal section (12°11.941'N/120°13.447) where samples HKU-PLWN02031201 to HKU-PLWN02031207 were collected consists of thinly bedded black-brown chert. The chert is recystallized. Bedding strikes 030 and dips steeply to the NW. Only two samples, HKU-PLWN02031201 and HKU-PLWN02031202, contain identifiable forms (Table 4). The assemblage is upper Norian-Rhaetian.

Mabintangin River (sample location 13)

Samples HKU-PLWN02031001 to HKU-PLWN02031017 were collected from a section along the Mabintangin River. The outcrop is composed of folded thinly to medium bedded black chert. Twelve samples yielded identifiable Upper Permian radiolarians. Table 5 shows the radiolarian assemblage in this section.

Table 5. Distribution of Upper Permian radiolarians within the Liminangcong Formation exposed along the Mabintangin River.

Sample Location	Mabintangin River											
Species / Sample Number	2031001	2031002	2031003	2031004	2031005	2031006	2031007	2031008	2031009	2031011	2031016	2031017
?*Latentifistula* sp. F sensu KUWAHARA AND YAO			x									
?*Stigmosphaerostylus* spp.						x			x			
Albaillella ? *protolevis* KUWAHARA											x	x
Albaillella cavitata KUWAHARA						x	x					
Albaillella sp.						x						
Albaillella triangularis ISHIGA, KITO & IMOTO								x				
broken *Follicucullus* spp.	x	x				x			x			
Copicyntra spp.			x		x	x						
Copicyntroides? sp. A sensu KUWAHARA AND YAO	x			x								
Copiellintra? sp. A sensu KUWAHARA AND YAO						x						
Deflandrella sp.			x									
Follicucullus porrectus RUDENKO	x	x										
Follicucullus sp. cf. *F. charveti* CARIDROIT & DE WEVER		x									x	x
Follicucullus sp. cf. *F. scholasticus* ORMISTON & BOBCOCK	x											
Foremanhelena musashiensis (SASHIDA & TONISHI)								x				x
Ishigaum nicolasensis TUMANDA							x					x
Ishigaum sp. cf. *I. trifustis* DE WEVER & CARIDROIT			x									
Ishigaum spp.		x	x	x						x		
Latentibifistula sp. cf. *L. asperspongiosa* SASHIDA & TONISHI	x											
Latentibifistula spp.			x		x		x		x			
Latentifistula sp. E sensu TUMANDA	x											
Latentifistula spp.			x	x		x	x			x		x
Nazarovella gracilis DE WEVER & CARIDROIT	x											x
Nazarovella sp.			x									
Polyfistula sp.									x			
Pseudotormentus sp.	x											
Ruzhencevispongus (?) sp. B sensu TUMANDA	x				x					x		
Spumellaria gen. et. sp. indet.	x	x	x	x	x	x	x		x	x		
Stauraxon F sensu YAO & KUWAHARA	x											
Stigmosphaerostylus sp. cf. *S. modesta* (SASHIDA & TONISHI)				x								x
Trilonche sp. cf. *T. cimelia* (NAZAROV & ORMISTON)		x			x	x	x					
Trilonche spp.							x		x			x
Range	**Upper Permian**											

Summary and Conclusion

This research has expanded the known geographic extent of studied sections and has contributed towards refinement of calibration of the Triassic in Northern Palawan. Samples collected from several sections on Busuanga and surrounding small islands yielded 173 species distributed among 92 genera in 45 families. Most of the samples are Triassic, spanning the time interval, Anisian to Rhaetian. Four of the sections also yield Jurassic faunas; three of which include Lower Jurassic radiolarians whereas two sites contain Bajocian-Bathonian? assemblages. Upper Permian radiolarians were recovered only from sample location 13 at the Mabintangin River in Busuanga. No other Permian outcrops were encountered on the surrounding islands investigated during this study. Lowermost Triassic (Griesbachian and Nammalian) sections are not documented herein and they have not been reported from earlier investigations. It remains unclear whether the absence of the lowermost Triassic outcrops is related to sampling density, poor radiolarian preservation or a possible hiatus coinciding with the worldwide latest Permian catastrophic event.

A previous study of Busuanga (Zamoras 2001; Zamoras & Matsuoka 2001) recognized three structural divisions: Northern Busuanga Belt (NBB), Middle Busuanga Belt (MBB) and Southern Busuanga Belt (SBB). Samples collected during this study were incorporated into the composite stratigraphy of the chert-clastic sequence described by Zamoras (2001). The sections from Calauit-Illultuk Bay, Sitio Tingil and Mabudyen are from the NBB. The Dabatonay, Mapadolo and Calindo sections lie within the MBB. The rest of the sections are from the SBB. Figure 3 shows the data from this study together with the Triassic data of Zamoras (2001) and Zamoras & Matsuoka (2001), and earlier results of Yeh (1992) and Tumanda (1991a, 1991b). In the SBB stratigraphy, Zamoras (2001) noted that the Upper Triassic cherts represent the lower part of the succession. However, the inclusion of data from the present study clearly demonstrates the presence of a Middle Triassic interval in the SBB. Tumanda (1991a, 1991b) also reported an upper Spathian assemblage in Decalangwang.

The chert-clastic succession in North Palawan documents almost 100 My of the history of a subducted oceanic plate.

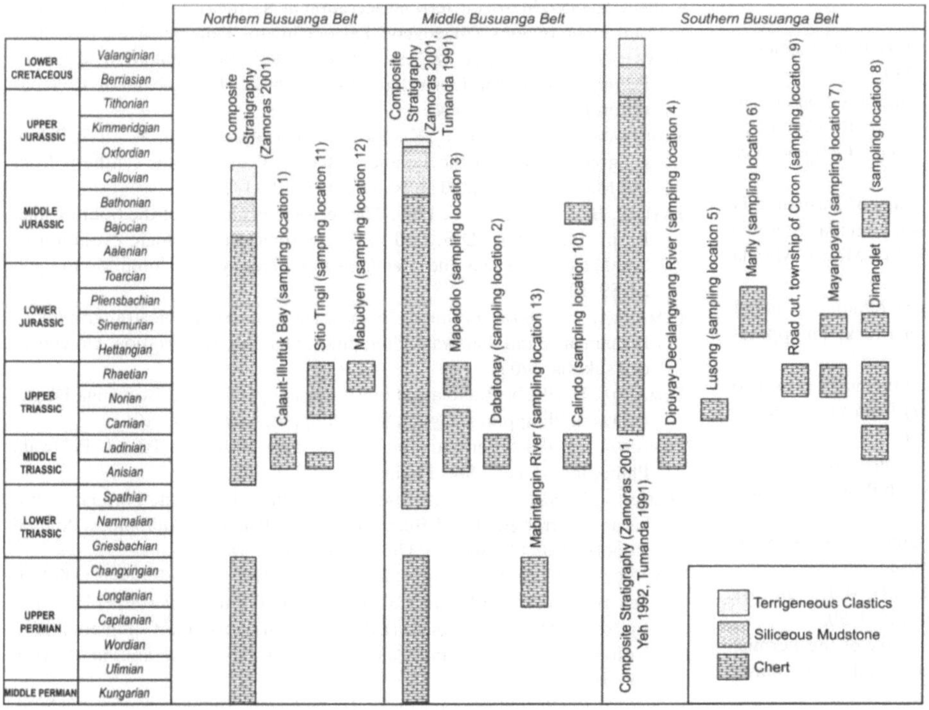

Fig. 3. Stratigraphic ranges of lithologies in the three accretionary belts of Busuanga as constrained by radiolarian faunas.

This plate traveled from a deep sea environment to a convergent margin, which developed along the eastern margin of Asia. However, the paleolatitudes of the block during this journey are still under investigation. Kiessling and Flügel (2000) suggested that the facies and ages of the carbonates from the islands of Malajon, Kalampisanan, Busuanga and Coron do not support a close paleogeographical connection between the NPB and South China Block. They suggested that during the Carboniferous and Permian, the NPB was part of the Indochina Block from which it separated during the Middle Permian. The NPB collided with the South China Block during the Late Cretaceous. Paleomagnetic studies carried out by Almasco et al. (2000) on the NPB show that the paleomagnetic directions fail regional fold tests and have alternating field (AF) demagnetisation features consistent with secondary magnetisation. However, they pointed out that the Cretaceous paleolatitude is still comparable to regions of pervasive Cretaceous remagnetisation in the South China borderland. They suggested that this may reflect similar remagnetisation and is consistent with the NPB's proposed South China origin.

Based on the tectonic model of Maruyama et al. (1997), Zamoras (2001) interpreted the NPB to have been part of the NNW-traveling Izanagi Plate from Late Permian to Late Jurassic-Early Cretaceous. He suggested that the Middle Carboniferous limestone described by Kiessling & Flügel (2000) probably originated from the Farallon Plate. Holloway (1982) noted that during the Middle Jurassic to Late Jurassic an episode of spreading commenced between Australia and a continent/continental fragments to the northwest. He added that this spreading episode produced a NNW-directed plate motion with the oceanic lithosphere being consumed at a NW-dipping subduction zone (East Asian Subduction Zone; Zamoras 2001) on the southeastern margins of the Indochina and South China blocks.

The pelagic sediments in Busuanga and surrounding islands records up to 100 My of deposition. The presence of benthonic and planktonic foraminifers at some localities probably indicates that not all of the chert was deposited below the CCD. Manganese deposits found on Busuanga and Dabatonay indicate an oceanic setting affected by a low average sedimentation rate. The limestone bodies presently found in fault contact with the chert-clastic succession can be interpreted as shallow water facies or carbonate build-ups on seamounts. The facies change from cherts through hemipelagic clastics to terrigenous clastics suggests that the ocean was being influenced by subduction zone related sedimentation by the Middle to Late Jurassic (Aalenian-Bajocian in the NBB, early-late Bathonian in the MBB and early-late Tithonian in the SBB).

Acknowledgements

The authors gratefully acknowledge the financial support of The University of Hong Kong-Committee on Research and Conference Grants towards this research. We also thank Dr. Jason R. Ali and Prof. Graciano P. Yumul, Jr. for their company and stimulating discussions while we were undertaking our field investigations. Valuable suggestions and comments by our reviewers, Dr. Spela Gorican and Dr. Martial Caridroit are greatly appreciated.

REFERENCES

ALMASCO, J.N., RODOLFO, K., FULLER, M. & FROST, G. 2000: Paleomagnetism of Palawan, Philippines. J. Asian Earth Sci. 18, 369–389.

BRIAIS, A., PATRIAT, P. & TAPPONIER, P. 1993: Updated interpretation of magnetic anomalies and sea floor spreading stages in the South China Sea: implications for the Tertiary tectonics of Southeast Asia. J. Geophys. Res. 98, 6299–6328.

BUREAU OF MINES & GEOSCIENCES. 1984: Geological maps of Busuanga Islands: Sheet nos. 2956 (I–II), 3055(IV), 3056 (I–IV). National Mapping and Resource Information Authority (NAMRIA), Metro Manila, 1:50,000.

CHENG, Y.-N. 1989: Upper Paleozoic and Lower Mesozoic radiolarian assemblages from the Busuanga Islands, North Palawan Block, Philippines. Bull. Natl. Mus. Nat. Sci. 1, 129–175.

CHENG, Y.-N. 1992: Upper Jurassic Pantanelliidae (Pantanelliinae Pessagno, 1977 and Vallupinae Pessagno and MacLeod, 1987) from the Busuanga Islands, Philippines. Bull. Natl. Mus. Nat. Sci. 3, 1–49.

FAURE, M. & ISHIDA, K. 1990: The mid-Upper Jurassic olistostrome of the west Philippines: a distinctive key-marker for the North Palawan Block. J. Asian Earth Sci. 8/1, 61–67.

FONTAINE, H. 1979: Note on the geology of the Calamian Islands, North Palawan, Philippines. ESAP-CCOP Newsletter 6, 40–47.

FONTAINE, H., DAVID, P., PARDEDE, R. & SUWARNA, R. 1983: Marine Jurassic in Southeast Asia. ESCAP-CCOP Technical Bulletin 16, 1–30.

FONTAINE, H., TIEN, N.D. & VACHARD, D. 1986: Discovery of the Permian limestone south of Tara Island in the Calamian Islands, Philippines. ESCAP-CCOP Technical Bulletin 18, 161–166.

HALL, R. 2002: Cenozoic geological and plate tectonic evolution of SE Asia and the SW Pacific: computer-based reconstructions, models and animations. J. Asian Earth Sci. 20, 353–431.

HAMILTON, W. 1979: Tectonics of the Indonesian Region. U.S. Geol. Surv. Prof. Paper 1078, 1–338.

HASHIMOTO, W. & SATO, T. 1973: Geological structure of North Palawan, and its bearing on the geological history of the Philippines. In: KOBAYASHI, T. & TORIYAMA, R. (Eds.). Geol. Paleont. Southeast Asia 13, 145–161.

HASHIMOTO, W., TAKIZAWA, S., BALCE, G.R., ESPIRITU, E.A. & BAURA, C.A. 1980: Discovery of Triassic conodonts from Malajon Island Group, Palawan Province, the Philippines and its geological significance. Proc. Jap. Acad. 56 (Ser. B), 69–73.

HOLLOWAY, N.H. 1982: North Palawan Block, Philippines – Its relation to Asian Mainland and role in evolution of South China Sea. Bull. Amer. Assoc. Petroleum Geol. 66, 1355–1383.

ISOZAKI, Y., AMISCARAY, E.A. & RILLON, A. 1988: Permian, Triassic and Jurassic bedded radiolarian cherts in North Palawan Block, Philippines: Evidence of late Mesozoic subduction-accretion. In: ICHIKAWA, K. (Ed.): Report No. 3 of the IGCP Project 224: Pre-Jurassic evolution of eastern Asia. Osaka, 99–115.

JAPAN INTERNATIONAL COOPERATION AGENCY-METAL MINING AGENCY OF JAPAN 1990: Report on the mineral exploration, mineral deposits and tectonics of two contrasting geologic environments in the Republic of the Philippines, terminal report. 121 p, Metal Mining Agency, Tokyo.

KIESSLING, W. & FLÜGEL, E. 2000: Late Paleozoic and Late Triassic limestones from North Palawan Block (Philippines): Microfacies and paleogeographical implications. Facies 43, 39–78.

KOJIMA, S. 1989: Mesozoic terrane accretion in Northeast China, Sikhote-Alin and Japan regions. Palaeogeog. Palaeoclimatol. Palaeoecol. 69, 213–232.

KUWAHARA, K., YAO, A. & YAMAKITA, S. 1998: Reexamination of Upper Permian radiolarian biostratigraphy. Earth Sci., J. Assoc. Geol. Collaboration Japan 52, 391–404.

MARUYAMA, S., ISOZAKI, Y., KIMURA, G. & TERABAYASHI, M. 1997: Paleogeographic maps of the Japanese Islands: Plate tectonic synthesis from 750 Ma to the present. The Island Arc 6, 121–142.

TAYLOR, B. & HAYES, D.E. 1980: The tectonic evolution of the South China Basin. In: HAYES, D.E. (Ed.): The tectonic and geologic evolution of Southeast Asian seas and islands. Amer. Geophysical Union Monograph 23, 89–104.

TUMANDA, F. 1991a: Permian to Jurassic radiolarian biostratigraphy of Busuanga Island, Palawan, Philippines. Doc. Sci. Dissertation, University of Tsukuba, 270 p.

TUMANDA, F. 1991b: Radiolarian biostratigraphy in central Busuanga Island, Palawan, Philippines. J. geol. Soc. Philippines *46*/1–2, 49–104.

TUMANDA, F. 1994: Permian radiolarian from Busuanga Island, Palawan, Philippines. J. geol. Soc. Philippines 49, 119–193.

TUMANDA, F., SATO, T. & SASHIDA, K. 1990: Preliminary Late Permian radiolarian biostratigraphy of Busuanga Island, Palawan, Philippines. Annual Reports, Inst. Geoscience, Univ. Tsukuba 16, 39–45.

TUMANDA-MATEER, F., SASHIDA, K. & IGO, H. 1996: Some Jurassic radiolarians from Busuanga Island, Calamian Island Group, Palawan, Philippines. In: NODA, H. & SASHIDA K. (Eds.): Prof. Hisayoshi Igo Commemorative Volume on Geology and Paleontology of Japan and Southeast Asia, 165–192.

WOLFART, R. 1984: Stratigraphy of Palawan Island, Philippines. Rep. Federal. Inst. Geosci. Nat. Resources, Archive no. 97074, 40 p.

WOLFART, R., CEPEK, P. GRAMANN, F., KEMPER, E. & PORTH, H. 1986: Stratigraphy of Palawan Island, Philippines. Newsl. Strat. 16/1, 19–48.

YEH, K.-Y. 1992: Triassic Radiolaria from Uson Island, Philippines. Bull. Natl. Mus. Nat. Sci. 3, 51–91.

YEH, K.-Y. & CHENG, Y.-N. 1996a: An Upper Triassic (Rhaetian) radiolarian assemblage from Busuanga Island, Philippines. Bull. Natl. Mus. Nat. Sci. 7, 1–43.

YEH, K.-Y. & CHENG, Y.-N. 1996b: Jurassic radiolarians from the northwest coast of Busuanga Island, North Palawan Block, Philippines. Micropaleontology 42/2, 93–124.

YEH, K.-Y. & CHENG, Y.-N. 1998: Radiolarians from the Lower Jurassic of the Busuanga Island, Philippines. Bull. Natl. Mus. Nat. Sci. 11, 1–65.

ZAMORAS, L. 2001: Jurassic-Early Cretaceous accretionary complex in the Calamian Islands, North Palawan Block (Philippines). Ph.D. Thesis, Niigata Univ., 120 p.

ZAMORAS, L.R. & MATSUOKA, A. 2000: Early Late Jurassic radiolarians from the clastic unit in Busuanga Island, North Palawan, Philippines. Sci. Rep., Niigata Univ. 15 (Ser. E, Geol.), 91–109.

ZAMORAS, L.R. & MATSUOKA, A. 2001: Malampaya Sound Group: A Jurassic-Early Cretaceous accretionary complex in Busuanga Island, North Palawan Block (Philippines). J. geol. Soc. Japan 107/5, 1–336.

Manuscript received January 2004
Revision accepted February 2005

Plate 1

Upper Permian radiolarians from the Mabintangin River (sample location 13), Busuanga, Philippines. Scale bars = 100 µm.

Figs. 1–5 from HKU-PLWN02031001, Upper Permian:
1. spherical radiolarian; 2. ?*Copicyntroides* sp. A sensu KUWAHARA & YAO; 3. ?*Nazarovella gracilis* DE WEVER & CARIDROIT; 4, 5. broken *Follicucullus*

Figs. 6–11 from HKU-PLWN02031002, Upper Permian:
6. *Trilonche* sp. cf. *T. cimelia* (NAZAROV & ORMISTON); 7, 8. *Follicucullus porrectus* RUDENKO; 9–11. *Follicucullus charveti* CARIDROIT & DE WEVER

Figs. 12–19 from HKU-PLWN02031003, Upper Permian:
12. spherical radiolarian; 13. ?*Copicyntra* sp.; 14. *Ishigaum* sp. cf. *I. trifustis* DE WEVER & CARIDROIT; 15. *Latentifistula* sp. F sensu KUWAHARA AND YAO?; 16. *Deflandrella* sp.; 17. *Nazarovella* sp.; 18, 19. *Latentibifistula* spp.

Figs. 20, 21 from HKU-PLWN02031004, Upper Permian:
20. *Copicyntroides*? sp. A sensu KUWAHARA AND YAO; 21. *Ishigaum* sp.

Figs. 22–27 from HKU-PLWN02031006, Upper Permian:
22. ?*Stigmosphaerostylus* sp.; 23. *Trilonche* sp. cf. *T. cimelia* (NAZAROV & ORMISTON); 24. *Copicyntra* sp.?; 25. *Copiellintra*? sp. A sensu KUWAHARA AND YAO; 26. *Albaillella* sp.; 27. *Albaillella* sp. cf. *A. cavitata* KUWAHARA

Figs. 28–31 from HKU-PLWN02031007, Upper Permian:
28, 29. *Albaillella cavitata* KUWAHARA; 30. *Ishigaum nicolasensis* TUMANDA; 31. *Latentibifistula* sp.

Figs. 32, 33 from HKU-PLWN02031008, Upper Permian:
32. *Foremanhelena musashiensis* (SASHIDA & TONISHI); 33. *Albaillella triangularis* ISHIGA, KITO & IMOTO

Figs. 34, 35 from HKU-PLWN02031009, Upper Permian:
34. ?*Stigmosphaerostylus* sp.; 35. *Polyfistula* sp.

Fig. 36 from HKU-PLWN02031011, Upper Permian:
36. *Ruzhencevispongus* (?) sp. B sensu TUMANDA

Fig. 37 from HKU-PLWN02031016, Upper Permian:
37. *Albaillella*? *protolevis* KUWAHARA

Figs. 38–41 from HKU-PLWN02031017, Upper Permian:
38. *Ishigaum nicolasensis* TUMANDA; 39. *Trilonche* sp.; 40. *Nazarovella gracilis* DE WEVER & CARIDROIT; 41. *Foremanhelena musashiensis* (SASHIDA & TONISHI)

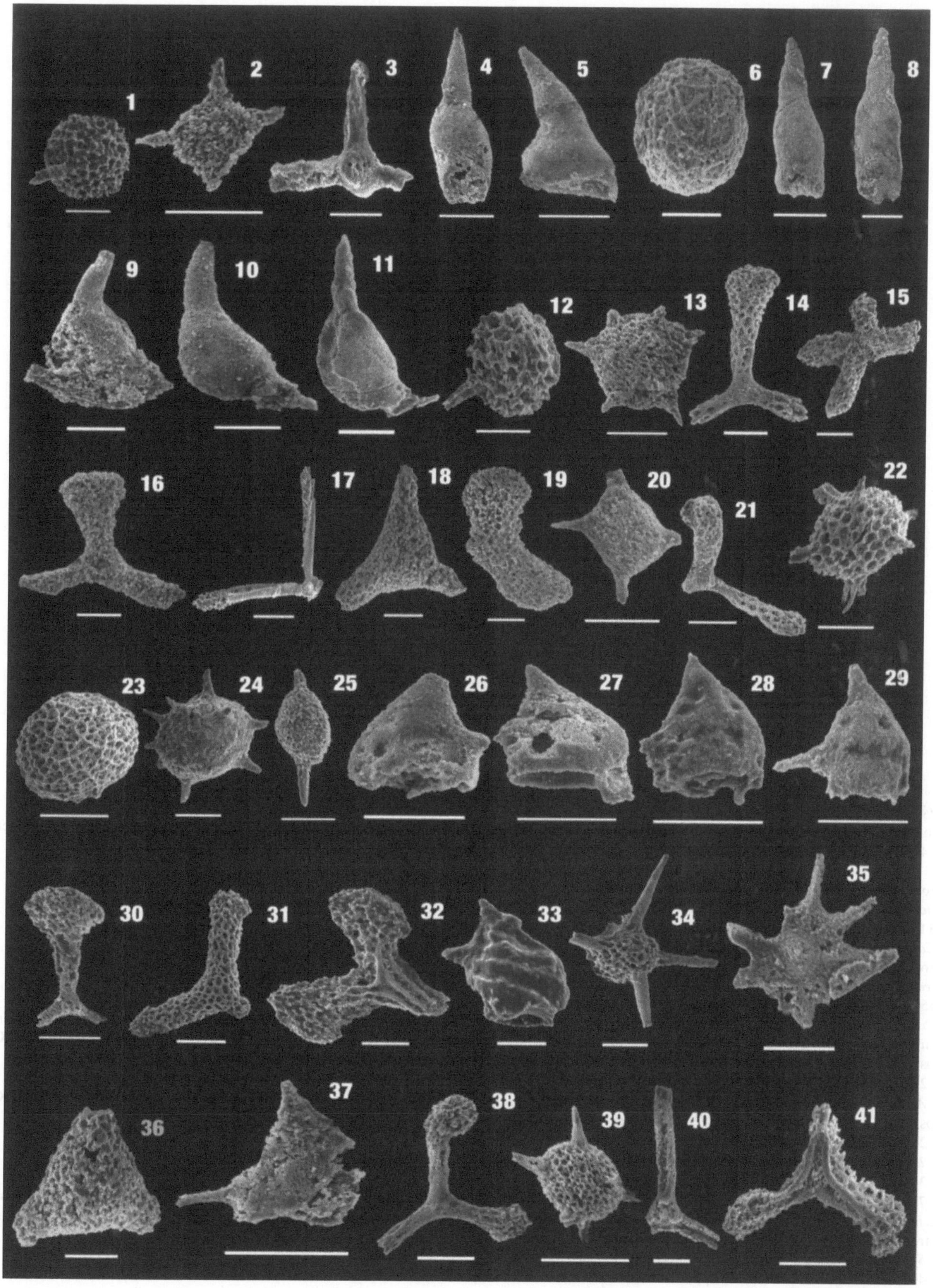
1
2
3
4
5
6
7
8
9
10
11
12
13
14
15
16
17
18
19
20
21
22
23
24
25
26
27
28
29
30
31
32
33
34
35
36
37
38
39
40
41

Plate 2

Middle to Upper Triassic radiolarians from Mapadolo (sample location 3) (figs. 1–4), Dabatonay (sample location 2) (figs. 5–35) and Mabudyen (sample location 12) (figs. 36–40), Philippines. Scale bars = 100 μm.

Fig. 1 from HKU-PLWN02031101, upper Anisian-Ladinian
Crucella sp.

Figs. 2, 3 from HKU-PLWN02031102, Carnian:
2. *Capnuchosphaera deweveri* KOZUR & MOSTLER, emend. BLOME; 3. *Capnuchosphaera missionensis* CORDEY

Fig. 4 from HKU-PLWN02031103, upper Norian-Rhaetian?
Haeckelicyrtium? sp.

Figs. 5–15 from HKU-PLWN02031109, uppermost Anisian-Ladinian:
5. *Pentactinocarpus fusiformis* DUMITRICA; 6. *Archaeocenosphaera* (?) sp.; 7. *Triassocampe* sp.; 8. *Spinotriassocampe hungarica* KOZUR; 9. *Eptingium manfredi* DUMITRICA; 10. *Spongostephanidium longispinosum* SASHID; 11. *Tiborella* sp.; 12. *Parasepsagon* sp.; 13. *Pseudostylosphaera longispinosa* KOZUR & MOSTLER; 14. *Pseudostylosphaera compacta* (NAKASEKO & NISHIMURA); 15. *Pseudostylosphaera magnispinosa* YEH

Figs. 16–22 from HKU-PLWN02031110, upper Anisian-Ladinian:
16. *Archaeocenosphaera* (?) sp.; 17, 18. *Poulpus* spp.; 19. *Pentactinocarpus fusiformis* DUMITRICA; 20. *Oertlispongus inaequispinosus* DUMITRICA, KOZUR & MOSTLER; 21. *Triassocampe deweveri* (NAKASEKO & NISHIMURA); 22. *Triassocampe coronata* BRAGIN

Figs. 23–26 from HKU-PLWN02031111, upper Anisian-Ladinian:
23. ?*Archaeocenosphaera* sp.; 24. *Pentactinocarpus fusiformis* DUMITRICA; 25. *Hozmadia* sp.; 26. *Pseudostylosphaera magnispinosa* YEH

Figs. 27–34 from HKU-PLWN02031114, upper Anisian-Ladinian:
27. *Triassocampe coronata* BRAGIN; 28. *Spinotriassocampe hungarica* KOZUR; 29. *Pseudostylosphaera japonica* (NAKASEKO & NISHIMURA); 30. *Pentactinocarpus tetracanthus* DUMITRICA; 31. *Pentactinocarpus fusiformis* DUMITRICA; 32. *Eptingium manfredi* DUMITRICA; 33. *Parasepsagon* sp.; 34. *Poulpus* sp.

Fig. 35 from HKU-PLWN 02031116, upper Anisian-Ladinian:
35. *Spinotriassocampe annulata* (NAKASEKO & NISHIMURA)

Figs. 36–40 from HKU-PLWN02031202, upper Norian-Rhaetian:
36. *Praemesosaturnalis huxleyensis* (CARTER); 37. *Praemesosaturnalis gracilis* (KOZUR & MOSTLER); 38–40. *Praemesosaturnalis* spp.

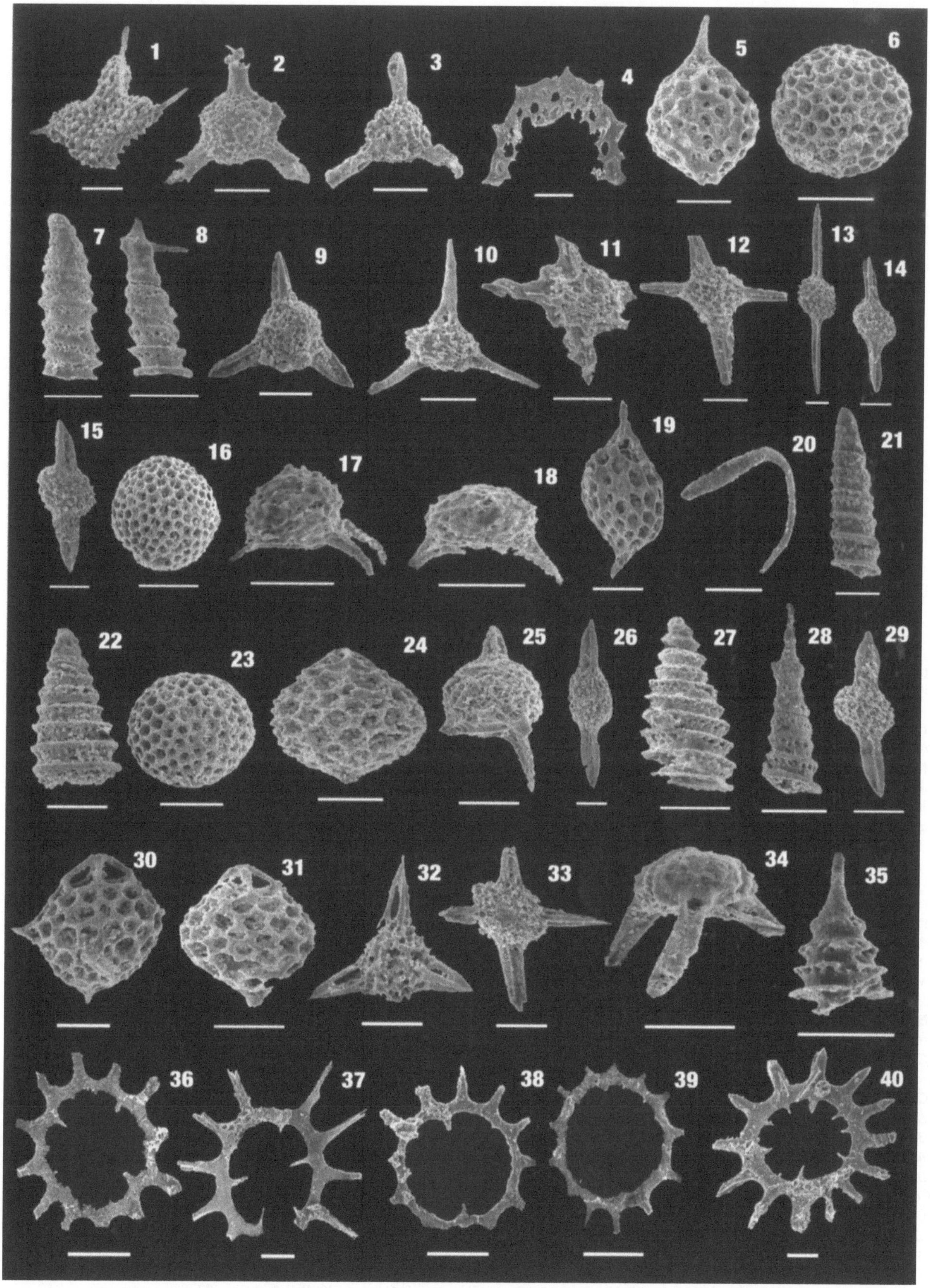
1
2
3
4
5
6
7
8
9
10
11
12
13
14
15
16
17
18
19
20
21
22
23
24
25
26
27
28
29
30
31
32
33
34
35
36
37
38
39
40

Plate 3

Middle Triassic to Lower Jurassic radiolarians from Mabudyen (sample location 12) (figs. 1, 2), Sitio Tingil (sample location 11) (figs. 3–6), Marily (Figs. 7–15) and Dimanglet (sample location 8) (figs. 16–37), Philippines. Scale bars = 100 μm.

Figs. 1, 2 from HKU-PLWN02031202, upper Norian-Rhaetian:
1. *Praemesosaturnalis rugosus* (YEH); 2. *Paleosaturnalis* sp. cf. *P. dotti* (BLOME)
Figs. 3–5 from HKU-PLWN02031210, upper Anisian-lower Ladinian:
3, 4. *Oertlispongus diacanthus* SUGIYAMA; 5. *Astrocentrus* sp. cf. *A. pulcher* KOZUR & MOSTLER
Fig. 6 from HKU-PLWN02031220, upper Carnian-upper Norian:
6. *Capnodoce* sp. cf. *C. anapetes* DE WEVER
Figs. 7–9, from HKU-PLWN02031302, Hettangian-Sinemurian?:
7, 8. *Parahsuum simplum* YAO; 9. *Sethocapsa* sp.;
Figs. 10, 11 from HKU-PLWN02031303, Hettangian-Sinemurian?:
10, 11. *Archicapsa* sp.
Figs. 12–15 from HKU-PLWN02031304, Hettangian-Sinemurian?:
12. *Pantanellium ultrasincerum* BLOME; 13. *Haliomma* sp.; 14. *Trilonche* sp.; 15. *Sethocapsa* sp.
Figs. 16–19 from HKU-PLWN02031402, Ladinian-lower Carnian:
16. *Napora* sp.; 17. *Silicarmiger* sp.; 18. *Muelleritortis* sp.; 19. *Pararuesticyrtium* sp.
Figs. 20–23 from HKU-PLWN02031405, upper Carnian-lower Norian:
20. *Haeckelicyrtium*? sp.; 21. *Latium* sp.; 22. *Capnuchosphaera lea* DE WEVER; 23. *Pachus multinodosus* TEKIN
Fig. 24 from HKU-PLWN02031407, upper Carnian-lower Norian
24. *Capnuchosphaera lea* DE WEVER
Fig. 25 from HKU-PLWN02031413, upper Carnian-Norian
25. *Capnodoce extenta* BLOME
Figs. 26–28 from HKU-PLWN02031414, upper Norian-Rhaetian:
26. *Betraccium maclearni* PESSAGNO & BLOME; 27. *Betraccium deweveri* PESSAGNO & BLOME; 28. *Praemesosaturnalis sandspitense* (BLOME)
Fig. 29 from HKU-PLWN02031418, upper Norian-Rhaetian:
29. *Betraccium deweveri* PESSAGNO & BLOME
Fig. 30 from HKU-PLWN02031419, upper Norian-Rhaetian:
30. *Canoptum* sp.
Figs. 31, 32 from HKU-PLWN02031420, upper Norian-Rhaetian:
31. *Acanthosphaera* sp.; 32. *Livarella magna* TEKIN
Figs. 34–37 from HKU-PLWN02031430, Sinemurian
33. *Archaeodictyomitra* sp.; 34. *Parahsuum* sp.; 35. *Katroma westermanni* WHALEN & CARTER; 36. *Podobursa* sp.; 37. *Sethocapsa* sp.

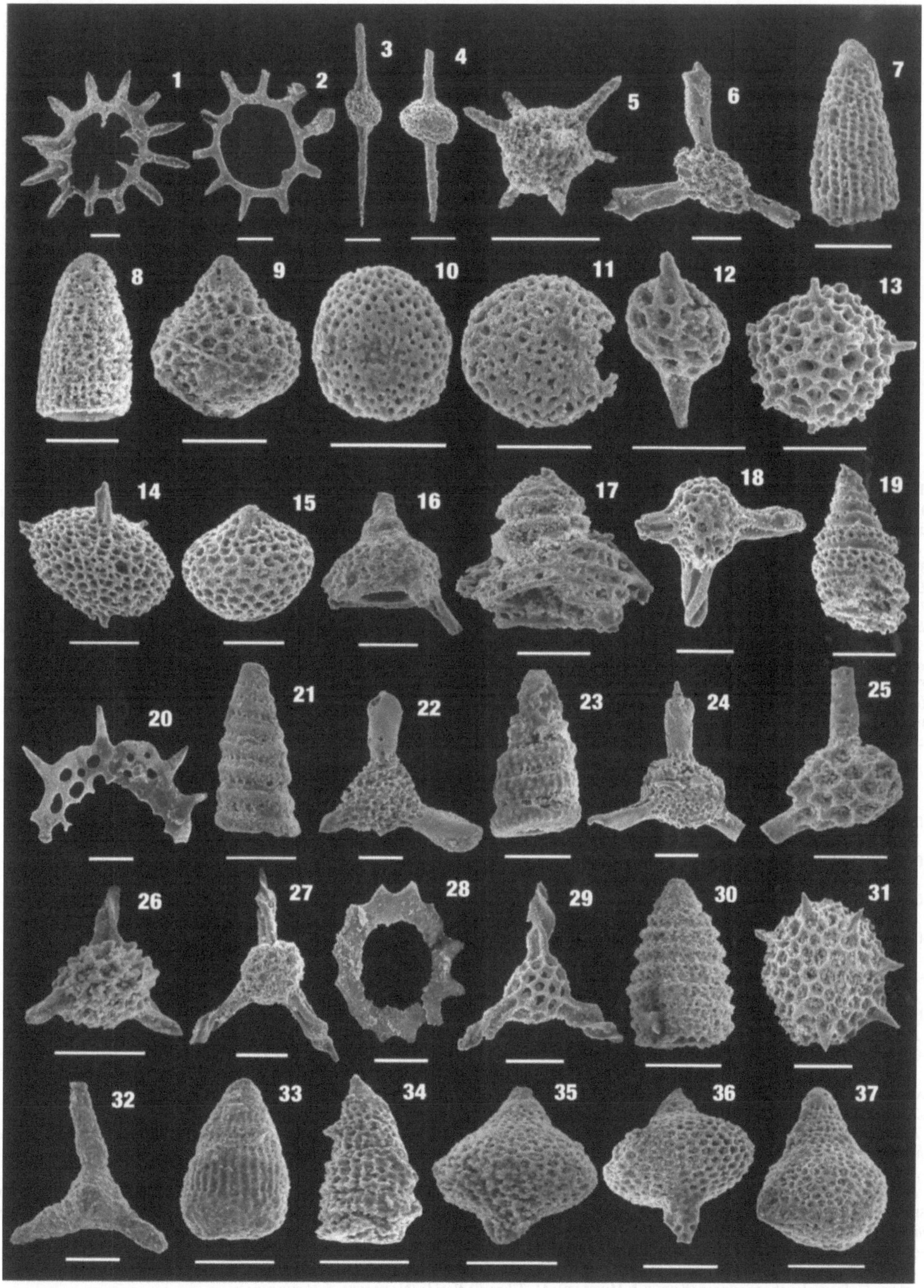

1
2
3
4
5
6
7
8
9
10
11
12
13
14
15
16
17
18
19
20
21
22
23
24
25
26
27
28
29
30
31
32
33
34
35
36
37

Plate 4

Middle Triassic to Lower Jurassic radiolarians from Dimanglet (sample location 8) (figs. 1–15), Mayanpayan (sample location 7) (figs. 16, 17), Lusong (sample location 5) (figs. 18, 19), Calauit-Illultuk Bay (sample location 1) (figs. 20–21) and Dipuyay-Decalangwang River, Busuanga (sample location 4) (figs. 22–44), Philippines. Scale bars = 100 µm.

Figs. 1–3 from HKU-PLWN02031701, upper Carnian-Norian:
1. *Eptingium* sp.; 2. *Japonocampe* sp.; 3. *Corum* sp. cf. *C. delgado* SUGIYAMA

Figs. 4–7 from HKU-PLWN02031702, Ladinian-lower Carnian:
4. *Busuanga* (?) sp.; 5. *Pseudostylosphaera compacta* (NAKASEKO & NISHIMURA); 6. *Vinassaspongus subsphaericus* KOZUR & MOSTLER; 7. *Muelleritortis cochleata* (NAKASEKO & NISHIMURA)

Fig. 8 from HKU-PLWN02031705, upper Norian-Rhaetian:
8. *Paroertlispongus* sp

Figs. 9, 10 from HKU-PLWN02031706, upper Norian-Rhaetian:
9, 10. *Praemesosaturnalis* sp. cf. *P. finchi* (Pessagno)

Figs. 11–15 from HKU-PLWN02031707, Bajocian?, could be younger:
11, 12. *Williriedellum* sp.; 13–15. *Parahsuum* sp.

Figs. 16, 17 from HKU-PLWN02031713, Sinemurian:
16. *Pleesus* sp. cf. *P. aptus* YEH; 17. *Katroma coliforme* (HORI)

Fig. 18 from HKU-PLWN02031802, upper Carnian-lower Norian:
18. *Canoptum farawayense* BLOME

Fig. 19 from HKU-PLWN02031803, upper Carnian-lower Norian:
19. *Xipotheca* sp. cf. *X. rugosa* BRAGIN, emend. TEKIN

Fig. 20 from HKU-PLWN02031904, upper Anisian-Ladinian:
20. *Rikivatella* sp.

Fig. 21 from HKU-PLWN02031905, upper Anisian- lower Ladinian:
21. *Triassocampe tulbuanensis* TUMANDA

Figs. 22–24 from HKU-PLWN02032013, upper Anisian-Ladinian:
22. *Paroertlispongus* sp.; 23. *Triassocampe* sp.; 24. *Triassocampe deweveri* (NAKASEKO & NISHIMURA)

Fig. 25 from HKU-PLWN02032018, upper Anisian-Ladinian:
25. *Praeconocaryomma* sp.

Figs. 26–28 from HKU-PLWN02032020, upper Anisian-Ladinian:
26. *Pseudostylosphaera timorensis* SASHIDA & KAMATA; 27. *Hindeosphaera spinulosa* (NAKASEKO & NISHIMURA); 28. *Spongostephanidium japonicum* (NAKASEKO & NISHIMURA)

Figs. 29–32 from HKU-PLWN02032021, upper Anisian-Ladinian:
29. *Pararuesticyrtium* sp.; 30. *Tiborella florida* (NAKASEKO & NISHIMURA); 31. *Eptingium nakasekoi* KOZUR & MOSTLER; 32. *Cryptostephanidium cornigerum* DUMITRICA

Figs. 33–39 from HKU-PLWN02032022, LADINIAN:
33. *Pentactinocapsa awaensis* NAKASEKO & NISHIMURA; 34. *Hozmadia rotunda* NAKASEKO & NISHIMURA; 35. *Spongostephanidium japonicum* (NAKASEKO & NISHIMURA); 36. *Pseudostylosphaera japonica* (NAKASEKO & NISHIMURA); 37. *Parasepsagon* sp.; 38. *Pseudostylosphaera compacta* (NAKASEKO & NISHIMURA); *Triassocampe coronata* BRAGIN

Figs. 40–44 from HKU-PLWN02032023, upper Anisian-Ladinian:
40. *Hindeosphaera spinulosa* (NAKASEKO & NISHIMURA); 41. *Pseudostylosphaera magnispinosa* YEH; 42. *Tiborella magnidentata* Dumitrica, KOZUR & MOSTLER; 43. *Staurolonche trispinosa* (KOZUR & MOSTLER); 44. *Tiborella florida* (NAKASEKO & NISHIMURA)

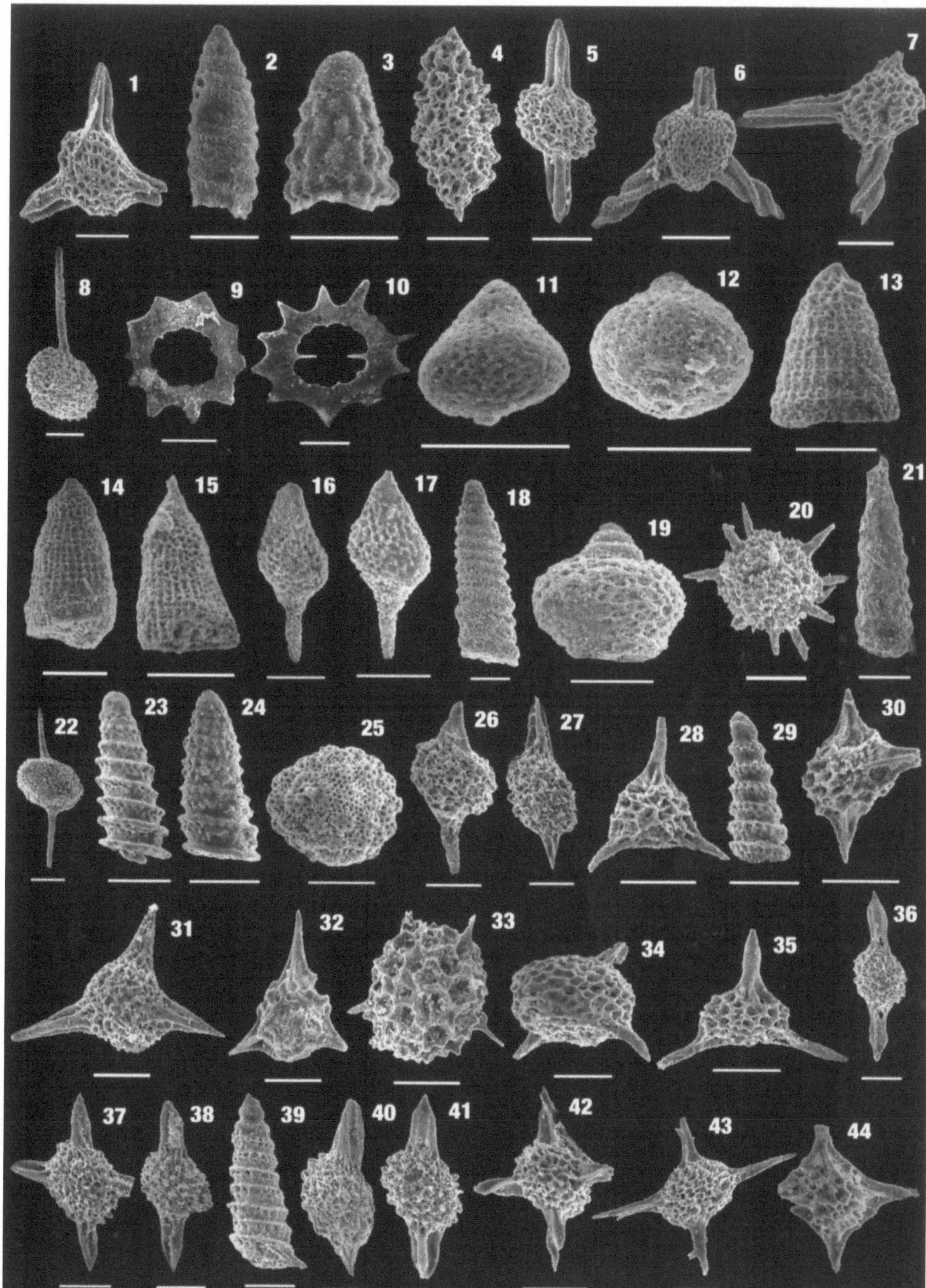

1
2
3
4
5
6
7
8
9
10
11
12
13
14
15
16
17
18
19
20
21
22
23
24
25
26
27
28
29
30
31
32
33
34
35
36
37
38
39
40
41
42
43
44

Plate 5

Middle to Upper Triassic radiolarians from Dipuyay-Decalangwang River (sample location 4), Busuanga, Philippines. Scale bars = 100 μm.

Figs. 1–4 from HKU-PLWN02032025, upper Anisian-Ladinian:
1. *Acanthosphaera* (?) *mocki* KOZUR & MOSTLER; 2. *Thaisphaera* sp.; 3. *Pararuesticyrtium* sp.; 4. *Eptingium ramovsi* KOZUR et al.

Figs. 5–9 from HKU-PLWN02032029, upper Anisian-Ladinian:
5. *Triassocampe coronata* BRAGIN; 6. *Staurolonche* sp.; 7. *Staurolonche trispinosa* (KOZUR & MOSTLER); 8. *Sarla* sp.; 9. *Triassistephanidium laticorne* DUMITRICA

Figs. 10–18 from HKU-PLWN02032027, upper Anisian-Ladinian:
10. *Pentactinocapsa awaensis* NAKASEKO & NISHIMURA; 11, 12. *Hozmadia reticulata* DUMITRICA, KOZUR & MOSTLER; 13. *Hexalonche* sp.; 14. *Pseudostylosphaera goricanae* KOZUR ET AL.; 15. *Epigondollela* sp.; 16. *Tiborella* sp.; 17. *Plafkerium abboti* PESSAGNO; 18. *Parasepsagon variabilis* NAKASEKO & NISHIMURA

Figs. 19–26 from HKU-PLWN02032031, upper Anisian-Ladinian:
19. *Pentactinocapsa awaensis* NAKASEKO & NISHIMURA; 20. *Hindeosphaera spinulosa* (NAKASEKO & NISHIMURA); 21. *Pseudostylosphaera timorensis* SASHIDA & KAMATA; 22, 23. *Pseudostylosphaera compacta* (NAKASEKO & NISHIMURA); 24. *Pseudostylosphaera japonica* (NAKASEKO & NISHIMURA); 25. *Triassocampe deweveri* (NAKASEKO & NISHIMURA); 26. *Triassocampe coronata* BRAGIN

Figs. 27, 28 from HKU-PLWN02032037, upper Anisian-Ladinian:
27. *Triassocampe scalaris* DUMITRICA, KOZUR & MOSTLER; 28. *Spinotriassocampe annulata* (NAKASEKO & NISHIMURA)

Figs. 29–31 from HKU-PLWN02032041, upper Anisian-Ladinian:
29, 30. *Hozmadia reticulata* DUMITRICA, KOZUR & MOSTLER; 31. *Napora* sp.

Figs. 32, 33 from HKU-PLWN02032049, upper Anisian-Ladinian:
32. *Cryptostephanidium cornigerum* DUMITRICA; 33. *Ladinocampe japonica* (NAKASEKO & NISHIMURA)

Fig. 34 from HKU-PLWN02032050, upper Anisian-Ladinian:
34. *Bulbocyrtium* sp.

Figs. 35–41 from HKU-PLWN02032052, upper Anisian-Ladinian:
35. *Sepsagon* sp.; 36, 38. *Hozmadia* sp.; 37. *Archaeocenosphaera* sp.; 39. *Hozmadia rotunda* (NAKASEKO & NISHIMURA); 40. *Tiborella* sp.; *Katorella bifurcata* KOZUR & MOSTLER

Figs. 42, 43 from HKU-PLWN02032057, upper Anisian-lower Ladinian:
42. *Archaeocenosphaera* sp.; 43. *Pentactinocorbis kozuri* DUMITRICA

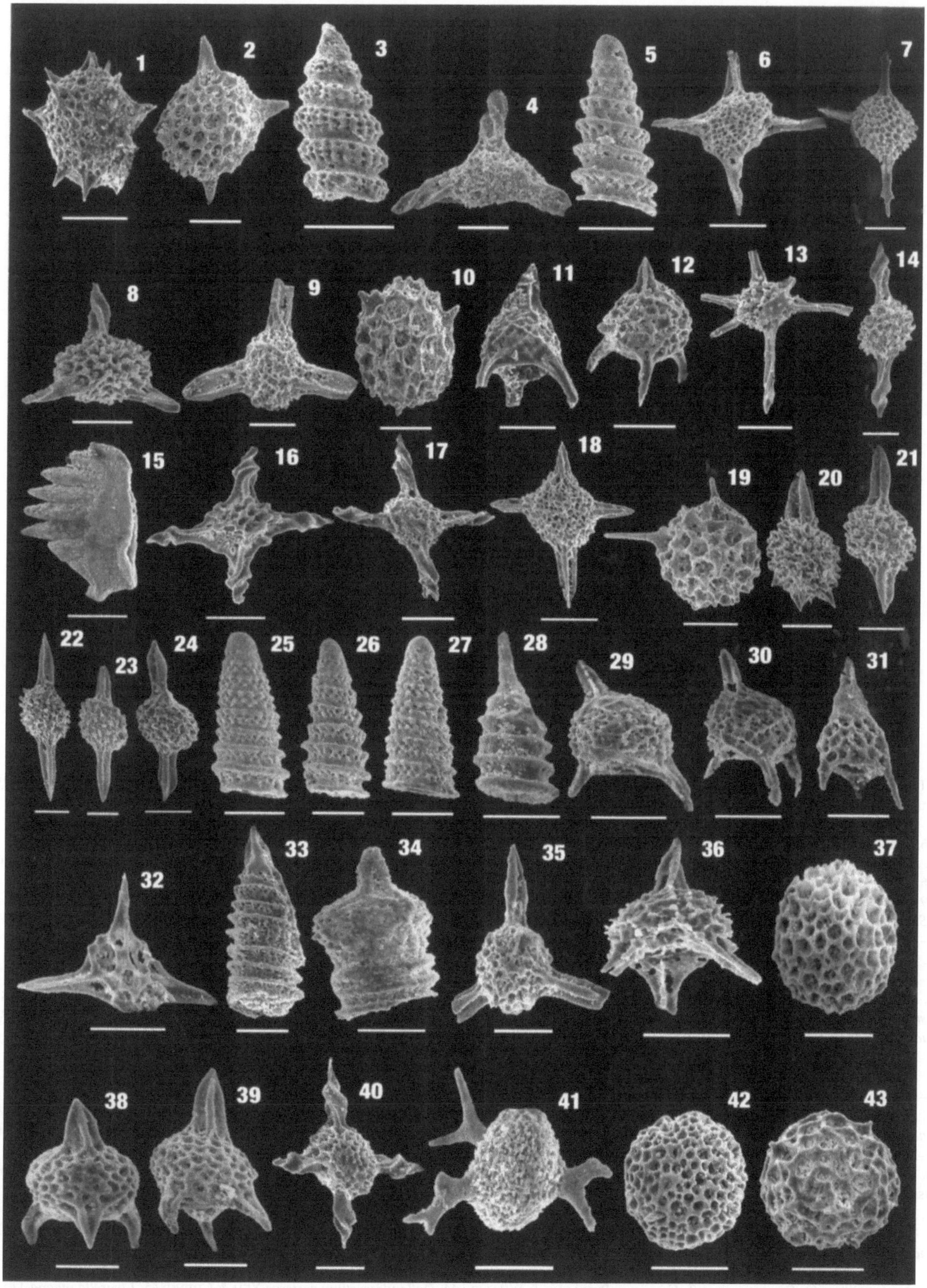
1
2
3
4
5
6
7
8
9
10
11
12
13
14
15
16
17
18
19
20
21
22
23
24
25
26
27
28
29
30
31
32
33
34
35
36
37
38
39
40
41
42
43

Plate 6

Middle Triassic to Middle Jurassic radiolarians from Dipuyay-Decalangwang River (sample location 4) (figs. 1–9), Calindo (sample location 10) (figs. 10–41) and a road cut, 3.5 kms east of the township of Coron (sample location 9) (figs. 42–45). Scale bars = 100 µm.

Figs. 1–6 from HKU-PLWN02032057, upper Anisian-lower Ladinian:
1a. *Hozmadia spinifera* Sugiyama; 1b. *Hindeosphaera spinulosa* (Nakaseko & Nishimura); 2, 3. *Paratriassocampe* sp.; 4. *Triassocampe coronata* Bragin; 5, 6. *Tiborella florida* (Nakaseko & Nishimura)

Figs. 7–9 from HKU-PLWN02032060, upper Anisian-Ladinian:
7. *Stauroloncha trispinosa* (Kozur & Mostler); 8. *Hozmadia rotunda* (Nakaseko & Nishimura); 9. *Hindeosphaera spinulosa* (Nakaseko & Nishimura)

Figs. 10–13 from HKU-PLWN02032101, upper Bajocian-uppermost middle Bathonian:
10. *Stichocapsa* sp.; 11. *Linaresia* sp.; 12. *Staurolonche* sp.; 13. *Transhsuum medium* Takemura

Figs. 14–22 from HKU-PLWN02032102, upper Anisian-Ladinian:
14. *Paratriassocampe* sp.; 15. *Pseudostylosphaera longispinosa* Kozur & Mostler; 16. *Pseudostylosphaera compacta* (Nakaseko & Nishimura); 17. *Spongostylus tricostatus* Kozur, Krainer & Mostler; 18. *Eptingium manfredi* Dumitrica; 19. *Hozmadia reticulata* Dumitrica, Kozur & Mostler; 20. *Hozmadia rotunda* (Nakaseko & Nishimura; 21. ?*Hexalonche* sp.; 22. *Staurolonche* sp.

Figs. 23–25 from HKU-PLWN02032106, Ladinian:
23. *Eptingium manfredi* Dumitrica; 24. *Stauracontium* sp.; 25. *Pentaspongodiscus* sp.

Figs. 26–32 from HKU-PLWN02032109, upper Anisian-Ladinian:
26. *Hozmadia rotunda* (Nakaseko & Nishimura); 27. *Spongosilicarmiger* sp.; 28. *Hindeosphaera spinulosa* (Nakaseko & Nishimura); 29. *Hozmadia rotunda* (Nakaseko & Nishimura); 30. *Triassocampe coronata* Bragin; 31. *Paratriassocampe* sp.; 32. *Parasepsagon variabilis* (Nakaseko & Nishimura)

Figs. 33–36, 38 from HKU-PLWN02032110, upper Anisian-Ladinian:
33. *Pentactinocapsa awaensis* Nakaseko & Nishimura; 34. ?*Trilonche* sp.; 35. *Celluronta* sp.; 36. *Bulbocyrtium* sp.; 38. *Tiborella florida* (Nakaseko & Nishimura)

Figs. 37, 39–41 from HKU-PLWN02032117, upper Anisian-Ladinian:
37. ?*Ladinocampe* sp.; 39. *Spinotriassocampe annulata* (Nakaseko & Nishimura); 40. *Spongostephanidium japonicum* (Nakaseko & Nishimura); 41. *Hozmadia* sp.

Figs. 42–45 from HKU-PLWN02032118, upper Norian-Rhaetian:
42. *Canoptum* sp.; 43. *Canoptum* sp. aff. *C. unicum* Pessagno & Whalen; 44. *Deflandrecyrtium carterae* Yeh & Cheng; 45. *Livarella magna* Tekin

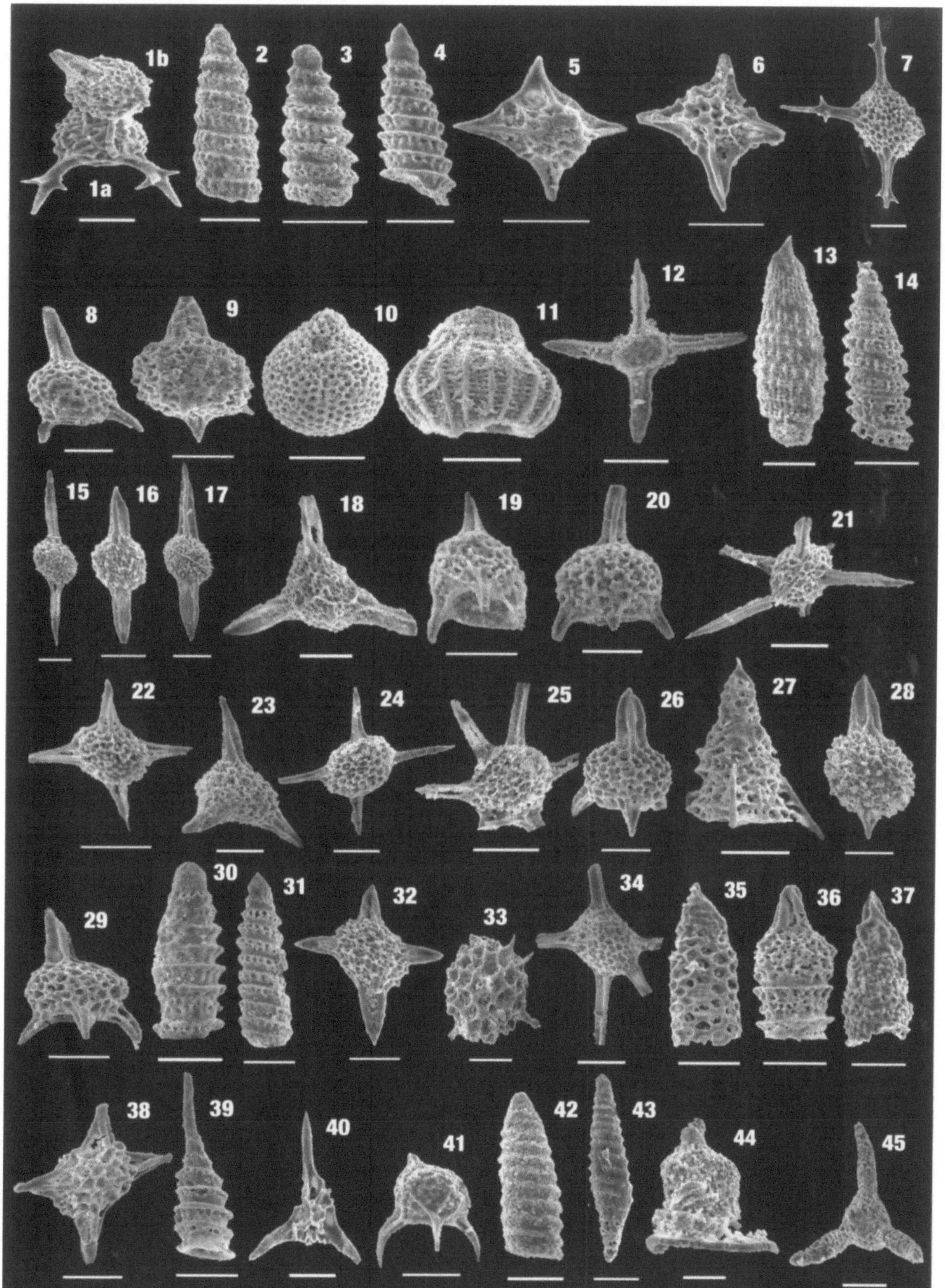
1b
1a
2
3
4
5
6
7
8
9
10
11
12
13
14
15
16
17
18
19
20
21
22
23
24
25
26
27
28
29
30
31
32
33
34
35
36
37
38
39
40
41
42
43
44
45

0012-9402/06/01S127-6
DOI 10.1007/s00015-006-0601-6
Birkhäuser Verlag, Basel, 2006

Eclogae geol. Helv. 99 (2006) Supplement 1, S127–S132

Influence of the Frasnian-Famennian event on radiolarian faunas

YU-JING WANG[1], HUI LUO[1] & JONATHAN C. AITCHISON[2]

Key words: Radiolaria, Frasnian, Famennian, extinction event, South China

ABSTRACT

The impact of the Frasnian-Famennian biotic mass extinction event (F-F event) on radiolarian faunas is examined in 13 sections, which belong to two types of cherty basin in South China. No appreciable decrease in radiolarian biodiversity is observed across the F-F boundary. Indeed, radiolarian faunas flourished during the Famennian. The F-F event may not have had any dramatic effect on the radiolarian faunas in deep water and the disappearance of Famennian radiolarians in shallow water platform sections may simply be a result of sea level change rather than any biotic crisis.

Introduction

The F-F event is one of the great biotic mass extinction events that occurred during the Phanerozoic. It took place between late Frasnian and Famennian time in the Late Devonian and its scale was of only marginally lesser magnitude than the largest mass extinction, which occurred at the P/T boundary. The F-F event caused a great extinction or rapid diminution in the abundance of marine invertebrate animals, such as stromatoporoids, brachiopods, rugose corals, tabulate corals, trilobites, ammonoids, tentaculitids, ostracods, conodonts and so on. Over 60% of total taxa of these animals disappeared at the end of the Frasnian (Xian et al. 1995). Based on Frasnian-Famennian radiolarian materials collected from South China and those reported from elsewhere, we discuss the effects of the F-F biotic mass extinction event on radiolarian faunas.

Stratigraphy and lithologic characteristics

Our radiolarian samples were gathered from thirteen stratigraphic sections containing cherty rocks distributed in Yunnan (6 sections), Guizhou (1 section) and Guangxi (6 sections) respectively in South China (Fig. 1). The general characteristics of these sections include an age range from Frasnian to Famennian, rocks composed of thin-bedded cherts associated with mudstone and shale, and some rich radiolarian faunas. According to the developmental background of the depositional basin and the characteristics of the fossil assemblages, these 13 stratigraghic sections may be divided into two basin types, namely the open-ocean cherty basin type and the cherty platform basin type (Fig. 2). These two cherty basins experienced different developmental histories during the F-F biotic mass extinction event.

The open-ocean cherty basin indicates deep-water basin deposition except for minor platform and the slope sediments. This is a basin of compensative lack that experienced the same deep-water anoxic conditions for a long time (possibly from Silurian to Permian or Triassic). The Devonian sections at the Shiti Reservoir, Bancheng, Qinzhou, Guangxi and in the Changning-Menglian terrane, west Yunnan are considered to be the typical representatives of this kind of basin. They contain continuous radiolarian zones composed of the Frasnian *Helenifore laticlavium* and *Helenifore robustum* zones and the Famennian *Holoeciscus foremanae* zone. The cherty platform basin developed as a result of siliceous pelagic sedimentation within a platform basin. This basin experienced slow sedimentation in an anoxic environment during periods of the platforms development (e.g. during the early or late Frasnian), owing to sea level rise caused by transgression in the basin. With the exception of the sections in the Qinzhou area, Guangxi and in Changning-Menglian terrane, Yunnan, the other stratigraphic sections cited-above belong to this basin type. In these sections, there are no continuous radiolarian zones. Generally, only one fossil zone, such as, the lower or

[1] Nanjing Institute of Geology and Palaeontology, Chinese Academy of Sciences, Nanjing 210008,China. Email: huiluo@nigpas.ac.cn
[2] Department of Earth Sciences, University of Hong Kong, Pokfulam Road, Hong Kong SAR, China

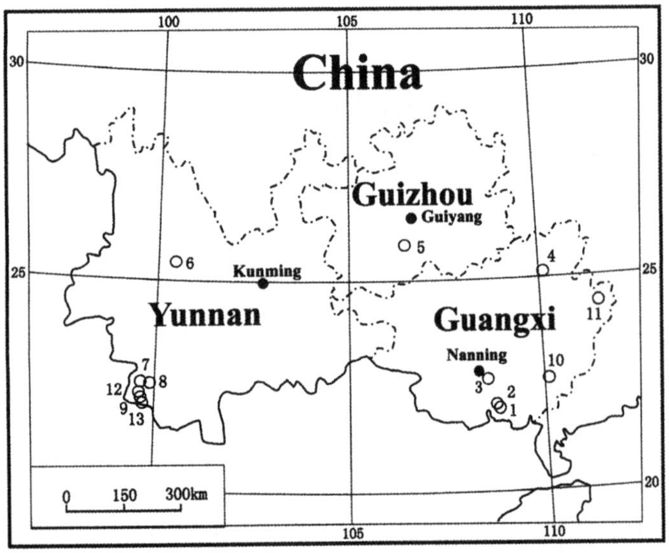

1. East section of Shiti Reservoir, Qinzhou, Guangxi
2. West section of Shiti Reservoir, Qinzhou, Guangxi
3. Wuxiangling section, Nanning, Guangxi
4. Yangdi section, Guilin, Guangxi
5. Bazhai section, Ziyun, Guangxi
6. Shaijingpo section, Xiangyun, Yunnan
7. Lila section, Lancang, Yunnan
8. Taierbu section, Lancang, Yunnan
9. Huiku section, Menglian, Yunnan
10. Niuyunling section, Yulin, Guangxi
11. Etang section, Hexian, Guangxi
12. Nanya section, Menglian, Yunnan
13. Ali section, Lancang, Yunnan

Fig. 1. Map showing the locations of sections studied through Upper Devonian cherty strata in South China.

upper Frasnian *Helenifore laticlavium* or *Helenifore robustum* zone can be distinguished. Moreover, Famennian radiolarians and radiolarian zones have never been discovered. From middle Frasnian to Famennian, the radiolarian-bearing cherts in the cherty platform basin were gradually replaced by phacoidal limestone due to regression. The timing of this facies change differed between areas. It occurred in the middle Frasnian in SE Guangxi and in the Famennian in SW Guizhou.

Age assignment

Upper Devonian radiolarian faunas

Our recent study suggests that the species *Helenifore laticlavium* identified by various scholars may be separated into two stratigraphically and morphologically distinct species (Wang et al. 2003). *Helenifore laticlavium* NAZAROV & ORMISTON (1983) is characterized by a shell with a thinner, nearly circular plate-ring, two spines (top and base spines) and an opening on one extremity of the ring (Plate 1, Figs. 1–5). This species occurs in the Gogo Formation, Canning Basin, western Australia (Nazarov & Ormiston 1983) and the Shiti Reservoir Formation, west bank of the Shiti Reservoir, Bancheng countryside, Qinzhou, Guangxi (Wang et al. 2003). It is considered to be lower Frasnian. *Helenifore robustum* (Boundy-Sanders & Murchey 1999) is characterized by an elliptical shell with a narrower and thicker ring (Plate 1, Figs. 6–10). This is an upper Frasnian species, which is distributed worldwide. It has been reported from eastern Australia (Ishiga 1988; Ishiga et al. 1988; Stratford & Aitchison 1997), Thailand (Sashida et al. 1998), Nevada (Boundy-Sanders & Murchey 1999) and Guangxi, Guizhou, Yunnan in China (Wang et al. 1998; Wang et al. 2003). Therefore, we subdivide the Upper Devonian radiolarian fauna into 3 zones: Two zones (*Helenifore laticlavium* zone and *H. robustum* zone) are Frasnian and one zone (*Holoeciscus foremanae* zone) is Famennian.

Correlation between Upper Devonian radiolarian and conodont zones

At present 15 typical Upper Devonian conodont zones are erected (7 zones belonging to Frasnian and 8 zones to Famennian, respectively) (Fig. 3) (Wang Cheng-yuang 2000). The *Helenifore laticlavium* zone from the Gogo Formation of western Australia is associated with conodonts that include *Polygnathus asymmetrica* and *Ancyrodella rotundilobata rotundilobata.* It is comparable to the lower-middle part of the *Polygnathus asymmetrica* zone (Nazarov & Ormiston 1983). The *Helenifore robustum* zone occurs in the Slaven Chert of the Shoshone Range, Nevada, USA (Boundy-Sanders and Murchey 1999) and in the Tarbu area, Lachang county, western Yunnan, China (Wang et al. 2000). It is associated with conodonts that include *Palmatolepis eureka, P. rhenana masuta, P. aubrecta, Polygnathus brevilaminus*, P. aff. *timanicus* (Boundy-Sanders and Murchey 1999) or *Palmatolepis rhenana rhenana, P. rhenana nasuta, P. hasii* and *P.* sp. (Wang et al. 2000). Consequently, the radiolarian *Helenifore robustum* zone may be correlated with the conodont *Palmatolepis hasii* zone to *Palmatolepis rhenana* zone. The *Holoeciscus foremanae* radiolarian zone from the Frankenwald area, northern Bavaria, Germany is associated with the conodonts *Palmatolepis rugosa, P. glabra lepta, P. gracilis gracilis, P. perlobata schindewolfii, Branmmehla bohlenana, Pseudopolygnathus granulasus* and occurs in the *Palmatolepis trachytera* zone to lower *P. postera* zone in the middle Famennian (Kiessling & Tragelehn, 1994). The same radiolarian zone from the Shiti Reservoir Formation in Bancheng, Qinzhou, Guangxi is also associated with the conodonts *Palmatolepis glabra acuta, P. perlobata schindewolfii.* It is correlated with the Famennian *Palmatolepis crepida* zone to *P. marginifera* zone (Wang et al. 1998). Therefore, the *Holoeciscus foremanae* zone can be correlated with the *Palmatolepis crepida* zone to *P. postera* zone (Fig. 3).

Upper Devonian radiolarian diversity

According to our Upper Devonian radiolarian materials collected from South China (Wang 1997; Wang et al. 2003) and

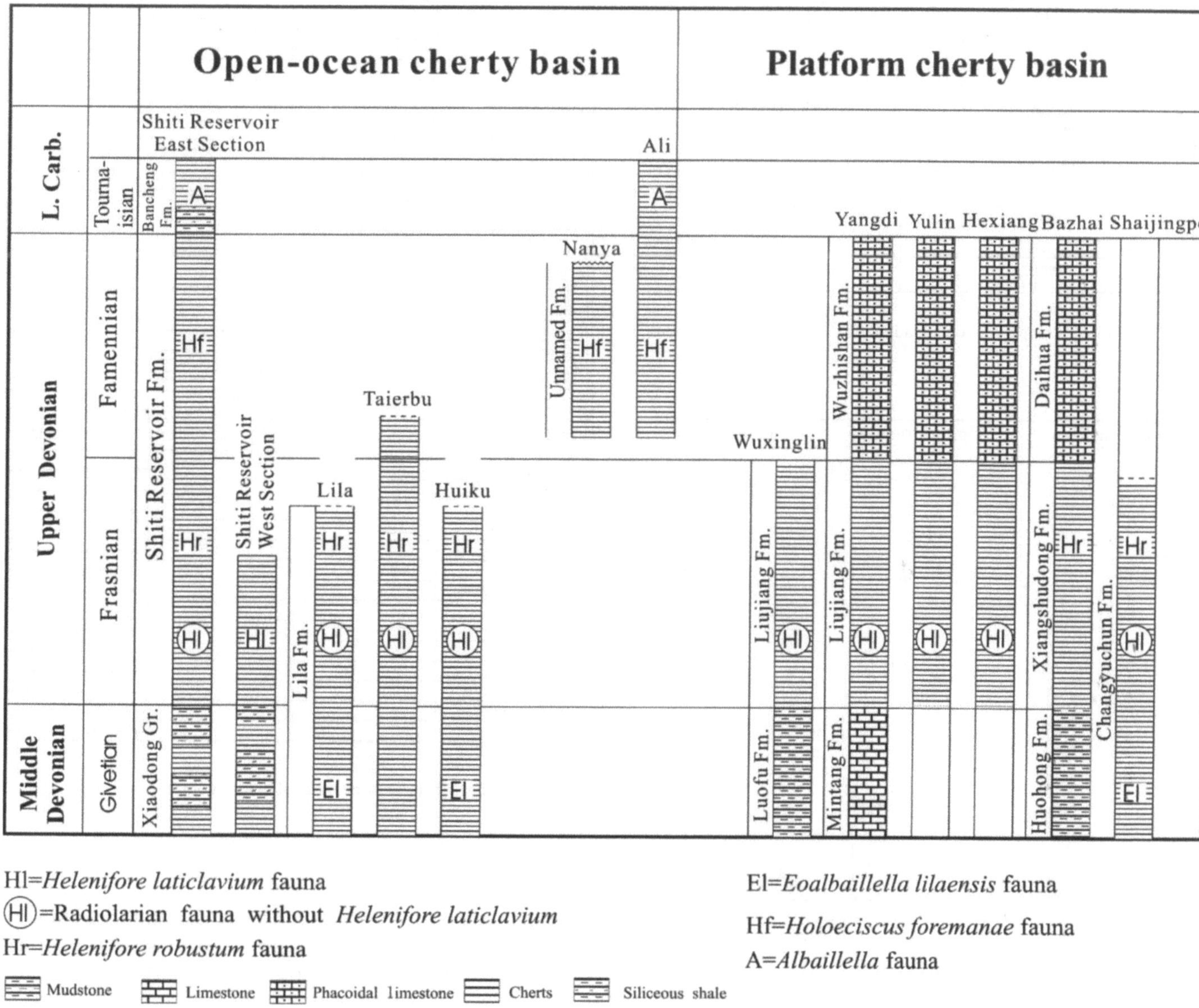

Fig. 2. Two different basin types containing cherty facies.

other published materials on Upper Devonian radiolarian faunas (Aitchison 1993; Boundy-Sanders et al. 1999; Cheng 1986; Foreman 1963; Holdsworth et al. 1978; Ishiga et al. 1988; Kiessling & Tagelehn 1994; Li & Wang 1991; Nazarov 1973, 1975; Nazarov & Ormiston 1983; Nazarov et al. 1982; Sashida et al. 1993; Schwartzapfel & Holdsworth 1996; Stratford & Aitchison 1997), we find that there are 77 and 35 species within the Frasnian *Helenifore laticlavium* zone and *H. robustum* zone respectively. They all belong to 15 genera and 8 families. Although these two faunas have the same number of families and genera, the number of species recognized in the *Helenifore robustum* zone is less half than that in the *H. laticlavium* zone. 10 genera and 12 species are same in both zones indicating a close affinity. However, in the Famennian *Holoeciscus foremanae* zone, 154 species, belonging to 28 genera and 10 families, are described. 18 species and 13 genera, as well as 8 families continued through from two Frasnian zones. 2 new families (Holoeciscidae and Pylentonemidae) and 14 new genera appear. The number of newborn genera is almost double the total number of radiolarian genera in the Frasnian. Although the species number decreased in Upper Frasnian compared to that of Lower Frasnian, the diversity of Famennian radiolarian families, genera and species are obviously higher than that of whole Frasnian (Fig. 4, Fig. 5).

	conodont zones	radiolarian zones
Famennian	*praesulcata*	
	expansa	
	postera	*Holoeciscus foremanae*
	trachytera	
	marginifera	
	rhomboidea	
	crepida	
	triangularis	
Frasnian	*linguiformis*	
	rhenana	*Helenifore robustum*
	jamieae	
	hassi	
	punctata	*Helenifore laticlavium*
	transitans	
	falsiovalis	

Fig. 3. Correlation of Upper Devonian radiolarian and conodont zonations.

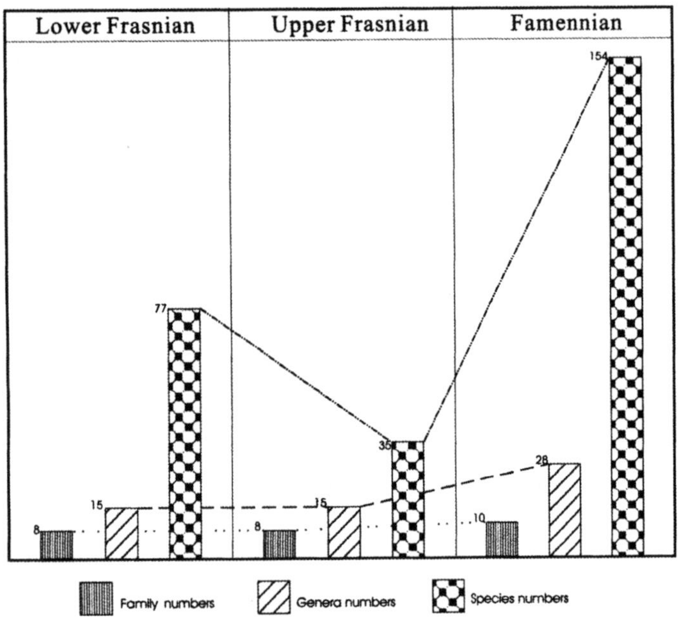

Fig. 4. Diversity changes of the radiolarian family, genera and species in Upper Devonian.

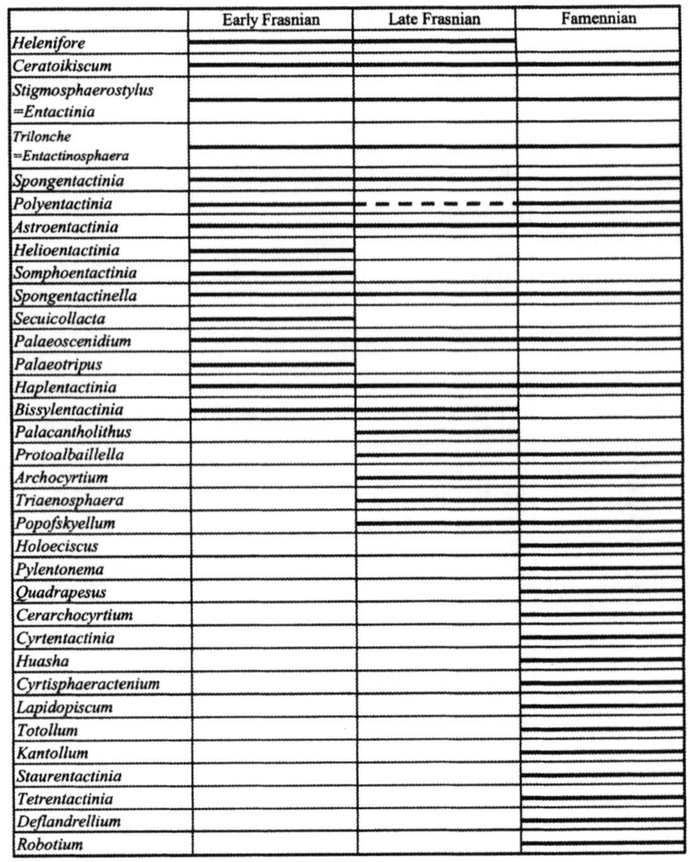

Fig. 5. Ranges of Upper Devonian radiolarian genera.

Influence of the F-F event on radiolarians

The materials mentioned above show that the F-F biotic mass extinction event did not have any marked negative effect upon the open ocean radiolarian fauna. In the open ocean cherty basin 8 upper Frasnian radiolarian families continue to flourish and only 2 out of 15 genera become extinct in the Famennian. No families and only 13% of genera become extinct. These figures are well below the minimum level of extinction (11% at family level and 20–25% of genera) regarded as indicative of a biotic mass extinction (Raup 1982). Elsewhere, in the cherty platform basin, the radiolarians belonging to *Helenifore laticlavium* and *H. robustum* zones thrived during the early and late Frasnian. However, as the radiolarian-bearing cherts were replaced by phacoidal limestone because of regression, no Famennian radiolarians are present. The disappearance of radiolarians from the platformal sections is obviously related to facies change rather than extinction. We note that in the Early Carboniferous, when deposition in the platform basin changed from limestone facies into a cherty facies again an albaillellid radiolarian fauna re-appeared in this platform basin.

Consequently, we conclude:

1. The F-F event may have only resulted in a major biotic mass extinction for biotas associated with shallow water reef facies and parts of the pelagic biota. It did not have a great influence on the radiolarian fauna in deep water and indeed radiolarians flourished during the Famennian.
2. The disappearance of radiolarians from shallow water platform sections may simply be a result of sea level change rather than any biotic crisis. In deeper water successions, radiolarians were less affected by sea level change. There was no appreciable decrease in radiolarian biodiversity across the F-F boundary.

In order to test this hypothesis we need to find and examine more Famennian platformal cherty successions.

Acknowledgements

We thank Dr. Martial Caridroit for his helpful review and comments on the manuscript. This study was supported by Chinese Academy of Sciences (Grant# KZCX2-SW-129), National Natural Science Foundation of China (Grant # 40172004), and the MOST-Pre 973 Project (2001CCA01800).

REFERENCES

AITCHISON, J.C. 1993: Devonian (Frasnian) radiolarians from the Gogo Formation, Canning Basin, Western Australia. Palaeontographica Abt. A 228(4–6), 105–128.

AITCHISON, J.C. & STRATFORD, J.M.C. 1997: Middle Devonian (Givetian) Radiolaria from Eastern New South Wales, Australia: a reassessment of the Hinde (1899) fauna. Neues Jahrbuch für Geologie und Paläontologie, Abhandlungen 203, 369–390.

AITCHISON, J.C., DAVIS, A.M., STRATFORD, J.M. & SPILLER, F.C. P. 1999: Lower and Middle Devonian radiolarian biozonation of the Gamilaroi terrane New England Orogen, eastern Australia. Micropaleontology 45, 138–162.

BOUNDY-SANDERS, S.Q., SANDBERG, C.A., MURCHEY, B.L. & HARRIS, A.G. 1999: A late Frasnian (Late Devonian) radiolarian, sponge spicule and conodont fauna from the Slaven Chert, northern Shoshone Range, Roberts Mountains allochthon, Nevada Micropaleontolgy 45(1), 62–68.

CHENG, Y.N. 1986: Taxonomic Studies on Upper Paleozoic Radiolaria. Nat. Mus. Nat. Sci. Spec. Publ. 1, 1–311.

FOREMAN, H.P. 1963: Upper Devonian Radiolaria from the Huron member of the Ohio shale. Micropaleontology 9(3), 267–304..

HOLDSWORTH, B.K., JONES, D.L. & ALLISON, C. 1978: Upper Devonian Radiolarians Separated from Chert of the Ford Lake Shale, Alaska. Jour. Research U. S. Geol. Survey 6(6), 775–788.

ISHIGA H., 1988: Paleontological study of radiolarians from the southern New England Fold Belt, Eastern Australia. In: IWASAKI, M., IIZUMI, S., WATANABE, T. & ISHIGA, H. (Eds.): Preliminary Report on the Geology of the New England Fold Belt, Australia. Co-operative Research Group of Japan and Australia, Matsue, Japan, 77–93.

ISHIGA H., LEITCH, E.C., NAKA, T., WATANABE, T. & IWASAKI, M. 1987: Late Devonian Paleoscenidiidae from the Hastings Block, New England Fold Belt, N.S.W., Australia. Earth Science (Chikyu Kagaku) 41(6), 297–302.

ISHIGA H., LEITCH, E.C., WATANABE, T., NAKA, T. & IWASAKI, M. 1988: Radiolarian and conodont biostratigraphy of siliceous rocks from the New England fold belt. Aust. J. Earth Sci. 35, 73–80.

KIESSLING, W.K. & TAGELEHN, H. 1994: Devonian Radiolarian Faunas of Conodont-Dated Localities in the Frankenwald (Northern Bavaria, Germany), Abhandlungen der geologisches Bundesanstalt in Wien, 50, 219–255.

LI, H. 1986: The Upper Paleozoic Radiolaria in Menlian County, Yunnan Province, Qinghai-Tibet Plateau. Research Symposium for the exploration in Hengduan Mountain 2, 8–15.

LI, Y.X., & WANG, Y.J. 1991: Upper Devonian (Frasnian) Radiolarian fauna from the Liukiang Formation, eastern and southeastern Guangxi. Acta Micropalaeontologica Sinica 8(4), 395–404.

NAZAROV, B.B. 1973: Pervye nakhodki radiolyarii Entactiniidae i Ceratoikiscidae v verkhnem devone yuzhnogo Urala (First occurrences of Radiolaria of the Entactiniidae and Ceratoikiscidae families in the upper Devonian of the Southern Urals). Doklady Akademii Nauk SSSR (Transactions of the Academy of Sciences, USSR) 210(3), 696–699.

NAZAROV, B.B. 1975: Radiolyarii nizhnego-srednego paleozoya Kazakhstana (metody issledovanii, sistematika, stratigraficheskoe znachenie (Lower and middle Paleozoic radiolarians of Kazakhstan (methods of investigation, systematics and stratigraphic significance)). In: RAABEN, M.E. (Ed.), Trudy Akademiya Nauk SSSR, Geologicheskii Institut (Transactions of the Academy of Sciences of the USSR, Geological Institute). Izdatelstvo Nauka, Moscow, USSR, 1–203.

NAZAROV, B.B. & ORMISTON, A.R. 1983: Upper Devonian (Frasnian) radiolarian fauna from the Gogo Formation, Canning Basin, Western Australia. Micropaleontology 29, 454–466.

NAZAROV, B.B., COCKBAIN, A.E. & PLAYFORD, P.E. 1982: Late Devonian Radiolaria from the Gogo Formation, Canning Basin,Western Australia. Alcheringa 6(3), 161–173.

RAUP, D.M. 1982: Biogeographic extinction: A feasibility test. In: SILVER, L. I. & SCHULTZ, P. H. (Eds.): Geological Implications of Impacts of Large Asteroids and Comets on the Earth. Geol. Soc. Amer. Spec. Pap. 190, 277–281.

SASHIDA, K. et al. 1993: Occurrences of Paleozoic and Mesozoic radiolarians from Thailand and Malaysia and its geologic significance (preliminary report). News of Osaka Micropaleontologists, Special Volume, 9, 1–17.

SCHWARTZAPFEL, J.A. & HOLDSWORTH, B.K. 1996: Upper Devonian and Mississippian radiolarian zonation and biostratigraphy of the Woodford, Sycamore, Caney and Goddard Formations, Oklahoma. Cushman Foundation Foram. Res. Spec. Publ. 33, 1–276.

SPILLER, F.C.P. & METCALFE, I. 1995: Late Paleozoic radiolarians from the Bentong-Raub, Jour. Southeast Asia Earth Sci. 11(3), 217–224.

STRATFORD, J. M. C. & AITCHISON, J. C. 1997: Lower to Middle Devonian radiolarian assemblages from the Gamilaroi terrane, Glenrock Station, NSW, Australia. Marine Micropaleontology 30, 225–250.

WANG, CHENG-YUANG 2000: Devonian. In: Nanjing Institute of Geology and palaeontology. In: ACADEMIA SINICA (Ed.): Stratigraphical studies in China (1979–1999). Hefei. Press of University of Science and Technology of China, 73–94.

WANG, Y.J. 1991: On progress in the study of Paleozoic radiolarians in China. Acta Micropalaeontologica Sinica (Wei Ti Ku Sheng Wu Hsueh Pao) 8(3), 237–251.

WANG, Y.J. 1994: Cherts and radiolarian assemblage zones of Qinzhou area, Guangxi. Chinese Science Bulletin 39(15), 1300–1304.

WANG, Y.J. 1997: An Upper Devonian (Famennian) radiolarian fauna from carbonate rocks, northern Xinjiang. Acta Micropalaeontologica Sinica 14(2), 149–160.

WANG, Y.J, & LUO, H., 2004: Impact of the Frasnian-Famennian Extinction Event on Radiolarian Faunas in South China. In: RONG J. Y. et al. (Eds.): Mass Extinction and Recovery – Evidences from the Palaeozoic and Triassic of South China. Hefei. Press of University of Science and Technology of China, 381–408.

WANG, Y.J. , LUO, H., KUANG, G.-D. & LI, J.-X. 1998: Late Devonian – Late Permian strata of cherty facies at Xiaodong and Bancheng counties of the Qinzhou area, SE Guangxi. Acta Micropalaeontologica Sinica,, 351–366.

WANG, Y.J. ,, FANG, Z.-J., YANG, Q., ZHOU, Z.-C., CHENG, Y.-N., DUAN, Y.-X. & XIAO, Y.-W. 2000: Middle-Late Devonian Strata of Cherty Facies and Radiolarian Faunas from west Yunnan. Acta Micropalaeontologica Sinica 17(3), 235–254.

WANG, Y.J., AITCHISON, J.C. & LUO, H. 2003: Devonian radiolarian faunas from South China. Micropaleontology 49(2), 127–145.

WU, H., XIAN, X. & KUANG, G. 1994: Late Paleozoic Radiolarian assemblages of southern Guangxi and its geological significance. Scientia Geologica Sinica 29(4), 339–344.

XIAN, S.Y., CHEN, J.R. & WANG, Z.Q. 1995: Devonian Ecostratigraphy, Sequence Stratigraphy and See-level Changes in Guanxi, Longmen Mountain Area, Sichuan. Sedimentary Facies and Palaeogeography 15(6), 1–47.

Manuscript received January 2004
Revision accepted February 2005

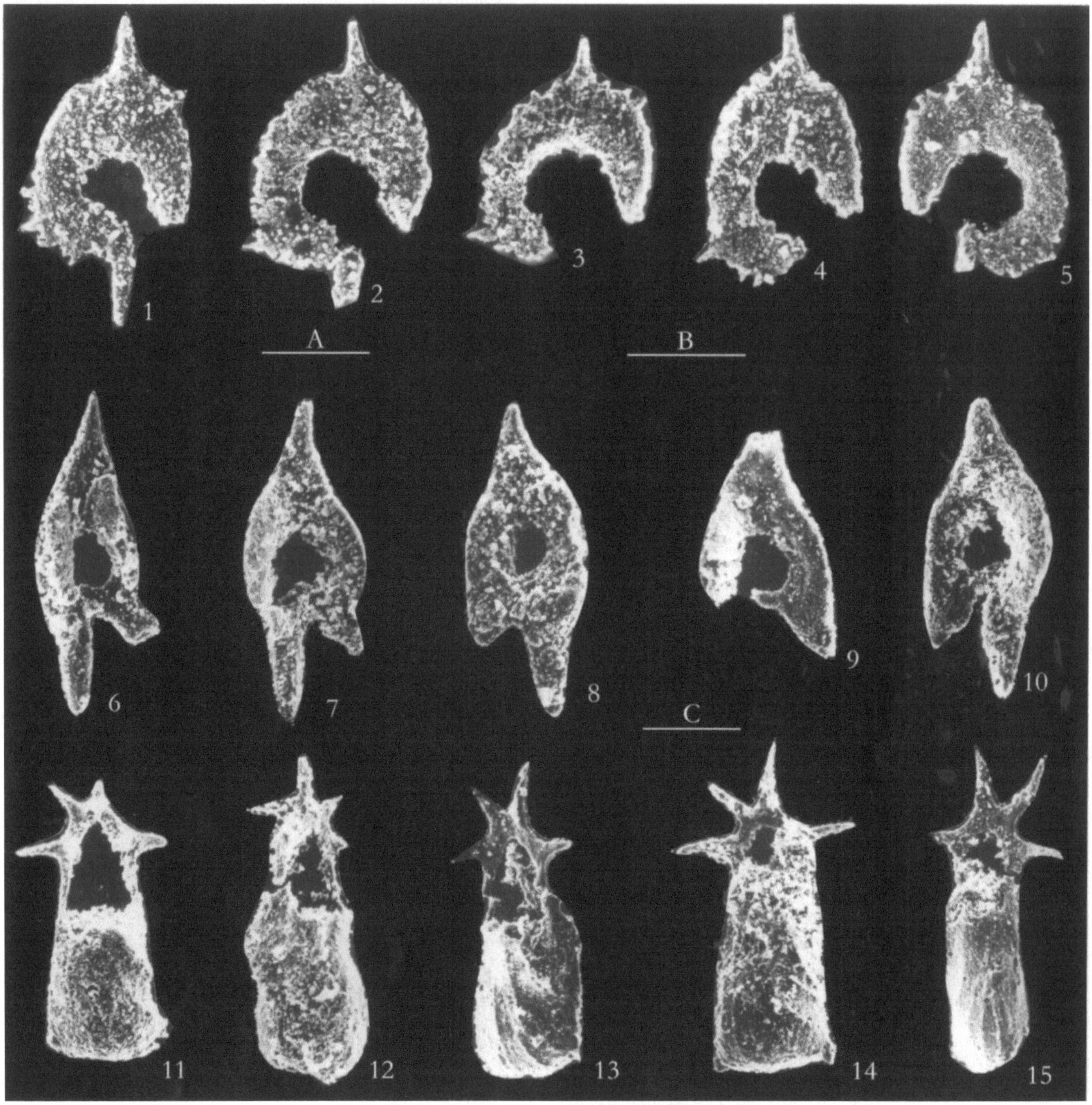

Plate 1

Index species of Late Devonian radiolarian zonations from South China. All scales = 100 μm: A for figure 1; B for figures 2–10; C for figures 11–15.

Figs. 1–5, *Helenifore laticlavium* NAZAROV & ORMISTON. 1–5 from sample 96shiti159, Shiti Reservoir Formation, Shiti, Bancheng, Guangxi.
Figs. 6–10, *Helenifore robustum* (BOUNDY-SANDERS & MURCHEY). 6–10 from sample 92J9, Shiti Reservoir Formation, Shija, Bancheng, Guangxi.
Figs. 11–13, *Holoeciscus foremanae* CHENG. 11–13 from sample 96shiti185, Bancheng Formation, Shiti, Bancheng, Guangxi.
Figs. 14–15, *Holoeciscus elongates* KIESSLING & TRAGELEHN. 14–15 from sample 96shiti185, Bancheng Formation, Shiti, Bancheng, Guangxi.

0012-9402/06/01S133-7
DOI 10.1007/s00015-006-0610-5
Birkhäuser Verlag, Basel, 2006

Eclogae geol. Helv. 99 (2006) Supplement 1, S133–S139

Middle to Upper Permian radiolarian faunas from chert blocks in Pai area, northwestern Thailand

NUTTHAWUT WONGANAN[1] & MARTIAL CARIDROIT[2]

Key words: Radiolaria, chert, biostratigraphy, Permian, northwestern Thailand
Mots clés: Radiolaire, «chert», biostratigraphie, Permien, Thaïlande du Nord-Ouest

ABSTRACT

Well-preserved Permian radiolarians have been discovered from chert blocks embedded in a unit in the Pai district, Mae Hong Son province, northwestern Thailand that was recently mapped as Carboniferous. Twenty-four taxa belonging to ten genera have been recognized and some significant forms of *Follicucullus* are present. Two upper Maokouan (Capitanian) to lower Wuchiapingian radiolarian assemblages (*Follicucullus monacanthus*, *Follicucullus charveti* – *F. porrectus* assemblages) are reported. *Follicucullus charveti* CARIDROIT & DE WEVER is interpreted as a possible provincialism rather than being endemic like other members of the same genus. These results provide more data to demonstrate that in northern Thailand, the Devonian to Triassic was of a zone of deep siliceous marine sedimentation. The resultant deposits are one of the longest witnesses of continuous deposition in an oceanic setting.

RESUME

Des assemblages de radiolaires bien préservés ont été extraits de blocks de «chert» présents dans la région de Pai (Province de Mae Hong Son, Thaïland du Nord), récemment cartographiés en tant que roches du Carbonifère. Vingt-quatre espèces appartentnant à 10 genres sont reconnues, comprenant plusieure Follicucullidae très significatifs pour la stratigraphie. Deux zones d'assemblages (*Follicucullus monacanthus* et *Follicucullus charveti-F. porrectus*) sont présentes et datent le Maokouien supérieur (Capitanien) et le Wuchiapingien inférieur; elles sont similairs à celles présentes en Chie du Sud et au Japon. La présence de *Follicucullus charveti* CARIDROIT & DE WEVER, espèce interprétée comme une forme d'eau chaude, est discutée. Ces résultas sont fondamentaux et sont une donnée supplémentaire pour prouver que l'ensemble des temps Dévonien à Trias supérieur est répresenté par des depôts siliceux de mer profonde; ceci étant un des plus longs témoins connus d'une sédimentation océanique continue.

1. Introduction

The study of Permian radiolarians in Thailand, has become a popular topic since radiolarian research started (Caridroit et al., 1990). Occurrences of Permian radiolarians are widespread and they have now been studied in many regions; northern Thailand (Chiang Mai: e.g. Caridroit 1991, 1993; Caridroit et al. 1992; Sashida et al. 1993; Wonganan et al. 2002), north-central Thailand (Sukhothai: Sashida & Nakornsri 1997), north-eastern Thailand (Loei: e.g. Sashida et al. 1993; Sashida & Igo 1999), eastern Thailand (Sra Khaew: e.g. Hada et al. 1999; Sashida et al. 2000). The goals of these studies are both to improve the biostratigraphy and to track radiolarites as remnants of oceanic palaeoenvironments. Only lower to middle Lower and uppermost Permian faunas have been reported from northern Thailand so far (e.g. Caridroit 1993; Sashida et al. 1993). The aim of this paper is to present details of a new upper Middle to lower Upper Permian radiolarian locality from Thailand. Radiolarian assemblage zones described from Japan and China are identified and briefly discussed. Possible provincialism is observed with the occurrence of *Follicucullus charveti* CARIDROIT & DE WEVER at few localities, compared with the endemic nature of other members of this genus.

2. General Geology and Occurrence of Radiolarians

The district of Pai is located in NW northern Thailand, approximately 110 km from Chiang Mai. According to Bunopas (1981) the main Palaeozoic sedimentary rocks widely exposed in the area is the Mae Hong Son Formation. Age assignments for this unit range from Siluro-Devonian (Bunopas 1981) to Carboniferous-Permian (Chuaviroj et al. 1985; Intawong et al. 1997). The formation is dominated by unmetamorphosed Palaeozoic sedimentary rocks, which are extensively distributed in northwest Thailand. It mainly consists of massive or

[1] Department of Geological Sciences, Faculty of Science, Chiang Mai University, Chiang Mai 50200 Thailand. E-mail: n-thawut@chiangmai.ac.th
[2] Laboratoire de Paléontologie et Paléogéographie du Paléozoïque (LP3), UMR 8014 et UFR 1818 du C.N.R.S.,
Sciences de la Terre: Université des Sciences et Technologies de Lille, F-59655 Villeneuve d'Ascq cedex France. E-mail: martial.caridroit@univ-lille1.fr

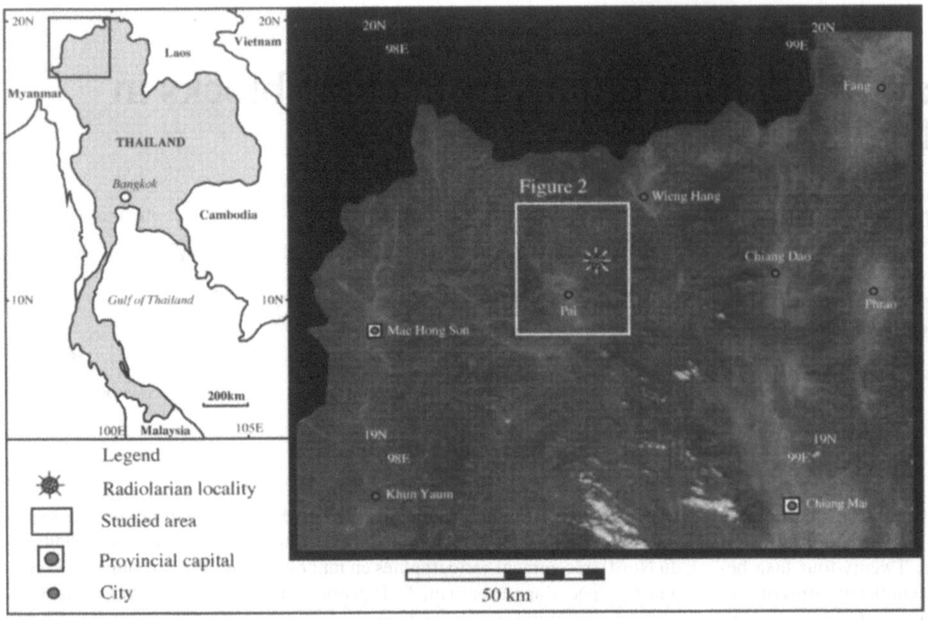

Fig. 1. Index map of northwestern Thailand showing the studied area and the radiolarian locality.

bedded sandstone, shale, chert and slate (Fig. 2). Limestones, forming isolated karstic hills are widespread and contain abundant Carboniferous and Permian foraminifers and other fossils. In general most Thai geologists correlate these limestones with the Ratburi Group. Limestones in this area are lithologically variable, but mainly consist of massive limestones, with local occurrences of dolomitic limestone with lenticular or nodular cherts. The relationship between limestone units and Palaeozoic siliceous sedimentary units is not clear. Field observations indicate the limestones unconformably overlie siliceous sediments, and where faulted the basal contact is locally marked by breccias.

The studied section is located approximately 7– 8 km north of Pai (Fig. 2), or at position 438455 on Thai topographic map sheet 4647 I Amphoe Pai, in the area of Ban Tan Jed Ton village. The section is exposed for approximately 850 m along a gravel road connecting Pai and Wieng Haeng that has been cut through a high mountainous area. The section consists of gray to greenish gray, partly dark gray, well-bedded chert with 4 to 10 cm thick beds intercalated with thin shale and clay (mm – few cm) layers. The main and lowermost exposed section is about 150 m long, before being obscured by strongly weathered yellow volcanic rocks for about 200 m. The second section, which is believed to be a separate block, is composed of greenish gray well-bedded chert and about 10 m long. Ten chert samples were collected from the main section (PAI-391 to PAI-400) and three more were taken from the second (PAI-401 to PAI-403). The next three samples (PAI-404 to PAI-406) were taken from a small chert outcrop exposed on the other side of 200 m of weathered tuffaceous material. Four chert samples were collected (PAI-407 to PAI-412) from the last section which ends at a summit approximately 150 m from the previous section. This chert has long been regarded as chert beds intercalated within the Upper Silurian to Carboniferous Mae Hong Son Formation (Bunopas 1981), or newly established Carboniferous-Permian of Chauviroj et al. (1985). According to field observations, no outcrops were found which expose the strata underlying the chert.

Under the microscope, the chert from lower part consists of mainly cryptocrystalline to microcrystalline quartz associated with very fine clay minerals. Radiolarians are rather well preserved and abundant and filled by chalcedony. Further up the lithologic column the cherts consist of cryptocrystalline quartz with more abundant fine clay mineral and fine volcanic grains, without any traces of fossils. This probably indicates that there was volcanic activity within or nearby the site of deposition area conditions did not favour the preservation of radiolarian skeletons.

Samples from the main chert section (PAI-391 through PAI-400) yielded the following Upper Permian radiolarian faunas: *Follicucullus ventricosus* Ormiston & Babcock, *F. scholasticus* Ormiston & Babcock, *F. orthogonus* Caridroit & De Wever, *F. porrectus* Rudenko, *F. charveti* Caridroit & De Wever, *F.* sp. cf. *F. bipartitus* Caridroit & De Wever, *Hegleria mammilla* (Sheng & Wang) and others. The association of *F. charveti* and *F. porrectus* Rudenko is diagnostic for assignment to the lower Wuchiapingian. Sample PAI-403 contains fewer radiolarians, however the occurrence of *F. monacanthus* Ishiga & Imoto is the diagnostic of the Middle Permian *F. monacanthus* Assemblage (Capitanian), therefore these chert beds are assigned to the Capitanian.

3. Radiolarian biostratigraphy

Twenty-two chert samples were collected, and nine samples yielded Permian radiolarians (Plate 1). The studied chert sec-

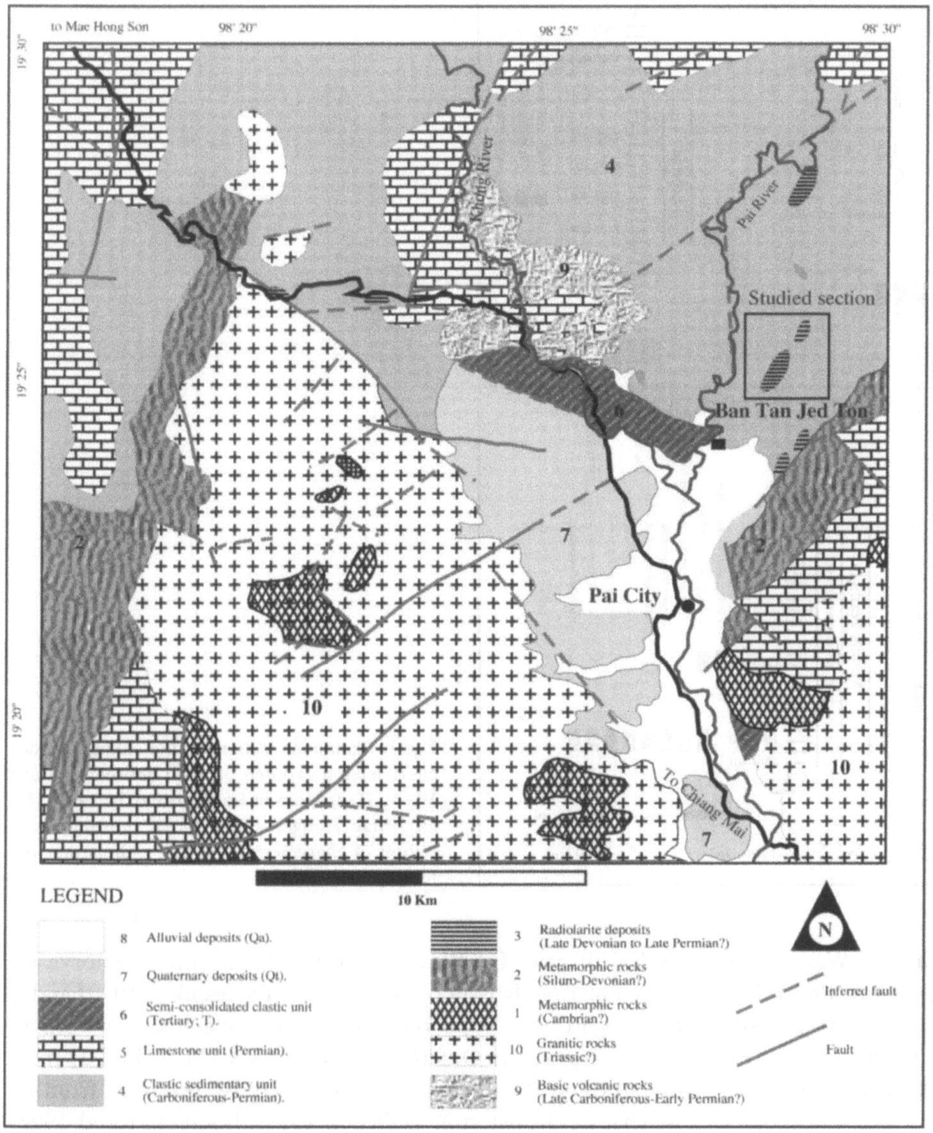

Fig. 2. Simplified geologic map of Pai area, Mae Hong Son province, northwestern Thailand (see Fig. 1 for location); shows the studied sections. (after Geologic map 1:50,000 Sheet Amphoe Pai Quadrangle, Geological Survey Division, Department of Mineral Resources, Bangkok, Thailand 1985; Intawong et al. 1997, modified)

tion extends over two radiolarian assemblage zones: the *Follicucullus monacanthus*, and *Follicuculllus charveti – F. porrectus* assemblage zones. These zones have been reported by Ishiga (1986, 1990) from Japan, Wang et al. (1994) from China and Caridroit in De Wever et al. (2001) and are known to occur in the middle Maokouan (Capitanian) to lower Wuchiapingian.

Follicucullus monacanthus assemblage zone.

This assemblage zone is found within sample PAI-403. It is characterized by the occurrence of *Follicucullus monacanthus* ISHIGA & IMOTO, which is the characteristic taxon. Other radiolarians species that occur in this assemblage include *F. scholasticus* ORMISTON & BABCOCK, *Pseudotormentus* sp. cf. *P. kamigoriensis* CARIDROIT & DE WEVER, *F. orthogonus* CARIDROIT & DE WEVER, *F. ventricosus* ORMISTON & BABCOCK, and *Ormistonella robusta* CARIDROIT & DE WEVER. The *Follicucullus monacanthus* assemblage is known from Japan (e.g. Ishiga 1986), Far East Russia (Rudenko & Panasenko 1990), Oregon in North America (Blome & Reed 1992), South China (Wang et al. 1994), and is also reported in studies by Caridroit (in De Wever et al. 2001, fig. 202, p. 313).

Follicuculllus charveti – F. porrectus assemblage zone

This assemblage zone is found within green chert samples PAI-393, 396, 398 & 399. It is defined by the co-occurrence of *Follicuculllus charveti* CARIDROIT & DE WEVER and *F. porrectus* RUDENKO. The range of this assemblage is considered as a total range of *F. charveti* CARIDROIT & DE WEVER. The other characteristic species are *Follicuculllus venticosus* ORMISTON & BABCOCK, *Follicucullus scholasticus* ORMISTON & BABCOCK, *Follicucullus* sp. cf. *F. bipatitus* CARIDROIT & DE WEVER, *Ishigaum similicutis* CARIDROIT & DE WEVER, *Triplanospongus musashiensis* SASHIDA & TONISHI, *Latentifistula texana* NAZAROV & ORMISTON, *Hegleria mammilla* (SHENG & WANG), and others. The assemblage contains the same radiolarian species reported from Japan (e.g. Ishiga 1990), North

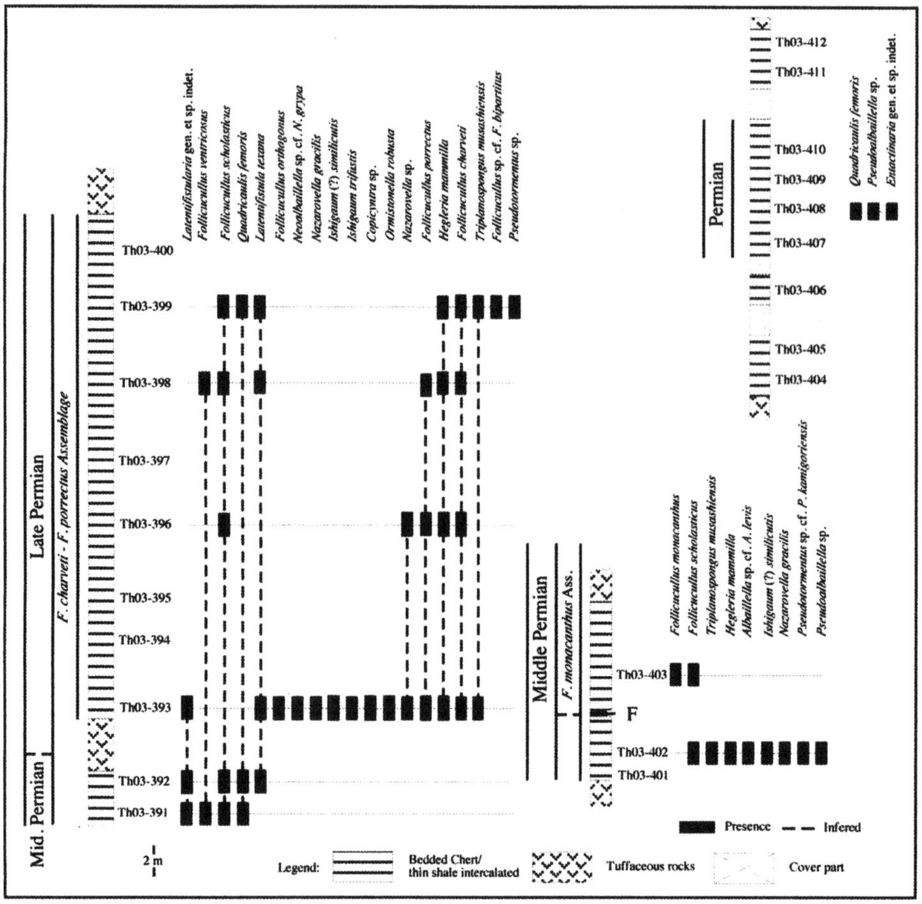

Fig. 3. Columnar sections of the studied bedded chert section exposed on road between Pai and Wieng Hang cities in Ban Tan Jed Ton area (Pai city, Mae Hong Son province), showing distribution of characteristic taxa of radiolarians obtained.

America (e.g. Murchey 1990), South China (e.g. Wang et al. 1994) and New Zealand (e.g. Caridroit & Ferriere 1988). It is comparable to the *F. scholasticus – F. venticosus* zone from China (Wang et at. 1994) and the *F. ventricosus – Ps. fusiformis* assemblage of North America (Murchey 1990).

4. Discussion: Follicucullus faunas and palaeobiogeography

Follicucullus charveti Caridroit & De Wever is characterized by a long apical cone, which is slightly curved to the ventral side (Pl. 1; Fig. 8). Specimens from the area we studied have a strongly inflated pseudothorax and a short pseudoabdomen with small, disk-like, rather short ventral flap compared to those reported from Japan and S. China (e.g. Caridroit & De Wever 1986; Wang & Li 1994). *F. scholasticus* Ormiston & Babcock m. I (e.g. Ishiga 1985; Blome & Reed 1992; Wang et al. 1994), and *F. scholasticus* Ormiston & Babcock m. II (e.g. Ishiga 1985) are considered as two different species. The first is the most abundant form and found in all levels of the section. This form has a smooth, undulating shell, however, somewhat obscure separated portions can be observed in some specimens. Small very short flaps, extending at apertural margin of the dorsal and ventral sides, are well preserved on most specimens. The second morphotype, on the other hand, has a bulbous shell characteristic similar to those reported from Japan by Caridroit & De Wever (1986) and from the Prymorye region, Far East Russia by Rudenko & Panasenko (1990) and is here regarded as *F. porrectus* Rudenko. Specimens of *Follicucullus ventricosus* Ormiston & Babcock (Pl. 1; Fig. 9) from the study area are closely similar to those reported from N. America (Ormiston & Babcock 1979; Blome & Reed 1992) and from S. China (Wang et al. 1994) having a slightly curved apical cone and big pseudothorax, with short pseudoabdomen; both ventral and dorsal flap are very short.

Materials presented in this study also contains some broken pieces that have been assigned to species such as *Gustefana obliqueannulata* Kozur (Pl. 1; Fig. 30) and *Nazarovella phlogidea* Wang (Pl. 1; Fig. 31). These forms in fact have little taxonomic value and in our collections all of these broken pieces appear to be the ending part of an arm of already defined radiolarian species.

It is not yet proven that siliceous skeleton zooplanktonic faunas are useful indicators for palaeo-ecology and/or palaeogeographic position. Nevertheless, the possible palaeogeographic implications of the occurrence of a particular taxon with distinctive form at a few localities merit some discussion. The Upper Permian *Follicucullus charveti – F. porectus* radiolarian assemblage recognized in this study is believed to be equivalent to the *F. bipartitus- F. charveti* assemblage known from southwest Japan (e.g. Caridroit & De Wever 1984, 1986),

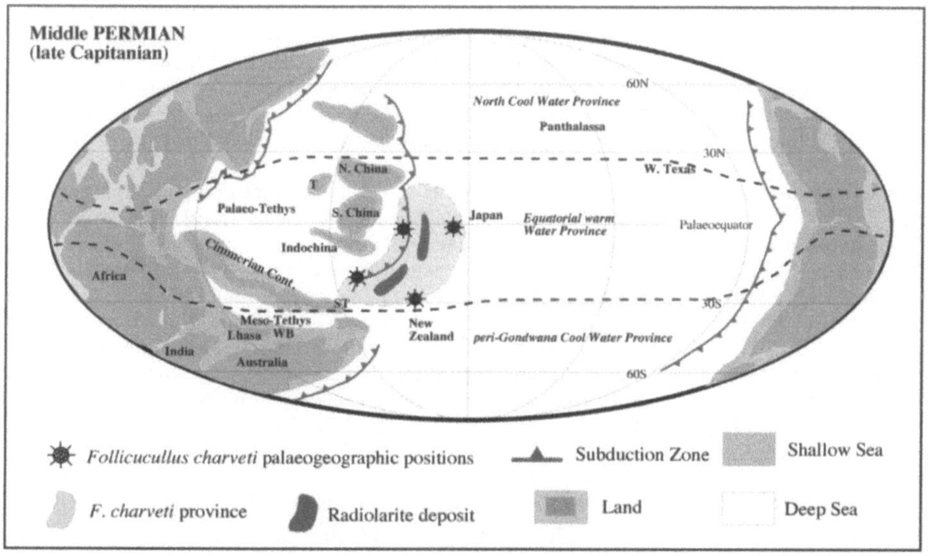

Fig. 4. Palaeogeographic reconstruction during Middle Permian (late Capitanian) showing relative palaeogeographic positions and possible *Follicucullus charveti* province discussed in the text. (Palaeogeographic map and terrane positions are based on Scotese 1997; Metcalfe 1996, and water province areas are after Henderson & Mei 2003, where ST: Shan-Thai; T: Tarim; WB: West Burma).

northern New Zealand (e.g. Caridroit & Ferrière 1988; Takemura et al. 1999), and south & southwest China (e.g. Wang & Li 1994). According to the known literature reported so far, this assemblage has only been discovered from the four different geographic areas mentioned above. Regarding *Follicucullus charveti* Caridroit & De Wever, a diagnostic taxon of this assemblage, it is not clear whether it occurs in a restricted area with short period (early Wuchiapigian). The occurrence of this fauna in a restricted area as plotted on a palaeogeographic map may suggest that its geographic range was probably confined to a special part in the tropical palaeoequator (low-latitude) realm within the equatorial warm water province. As this species can likely be defined as a warm-water taxon the palaeogeographic positions of terranes containing radiolarites with this fauna were probably at low latitudes not far from one another (Fig. 4).

Acknowledgements

Our field survey in January 2003 was financial supported by the Chiang Mai University (Thailand), the Laboratoire de Paléontologie et Paléogeographie du Paléozoïque (*CNRS-UMR 8014*) and the Université des Sciences et Technologies de Lille (France). The Faculty of Science, Chiang Mai University for supporting field-working facilities is credited here. Fabrice Corday and Frances Spiller are thanked for reviews that helped to improve the manuscript.

REFERENCES

Blome, C. D. & Reed, K. M. 1992: Permian and Early (?) Triassic radiolarian faunas from the Grindstone terrane, central Oregon. Journal of Paleontology 66, 351–383.

Bunopas, S. 1981: Palaeogeographic history of western Thailand and adjacent parts of South-east Asia: a plate tectonic interpretation. Geological Survey paper vol. 5, Department of Mineral Resources, Bangkok, Special Issue, 810.

Caridroit, M. 1991: Taxonomic study on Carboniferous and Permian Radiolaria from NW Thailand. Paleontologic, stratigraphic and tectonic significances. Abstracts. Sixth Meeting, International Association of Radiolarian Paleontologists (INTERRAD VI), p. 21.

Caridroit, M. 1993: Permian Radiolarian from NW Thailand. In: Thanasuthipitak, T. (Ed.): Proceedings of the International Symposium on Biostratigraphy of Mainland Southeast Asia Facies and Paleontology 1, Chiang Mai, Thailand, 83–96.

Caridroit, M. & De Wever, P. 1984: Description de quelques nouvelles espèces de Follicucullidae et d'Entactinidae (Radiolaires polycystines) du Permien du Japon (Description of some new species of Follicucullidae and Entactinidae (Polycystin radiolarians) from the Permian of Japan). Geobios 17, 639–644.

Caridroit, M. & De Wever, P. 1986: Some Late Permian Radiolarians from pelitic rocks of the Tatsuno Formation (Hyogo Prefecture), Southwest Japan. Marine Micropaleontology 11, 55–90.

Caridroit, M. & Ferriere, J. 1988: Premieres datations precises du Paleozoique par radiolaires en Nouvelle Zealande: interet geologique et paleontologique. Compte rendus Academie des Sceances de l'Academie des Sciences (Paris), series II 306, 321–326.

Caridroit, M., Fontaine, H., Jongkanjamasoontorn, Y., Suteethorn, V. & Vachard, D. 1990: First results of a paleontological study of Northwest Thailand. CCOP Technical Secretariat, 337–350.

Caridroit, M., Vachard, D. & Fontaine, H. 1992: Datations par radiolaires (Carbonifère, Permien et Trias) en Thailande nord-occidentale. Mise en évidence de nappe de charriage et d'olisostromes. Academie des Sciences Compte Rendu Paris 314, 515–520.

Chauviroj, S., Chaturongkawanich, S. Leevongcharoen, S. & Sopon-pongpipat, P. 1985: Hydrothermal Research Project, Geological Survey Division, Department of Mineral Resources, Bangkok, Thailand. 66.

De Wever, P., Dumitrica, P., Caulet, J.P., Nigrini, C. & Caridroit, M. 2001: Radiolarians in the Sedimentary Record. 533 p., Gordon and Breach Science Publishers.

Hada, S., Bunopas, S., Ishii, K. & Yoshikura, S. 1999 Rift-drift history and the amalgamation of Shan-Thai and Indochina/East Malaya blocks In: Metcalfe, I. (Ed.): Gondwana Dispersion and Asian Accretion. IGCP 321, Sp. Publ., 67–88, A.A. Balkema.

Henderson, C.M. & Mei, S. 1990: Stratigraphic versus environment significance of Permian serrated conodonts around the Cisuralian-Guadalupian boundary: new evidence from Oman. Palaeogeography, Palaeoclimatology, Palaeoecology 191, 301–328.

Intawong, A., Kunmanee, J. & Wonganan, N. 1997: Geological report in the area of Ban Tan Jed Ton, Tambol Wieng Nue, Amphoe Pai, Chang Wat Mae Hong Son. Department of Geological Sciences, Chiang Mai University, 82.

Ishiga, H. 1985: Discovery of Permian radiolarians from Katsumi and Oi Formation along south of Maizuru Belt, Southwest Japan and its significance. Ibid 39, 175–185.

ISHIGA, H. 1986: Late Carboniferous and Permian radiolarian biostratigraphy of southwest Japan. Journal of Geoscience, Osaka City University 29, 89–100.

ISHIGA, H. 1990: Paleozoic radiolarians. In: ICHIKAWA, K., MIZUTANI, S.,HARA, I., HADA, S. & YAO, A. (Eds.): Pre-Cretaceous terranes of Japan. Publication of IGCP Project No. 224: Pre-Jurassic Evolution of Eastern Asia, Nippon Insatsu Shuppan, Co. Ltd, Osaka, 285–295.

METCALFE, I. 1996: Gondwana dispersion, Asian accretion and evolution of Eastern Tethys. Australian Journal of Earth Sciences 43, 605–623.

MURCHEY, B.L. 1990: Age and Depositional setting of siliceous sediments in the Upper Paleozoic Havallah sequence near Battle Mountain, Nevada: implications for the paleogeography and structural evolution of the western margin of North America. Geological Society of America, Special paper 255, 137–155.

ORMISTON, A. & BABCOCK, L. 1979: Follicucullus, new radiolarian genus from the Gaudalupian (Permian) Lamar Limestone of the Delaware Basin. Journal of Paleontology 53, 328–334.

RUDENKO, V. & PANASENKO, E.S. 1990: A new findings of the Upper Permian radiolarians in Prymorye region. In: ZAKHAROV, Y.D., BELYAEVA, G.V. & NIKITINA, A.P. (Eds.): New data on Paleozoic and Mesozoic biostratigraphy of the South Far East. Far Eastern Branch of the USSR Acad. of Sci., Vladivostok, 117–124.

SASHIDA, K. & IGO, H. 1999: Occurrence and tectonic significance of Paleozoic and Mesozoic Radiolaria in Thailand and Malaysia. In: METCALFE, I. (Ed.): Gondwana Dispersion and Asian Accretion Final Results Volume for IGCP Project 321, A.A. Balkema, Rotterdam, 175–196.

SASHIDA, K. & NAKORNSRI, N. 1997: Lower Permian radiolarian faunas from the Khanu Chert Formation distributed in the Sukhothai Area, northern central Thailand. In DHEERADILOK, P., C. HINTHONG, P. CHAODUMRONG, P. PUTTHAPIBAN, W. TANASTHIEN, C. UTHA-AROON, N. SATTYARAK, N. NUCHANONG & T. TECHAWAN (Eds.): Proceedings of the International Conference on Stratigraphy and Tectonic Evolution of Southeast Asia and the South Pacific, Bangkok, Thailand, 101–108.

SASHIDA, K., IGO, H., HISADA, K., NAKORNSRI, N. & AMPORNMAHA, A. 1993: Occurrence of Paleozoic and early Mesozoic Radiolaria in Thailand (preliminary report). Journal of Southeast Asian Earth Science 8, 97–108.

SASHIDA, K., IGO, H., UENO, K., NAKORNSRI, N. & SARDSUD, A. 1998: Late Paleozoic radiolarian fauna from northern and northeastern Thailand. Science Reports of the Institute of Geoscience, University of Tsukuba, Section B 19, 1–27.

SASHIDA, K., IGO, H., ADASHI, S., UENO, K., KAJIWARA, Y., NAKORNSRI, N. & SARDSUD, A. 2000: Late Permian and Middle Triassic radiolarian faunas from northern Thailand. Journal of Paleontology 74, 789–811.

SCOTESE, K. 1997: Paleogeographic Atlas Paleomap Progress Report 90-0497. Paleomap Project ed., Univ. of Texas at Arlington, Arlington, Texas, 45.

TAKEMURA, A., MORIMOTO, T., AITA, Y., HORI, R.S., HIGUCHI, Y., SPORLI, K.B., CAMPBELL, H.L., KODAMA, K. & SAKAI, T. 1999: Permian Albaillellaria (Radiolaria) from a limestone lens at the Arrow Rocks in the Waipapa Terrane, Northland, New Zealand. In: DE WEVER, P. & CAULET, J.-P. (Eds.): InterRad VII, Paris/Bierville, September 1997. Geodiversitats 21(4), 789–811.

WANG, Y. & LI, J. 1994: Discovery of the *Follicucullus bipartitus –F. charveti* radiolarian assemblage zone and its geological significance. Acta Micropalaeontologica Sinica 11, 201–212.

WANG, Y., CHENG, Y.N. & YANG, Q. 1994: Biostratigraphy and systematics of Permian radiolarians in China. Paleoworld 4, 172–202.

WONGANAN, N., CARIDROIT, M. & RANDON, C. 2002: Radiolarians and conodonts from radiolarites in NW-Thailand; witnesses of a 140 my (at least) oceanic realm. In: The 64th Palaeontological Association Annual Meeting, Cambridge, abstract. The Palaeontology Newsletter 51, 73.

Manuscript received January 2004
Revision accepted March 2005

Plate 1

Permian radiolarians from Pai area, northwestern Thailand. All figures are scanning electronic micrographs. Scale bar 100 µm applies to all specimens. 1–6. *Follicucullus scholasticus* ORMISTON & BABCOCK, PAI-421; 7. *Follicucullus porrectus* RUDENKO, PAI-421; 8–10. *Follicucullus charveti* CARIDROIT & DE WEVER, PAI-421; 11. *Follicucullus monacanthus* ISHIGA & IMOTO, PAI-421; 12–13. *Follicucullus ventricosus* ORMISTON & BABCOCK, PAI-419; 14. *Follicucullus orthogonus* CARIDROIT & DE WEVER, PAI-421; 15–17. *Pseudoalbaillella* sp. A, PAI-421; 18. *Neoalbaillella* sp. cf. *N. grypa* ISHIGA ET AL., PAI-421; 19. *Albaillella* sp. aff. *A. levis* ISHIGA ET AL., PAI-421; 20. *Nazarovella gracilis* DE WEVER & CARIDROIT, PAI-402; 21. *Quadricaulis femoris* CARIDROIT & DE WEVER, PAI-393; 22. *Ishigaum* (?) *similicutis* CARIDROIT & DE WEver, PAI-393; 23. *Ishigaum trifustis* DE WEVER & CARIDROIT, PAI-393; 24. *Trianospongos musashiensis* SASHIDA & TONISHI, PAI-393; 25. *Latentifistula texana* NAZAROV & ORMISTON, PAI-398; 26. *Ormistonella robusta* DE WEVER & CARIDROIT, PAI-393; 27. *Pseudotormentus* sp. cf. *P. kamigoriensis* DE WEVER & CARIDROIT, PAI-402; 28. *Copicyntra* sp., PAI-393; 29. *Hegleria mammila* (SHENG & WANG), PAI-393; 30–31. broken part of latentifistulinid (30: cf. *Gustefana obliqueannulata* KOZUR; 31: cf. *Nazarovella phlogidea* WANG), PAI-393.

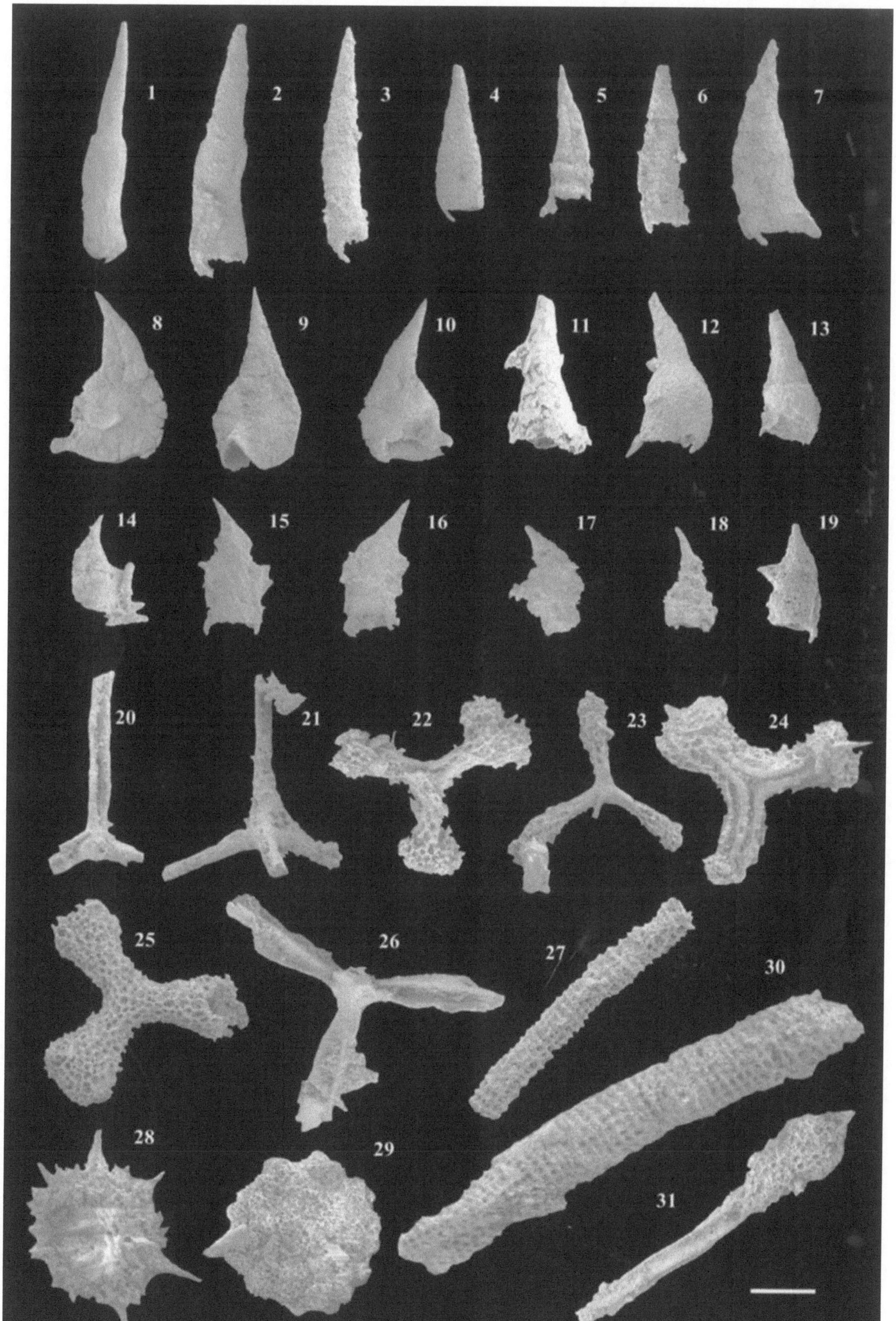
1
2
3
4
5
6
7
8
9
10
11
12
13
14
15
16
17
18
19
20
21
22
23
24
25
26
27
28
29
30
31

MIX
Papier aus verantwortungsvollen Quellen
Paper from responsible sources
FSC® C105338
FSC
www.fsc.org

If you have any concerns about our products,
you can contact us on
ProductSafety@springernature.com

In case Publisher is established outside the EU,
the EU authorized representative is:
Springer Nature Customer Service Center GmbH
Europaplatz 3, 69115 Heidelberg, Germany

Printed by Libri Plureos GmbH
in Hamburg, Germany